SUSTAINABLE AGRICULTURAL AND FOOD SYSTEMS ENGINEERING

New Research, Practices, and Developments

SUSTAINABLE AGRICULTURAL AND FOOD SYSTEMS ENGINEERING

New Research, Practices, and Developments

Jose Juan Buenrostro-Figueroa, PhD

Julio César Tafolla-Arellano, PhD

Leonardo Sepúlveda Torre, PhD

Porteen Kannan, PhD

A. K. Haghi, PhD

Editors

AAP | APPLE ACADEMIC PRESS

First edition published 2025

Apple Academic Press Inc.
1265 Goldenrod Circle, NE,
Palm Bay, FL 32905 USA

760 Laurentian Drive, Unit 19,
Burlington, ON L7N 0A4, CANADA

CRC Press
2385 NW Executive Center Drive,
Suite 320, Boca Raton FL 33431

4 Park Square, Milton Park,
Abingdon, Oxon, OX14 4RN UK

Library and Archives Canada Cataloguing in Publication

CIP data on file with Canada Library and Archives

Library of Congress Cataloging-in-Publication Data

CIP data on file with US Library of Congress

ISBN: 978-1-77491-729-9 (hbk)
ISBN: 978-1-77491-730-5 (pbk)
ISBN: 978-1-00350-421-4 (ebk)

About the Editors

Jose Juan Buenrostro-Figueroa, PhD
Researcher, Research Center for Food and Development, A.C., Mexico

Jose Juan Buenrostro-Figueroa, PhD, is a Researcher at the Research Center for Food and Development, A.C., Mexico. Dr. Buenrostro has experience in bioprocess development, including microbial processes for enzyme production and recovery of bioactive compounds; valorization of agroindustrial by-products, and extraction and characterization of bioactive compounds from plants and agroindustrial wastes. He has published 35 papers in indexed journals, 19 book chapters, three books as editor, four patent requests, and more than 65 contributions at scientific meetings. Dr. Buenrostro has been a member of S.N.I. (National System of Researches, Mexico), the Mexican Society of Biotechnology and Bioengineering (SMBB), and the Mexican Society for Food Protection affiliate of the International Association for Food Protection.

He became a Food Engineer at the Antonio Narro Agrarian Autonomous University. He earned his MSc and PhD degrees in Food Science and Technology from the Autonomous University of Coahuila, México, where he worked on the development of bioprocesses for the valorization of agroindustrial by-products. He also worked at in the Biotechnology Department of the Metropolitan Autonomous University, Mexico City, Mexico.

Julio César Tafolla-Arellano, PhD
Research Professor, Department of Basic Sciences,
Autonomous Agrarian University Antonio Narro, Coahuila, Mexico

Julio César Tafolla-Arellano, PhD, is a Full-Time Research Professor of the Department of Basic Sciences of the Autonomous Agrarian University Antonio Narro in Coahuila, Mexico. He is the Head of the Biotechnology and Molecular Biology Laboratory and a member of the Mexican Association of Food Engineering and Biotechnology. He is editor in-chief of the journals *Universitas Agri* and *Revista Agraria*. Dr. Tafolla-Arellano is a member of the Mexican National System of Researchers (level I), and his research is focused on the study of the biotechnological use of plants through an interdisciplinary omics approach, biocatalysis, bioinformatics, and food biotechnology. He became an engineer in food science and technology at the Autonomous Agrarian University Antonio Narro. He earned his PhD degree in Physiology and Plant Molecular Biology at the Center for Research and Food and Development A.C. of Sonora, Mexico. Also, he performed postdoctoral research on the postharvest shelf life of fruit and transcriptomics analysis. He has published several books chapters and articles in indexed journals and developed eight research projects. He has been advisor of several PhD, MSc, and BSc theses.

Leonardo Sepúlveda Torre, PhD
Professor, School of Chemistry at the Universidad Autónoma de Coahuila, Saltillo, Coahuila, Mexico

Leonardo Sepúlveda Torre, PhD, is a full Professor at the School of Chemistry at the Universidad Autónoma de Coahuila, Saltillo, Coahuila, México, where he is also a leader of the Bioprocesses and Microbial Biochemistry Group. His postgraduate studies were related to topics on food biotechnology in the Food Research Department of the UAdeC. In 2011 he made a research stay at the Institute of Biotechnology and Bioengineering at the University of Minho (Uminho), Braga, Portugal. He also worked as a collaborator at the Center of Biological Engineering in Uminho, Braga, Portugal, with the link project "Biotechnologies for Regional Food Biodiversity in Latin America." He participated as author and co-author in 35 scientific articles in indexed journals, several books, and 24 book chapters, and he holds two national patents.

Porteen Kannan, PhD
Assistant Professor, Department of Veterinary Public Health,
Madras Veterinary College, Tamil Nadu Veterinary and Animal Sciences University, India

Porteen Kannan, PhD, is an Assistant Professor in the Department of Veterinary Public Health at Madras Veterinary College, Tamil Nadu Veterinary and Animal Sciences University, India. The research activities of Dr. Kannan include food safety and antimicrobial resistance. He performed his postdoctoral studies at the US Department of Agriculture, Maryland,

USA, with a specialization in foodborne pathogens. He has published his work in both national and international journals. He is actively involved in mentoring both MVSc and PhD students.

A. K. Haghi, PhD
Research Associate, Department of Chemistry, University of Coimbra, Portugal

A. K. Haghi, PhD, is a retired professor and has written, co-written, edited or co-edited more than 1000 publications, including books, book chapters, and papers in refereed journals with over 3800 citations and h-index of 32, according to the Google Scholar database. He is currently a research associate at university of Coimbra, Portugal. Professor Haghi has received several grants, consulted for several major corporations, and is a frequent speaker to national and international audiences. He is Founder and former Editor-in-Chief of *the International Journal of Chemoinformatics and Chemical Engineering* and *Polymers Research Journal*. Professor Haghi has acted as an editorial board member of many international journals. He has served as a member of the Canadian Research and Development Center of Sciences & Cultures. He has supervised several PhD and MSc theses at the University of Guilan (UG) and co-supervised international doctoral projects. Professor Haghi holds a BSc in urban and environmental engineering from the University of North Carolina (USA) and holds two MSc, one in mechanical engineering from North Carolina State University (USA) and another one in applied mechanics, acoustics, and materials from the Université de Technologie de Compiègne (France). He was awarded a PhD in engineering sciences at Université de *Franche-Comté* (France). He is a persistent reviewer of leading international journals.

Contents

Contributors

P. Álvarez-Vazquez
Natural and Renewable Resources Department, UAAAN, Saltillo, Mexico

Juan Alberto Ascacio-Valdés
Department of Food Research, School of Chemical Sciences, Autonomous University of Coahuila, Blvd. V. Carranza e Ing. Jose Cardenas, Saltillo, Coahuila, Mexico

Claudia Magdalena López Badillo
Facultad de Ciencias Químicas, Universidad Autónoma de Coahuila, Saltillo, Mexico

D. Belleza-Pacheco
Food Science and Technology Department, Universidad Autónoma Agraria Antonio Narro, Calzada Antonio Narro, Buenavista, Saltillo, Coahuila, C.P., Mexico

Sonia Yesenia Silva Belmares
Department of Food Research, Faculty of Chemistry, Autonomous University of Coahuila, Blvd. Venustiano Carranza S/N, Saltillo, Coahuila, Mexico

Juan C. Bustillos-Rodríguez
National Technological of Mexico, Technological Institute of Cd. Cuauhtemoc, Chihuahua, Mexico

N. Camposeco-Montejo
Plant Breeding Department, UAAAN, Saltillo, Mexico

Elda Patricia Segura Ceniceros
Department of Food Research, Faculty of Chemistry, Autonomous University of Coahuila, Blvd. Venustiano Carranza S/N, Saltillo, Coahuila, Mexico

J. D. Corona-Flores
Planning Direction, UAAAN, Saltillo, Mexico

Ndahita De Dios-Avila
School of Agrotechnological Sciences, Autonomous University of Chihuahua, Chihuahua, Mexico

Domínguez-Rosas Edmundo
Estudiante de Doctorado, Coordinación de Tecnología de Alimentos de Origen Vegetal, Centro de Investigación en Alimentación y Desarrollo A.C. Carretera Gustavo Enrique Astiazarán Rosas No. 46, Hermosillo, Sonora, Código, Mexico

P. C. Espinoza-López
Coordinación de Desarrollo Regional, Centro de Investigación en Alimentación y Desarrollo, Calle Gustavo Enrique Astiazarán Rosas, Apartado, Hermosillo, Sonora, Mexico

S. Ezhilvalavan
Tamil Nadu Veterinary and Animal Sciences University, Chennai, Tamil Nadu, India

D. Félix-Gurrola
Programa de Doctorado en Desarrollo Regional, Centro de Investigación en Alimentación y Desarrollo Regional, A. C. Carretera Gustavo Enrique Astiazarán Rosas, C. P., Hermosillo, Sonora, Mexico

Adriana Carolina Flores-Gallegos
Department of Food Research, School of Chemical Sciences, Autonomous University of Coahuila, Blvd. V. Carranza e Ing. Jose Cardenas, Saltillo, Coahuila, Mexico

A. Flores-Naveda
Plant Breeding Department, UAAAN, Saltillo, Mexico

L. O. Fuentes-Lara
Animal Nutrition Department, UAAAN, Saltillo, Mexico

J. A. García-Figueroa
Programa de Doctorado en Desarrollo Regional, Centro de Investigación en Alimentación y Desarrollo, Calle Gustavo Enrique Astiazarán Rosas, Apartado, Hermosillo, Sonora, Mexico

J. I. García-López
Plant Breeding Department, UAAAN, Saltillo, Mexico

Luis Carlos Mata Gómez
Facultad de Ciencias Químicas, Universidad Autónoma de Coahuila, Saltillo, Mexico

Iván de Jesús Zapata González
Centro de Investigación en Química Aplicada, Saltillo, Mexico

Sandra Cecilia Esparza González
School of Odontology, Autonomous University of Coahuila, Dra Cuquita Cepeda de Dávila, Saltillo, Coahuila, Mexico

Anna Iliná
Department of Food Research, Faculty of Chemistry, Autonomous University of Coahuila, Blvd. Venustiano Carranza S/N, Saltillo, Coahuila, Mexico

Robles-Parra Jesús-Martín
Profesor-Investigador, Coordinación de Desarrollo Regional, Centro de Investigación en Alimentación y Desarrollo A.C. Carretera Gustavo Enrique Astiazarán Rosas No. 46, Hermosillo, Sonora, Código, Mexico

Edmundo Juárez-Enriquez
School of Chemistry, Autonomous University of Chihuahua, Chihuahua, Mexico

G. N. Juárez-Luna
Departamento de Ingeniería Química, Industrial y de Alimentos, Universidad Iberoamericana, Prolongación Paseo de Reforma 880, Lomas de Santa Fe, Ciudad de Mexico, C.P., Mexico

Tafolla-Arellano Julio-Cesar
Profesor Investigador Titular C. Laboratorio de Biotecnología y Biología Molecular, Departamento de Ciencias Básicas, Universidad Autónoma Agraria Antonio Narro, Calzada Antonio Narro, Buenavista, Saltillo, Código, Mexico

Porteen Kannan
Tamil Nadu Veterinary and Animal Sciences University, Chennai, Tamil Nadu, India

N. López-Ramírez
División de Energías Renovables, Tecnológico Nacional de Mexico/TES Chicoloapan, Camino al Monte s/n, Esquina con Camino a Tlalmimilolpan, Loma de Guadalupe, Ejido de Chicoloapan, Estado de Mexico, C.P., Mexico

Tiznado-Hernández Martín-Ernesto
Profesor-Investigador, Coordinación de Tecnología de Alimentos de Origen Vegetal, Centro de Investigación en Alimentación y Desarrollo A.C. Carretera Gustavo Enrique Astiazarán Rosas No. 46, Hermosillo, Sonora, Código, Mexico

Hernández-Oñate Miguel-Ángel
Profesor-Investigador, CONACYT-Coordinación de Tecnología de Alimentos de Origen Vegetal, Centro de Investigación en Alimentación y Desarrollo, A.C. Carretera Gustavo Enrique Astiazarán Rosas No. 46. Colonia La Victoria, Hermosillo, Sonora, Código, Mexico

I. Morales-Ibarra
Food Science and Technology Department, Universidad Autónoma Agraria Antonio Narro, Calzada Antonio Narro, Buenavista, Saltillo, Coahuila, C.P., Mexico

Andrea Elizabeth Cortes Morán
Department of Food Research, Faculty of Chemistry, Autonomous University of Coahuila, Blvd. Venustiano Carranza S/N, Saltillo, Coahuila, Mexico

David Neder-Suárez
School of Chemistry, Autonomous University of Chihuahua, Chihuahua, Mexico

Magali Ordóñez-García
National Technological of Mexico, Technological Institute of Cd. Cuauhtemoc, Chihuahua, Mexico

Ernesto Oyervides-Muñoz
Facultad de Ciencias Químicas, Universidad Autónoma de Coahuila, Saltillo, Mexico

J. L. Paz-Luna
Estudios Sociales, Revista de Alimentación Contemporánea y Desarrollo Regional, Centro de Investigación en Alimentación y Desarrollo Regional, A. C. Carretera Gustavo Enrique Astiazarán Rosas, C. P., Hermosillo, Sonora, Mexico

Luis Enrique Cobos Puc
Department of Food Research, Faculty of Chemistry, Autonomous University of Coahuila, Blvd. Venustiano Carranza S/N, Saltillo, Coahuila, Mexico

M. Nithya Quintoil
Rajiv Gandhi Institute of Veterinary Education and Research, Puducherry, Tamil Nadu, India

O. N. Rebolloso-Padilla
Animal Production Department, UAAAN, Saltillo, Mexico

J. M. Robles-Parra
Coordinación de Desarrollo Regional, Centro de Investigación en Alimentación y Desarrollo Regional, A. C. Carretera Gustavo Enrique Astiazarán Rosas, C. P., Hermosillo, Sonora, Mexico

Raúl Rodriguez-Herrera
Department of Food Research, School of Chemical Sciences, Autonomous University of Coahuila, Blvd. V. Carranza e Ing. Jose Cardenas, Saltillo, Coahuila, Mexico

María J. Rodríguez-Roque
School of Agrotechnological Sciences, Autonomous University of Chihuahua, Chihuahua, Mexico

R. Rojas-Molina
Agronomy Faculty, Universidad Autónoma de Nuevo Léon, Francisco I. Madero, ExHacienda el Cañada, General Escobedo, Nuevo Léon, C.P., Mexico

X. Ruelas-Chacón
Food Science and Technology Department, Universidad Autónoma Agraria Antonio Narro, Calzada Antonio Narro, Buenavista, Saltillo, Coahuila, C.P., Mexico

Laura María Solís Salas
Department of Basic Sciences, Antonio Narro Agrarian University, Calz. Antonio Narro, Saltillo, Coahuila, Mexico

María del Carmen Rodríguez Salazar
Department of Food Research, Faculty of Chemistry, Autonomous University of Coahuila, Blvd. Venustiano Carranza S/N, Saltillo, Coahuila, Mexico

Karla Segura-Almaraz
Department of Food Research, School of Chemical Sciences, Autonomous University of Coahuila, Blvd. V. Carranza e Ing. Jose Cardenas, Saltillo, Coahuila, Mexico

Sonia Yesenia Silva-Belmares
Department of Food Research, School of Chemical Sciences, Autonomous University of Coahuila, Blvd. V. Carranza e Ing. Jose Cardenas, Saltillo, Coahuila, Mexico

K. Terán-Samaniego
Coordinación de Desarrollo Regional, Centro de Investigación en Alimentación y Desarrollo Regional, A. C. Carretera Gustavo Enrique Astiazarán Rosas, C. P., Hermosillo, Sonora, Mexico

Juan Manuel Tirado-Gallegos
School of Animal Sciences and Ecology, Autonomous University of Chihuahua, Chihuahua, Mexico

Juan Alberto Ascacio Valdés
Department of Food Research, Faculty of Chemistry, Autonomous University of Coahuila, Blvd. Venustiano Carranza S/N, Saltillo, Coahuila, Mexico

A. Varun
Rajiv Gandhi Institute of Veterinary Education and Research, Puducherry, Tamil Nadu, India

Preface

This book provides valuable research on the latest topics and research surrounding sustainable agriculture and food systems engineering.

It covers key concepts on sustainable agriculture, food systems engineering for sustainable development, and agroecological transitions as an interdisciplinary field that comprises the concepts of agronomy, economics, and livestock farming sciences. It also reports important techniques in transforming new concepts involved in both agricultural production and food consumption choices. This new research-oriented book addresses a range of physical and engineering properties of foods and food processing systems applicable to many areas of the food industry.

With a food systems strategy for health and sustainability analyzed as a priority area for public health, this research book reports core knowledge for a range of disciplines including food and nutrition sciences, public health, and environmental science and will be of great interest to scientists, researchers, and postgraduate students working on sustainable food systems engineering.

This book:

- explores case studies on food processing systems and highlights the key concepts and applications.
- includes engineering and physical properties of foods applicable to many areas of the food industry.
- presents key strategies and advanced technologies used in the improvement of healthy and sustainable foods in a multidisciplinary manner.

CHAPTER 1

Utilization of DNA Recombinant Technology for the Sustainability of Agricultural Production Systems

DOMÍNGUEZ-ROSAS EDMUNDO,[1] TAFOLLA-ARELLANO JULIO-CESAR,[2] HERNÁNDEZ-OÑATE MIGUEL-ÁNGEL,[3] ROBLES-PARRA JESÚS-MARTÍN,[4] and TIZNADO-HERNÁNDEZ MARTÍN-ERNESTO[5]

[1]*Estudiante de Doctorado, Coordinación de Tecnología de Alimentos de Origen Vegetal, Centro de Investigación en Alimentación y Desarrollo A.C. Carretera Gustavo Enrique Astiazarán Rosas No. 46, Hermosillo, Sonora, Código, Mexico*

[2]*Profesor Investigador Titular C. Laboratorio de Biotecnología y Biología Molecular, Departamento de Ciencias Básicas, Universidad Autónoma Agraria Antonio Narro, Calzada Antonio Narro, Buenavista, Saltillo, Código, Mexico*

[3]*Professor-Investigador, CONACYT-Coordinación de Tecnología de Alimentos de Origen Vegetal, Centro de Investigación en Alimentación y Desarrollo, A.C. Carretera Gustavo Enrique Astiazarán Rosas No. 46. Colonia La Victoria, Hermosillo, Sonora, Código, Mexico*

[4]*Professor-Investigador, Coordinación de Desarrollo Regional, Centro de Investigación en Alimentación y Desarrollo A.C. Carretera Gustavo Enrique Astiazarán Rosas No. 46, Hermosillo, Sonora, Código, Mexico*

[5]*Professor-Investigador, Coordinación de Tecnología de Alimentos de Origen Vegetal, Centro de Investigación en Alimentación y Desarrollo A.C. Carretera Gustavo Enrique Astiazarán Rosas No. 46, Hermosillo, Sonora, Código, Mexico*

Sustainable Agricultural and Food Systems Engineering: New Research, Practices, and Developments.
Jose Juan Buenrostro-Figueroa, Julio César Tafolla-Arellano, Leonardo Sepúlveda Torre,
Porteen Kannan, A. K. Haghi (Eds.)

ABSTRACT

There are environmental factors, known as stresses, that can reduce the quality and yield of agricultural products. The traditional genetic improvement programs allowed for the selection of varieties with greater resistance to these different stresses. However, this experimental approach is unpredictable and requires many years to develop improved varieties. The discovery and study of DNA have made it possible to develop DNA recombinant technology, which allows for the introduction of specific genes into the genome to induce specific phenotypic characteristics or to induce changes in the gene sequence through the protocol known as genomic editing.

Based on the above, this technology allows for the manipulation of DNA, making it possible to develop improved varieties using a stepwise approach, thus reducing the time required to obtain a variety with new and desirable characteristics. In this way, it is possible to design varieties with better resistance to different stresses, thereby increasing yield and food security. This chapter will describe the application of DNA recombinant technology in agricultural systems with the goal of improving the resistance of plants to various stresses, aiming to provide a comprehensive understanding of the possibilities of this technology in enhancing agricultural sustainability.

1.1 INTRODUCTION

At the present time, there is concern about the sustainability of the economic system, mainly due to the compromises signed at the global level regarding the environment. In this regard, agricultural activities often deplete freshwater resources, negatively impacting human health and genetic diversity. Altogether, this is a challenge that science must face with the goal of providing viable solutions (Sands & Podmore, 2000).

In this context, recombinant DNA technology can help mitigate the reduction in agricultural yield caused by both biotic and abiotic stresses, thereby improving the environmental sustainability of agricultural systems (Gimenez et al., 2018).

Among the abiotic stresses that can be mentioned are large visible light irradiation, high temperatures, high concentrations of salts and heavy metals, as well as hydric stress (Gong et al., 2020).

In the case of some states in the northwest part of Mexico such as Sonora, Sinaloa, Nayarit, Baja California Norte, and Baja California Sur, one of the main problems is hydric stress (Díaz-Caravantes & Wilder, 2014; FAO and UN Water, 2021).

Moreover, current predictions suggest that these stress conditions will increase due to global warming (Dobler-Morales & Bocco, 2021).

1.2 HYDRIC STRESS EFFECT ON AGRICULTURAL YIELD

An experiment carried out at CIMMYT-CIANO located in the city of Obregon, Sonora, Mexico, with 75 bread wheat varieties found a yield reduction from 4,560 to 1,090 kg per hectare due to hydric stress (Sayre et al., 1995). Moreover, eight varieties of wheat evaluated at the same place showed a yield reduction between 33% and 50% compared with the control when they were treated under moderate to severe hydric stress, respectively (Pask & Reynolds, 2013).

The development of varieties more resistant to hydric stress through the utilization of the genetic improvement program using the traditional Mendelian approach was an alternative utilized in the past because it was the only experimental approach available. This experimental approach has several disadvantages such as the need for long periods of time, the loss of resistance genes naturally present in the genome, and the inability to induce a precise phenotypic change without genetic exhaustion (Calhoun et al., 1994). Furthermore, this experimental approach requires the availability of varieties showing a resistance phenotype (Reynolds et al., 2007).

However, with the discovery of DNA and the development of DNA recombinant technology as a consequence, it is possible to develop a genetic improvement program assisted by markers. Furthermore, with this approach, the time to develop new varieties is shorter, specific phenotypic changes can be induced, and there is a reduction in the random changes induced by traditional Mendelian genetic programs (Ahmad et al., 2021).

In this chapter, it will be described how DNA recombinant technology can reduce the negative effects on agricultural yield due to water stress only. Other approaches to improve plant resistance will also be analyzed. However, other abiotic stresses will not be considered. In this way, the positive effect of DNA recombinant technology on agricultural sustainability will be better understood.

1.3 HISTORY AND PROGRESS OF PLANT GENETIC IMPROVEMENT

The change in humankind from hunter-gatherers to communities living in cities was possible due to agricultural development. This, in turn, brought the need to start genetic improvement programs using wild-type species with the goal of developing plants with better organoleptic characteristics (Doebley et al., 2006).

The experimental approach just mentioned was utilized for approximately 10,000 years, and basically, the selection was carried out based on size, color, vigor, taste, and better resistance to abiotic stresses. As a result of this work, one of the most important examples is the improvement of maize from teosinte about 7,500 years ago. A negative effect of this improvement is the fact that modern maize is unable to propagate naturally (Doebley et al., 2006).

Later, it was discovered that the interchange of DNA regions between homologous chromosomes made it possible to quantify genetic recombination in Morgan units, which in turn made it possible to map chromosome regions related to phenotypic characteristics. It is important to mention that Morgan units quantify genetic recombination and are not related to the actual distance in base pairs between genes.

However, this knowledge allowed the development of quantitative genetics, which can identify quantitative trait loci that are associated with phenotypic characteristics related to the interaction of many genes (Chakravarti, 1991). This is the case of the plant response to hydric stress, heat stress, and pathogen attack, among others (Wang et al., 2020). All the mentioned discoveries allowed the creation and implementation of genetic improvement programs, giving rise to new improved varieties such as wheat (Ortiz et al., 2008). However, the development of DNA recombinant technology, which will be described in the next section, made it possible to create successful genetic programs with better direction and in a shorter period of time to obtain desirable characteristics of a given species.

1.4 DNA RECOMBINANT TECHNOLOGY

It is important to comment that the DNA recombinant technology has been under extensive discussion and is still ongoing. This is mainly due to the lack of agreement among groups that either favor or oppose the technology, as well as the dissemination of false information on social networks by individuals who are not experts in the field. One of the authors of this chapter conducted a literature review to assess the various negative effects associated with transgenics, such as the transfer of DNA constructs from transgenic organisms to the human genome, horizontal transfer of transgenes from transgenics to soil bacteria, the uncontrollable release of transgenes through pollen, and the presence of chimeric proteins with potential toxic characteristics, including allergens.

Most of the negative effects mentioned above are not real, although horizontal gene transfer and pollen-mediated gene release can occur, albeit with low probability. From my perspective, the release of transgenic organisms into the biosphere is an event with unpredictable consequences, and it is important to take precautionary measures. On the other hand, it is important to note that DNA recombinant technology is highly beneficial for producing recombinant proteins at a low cost and safely, such as insulin for diabetic individuals, blood coagulation factors for people with hemophilia, growth hormones, and vaccines to combat infections by various strains of the COVID virus, among others. There are other applications for this technology that will not be described here as they are not the main focus of this chapter.

The development of DNA recombinant technology was created from the scientific knowledge developed from the study of the DNA molecule, such as the enzymes playing a role in DNA replication, repair, degradation, and ligation, among others. Furthermore, the knowledge of the DNA genetic code, protein synthesis directed by DNA, presence of responsive elements located in gene promoters, and the role of different genetic elements known as non-coding RNA. Besides that, there was the creation of protocols to introduce a gene construct into the plant genome. Moreover, it was necessary to develop protocols to induce a complete plant from tissues such as leaves, epicotyls, hypocotyls, and cotyledons, among others.

The knowledge of the DNA sequence was made possible with the development of the first generation of DNA sequencing technology known as Sanger-dideoxy (Sanger et al., 1977). With this tool and the availability of enzymes such as restriction enzymes, DNA exonucleases, ligases, phosphotransferases, and DNA polymerases, among others, it was possible to create the first DNA recombinant molecule (Cohen et al., 1973). Furthermore, the sequence of the first genome was known (Frederick Sanger et al., 1977), which was the sequence of a virus with a very small genome. In addition to the above-mentioned, the protocols to carry out plant organogenesis were developed more than 60 years ago (Sussex, 2008).

With all the knowledge mentioned, the creation of the first transgenic plants was carried out more than 40 years ago (Herrera-Estrella et al., 1983) using the Agrobacterium tumefaciens, which is a soil bacterium that can naturally introduce a DNA fragment into the plant genome (Zupan et al., 2000).

Besides the utilization of *A. tumefaciens*, there are other protocols to introduce a DNA fragment into the plant genome such as bioballistics (Lorence & Verpoorte, 2004), utilization of protoplasts, electroporation, microinjection, injection into the pollen tube, utilization of fibers of different compositions (Davey et al., 1989). The different protocols mentioned will not be described in detail because it is not the main objective of the chapter.

Besides the protocols mentioned above, there are other methodologies to change the plant genome that can be classified into two groups. One of them is the protocols for genome editing such as oligonucleotide-directed mutagenesis, zinc finger nucleases, nucleases containing DNA genomic binding domains, and clustered regularly interspaced short palindromic repeats associated protein nine (Zhang et al., 2017). In the other group, there are technologies such as seed production, DNA methylation-dependent on RNA, reverse genetic improvement, cisgenesis, intragenesis, and the utilization of grafting into genetically modified rootstock (Limera et al., 2017).

The technologies will not be described in detail, although it is important to mention that the genomic edition by the protocol of clustered regularly interspaced short palindromic repeats associated with protein nine is one of the most important and it has been utilized a lot with good results.

Besides these new technologies, there are protocols of massive DNA sequencing for the next generation of second generation such as Illumina, Solid, Roche 454, Ion Torrent, Helicos, PacBio, and Nanopore (Paajanen et al., 2019). The creation of the technologies mentioned was possible due to the development of high-performance computers and information storage systems with high capacity (Schmidt & Hildebrandt, 2017). With the utilization of the technologies just mentioned, DNA sequencing is cheaper, and consequently, there are genomes of different organisms completely sequenced.

1.5 UTILIZATION OF DNA RECOMBINANT TECHNOLOGY FOR THE SUSTAINABILITY OF AGRICULTURAL SYSTEMS

1.5.1 MOLECULAR MARKERS

The utilization of molecular markers has been very useful for genetic improvement programs, which is going to be the main topic of this part of the chapter.

There are several molecular markers such as restriction fragment length polymorphism (RFLP), random amplified polymorphic DNA (RAPD), inter simple sequence repeats (ISSR), amplified fragment length polymorphism (AFLP), resistance gene analog polymorphism (RGAP), repetitive element palindromic PCR (rep-PCR), simple sequence repeats (SSR), single nucleotide polymorphism, among others. Moreover, there are different experimental approaches for the basic protocol of molecular markers. In this way, there are six protocols with differences in the case of random amplified polymorphic DNA and seven protocols with differences in the case of inter-simple sequence repeats (Grover & Sharma, 2016).

The utilization of molecular markers allows us to know if the genes related to a phenotypic characteristic are present in the genome, even in young individuals. This way, it is possible to determine which individuals of the progeny should be chosen to continue the genetic improvement program. Consequently, the time to develop a new variety is shorter. In the next part of the section, several experiments using molecular markers will be described.

Using RAPDs, it was possible to estimate the genetic diversity of 38 cultivars of three varieties of tea plants (Wachira et al., 1995). Also, with the same molecular markers, it was possible to obtain the genetic diversity of 24 species of the genus *Camellia* (Chen & Yamaguchi, 2002). Moreover, with the utilization of AFLP markers, the genetic diversity of 49 cultivars of the tea plant from India was obtained (Balasaravanan et al., 2003). Besides, the use of ISSR markers allows the establishment of the genetic diversity of six cultivars of tea plants from China (Yao et al., 2005).

Using SSR markers, it was possible to create a maize variety with many genes related to resistance to the maize Rayado Fino virus (Prasanna & Hoisington, 2003).

Using RFLP markers in a genetic improvement program with 125 cultivars of rice, it was possible to map genes related to different phenotypic characteristics such as number of grains, grain weight, yield, and resistance to hydric stress. With all this data, it was possible to create rice varieties with improved phenotypic characteristics (Hittalmani et al., 2002).

Eight molecular markers including five from the chloroplast genome and three from the nuclear genome were used to identify lima bean (*Phaseolus lunatus* L.) varieties with a phenotype of resistance to hydric stress. This information was utilized to develop bean varieties with high resistance to hydric stress (Martínez-Nieto et al., 2020).

1.5.2 TRANSGENIC PLANTS

Tobacco (*Nicotiana tabacum*) transgenic plants were created expressing a transcription factor WRKY type, designated as BoWRKY10, isolated from *Brassica oleracea.* Furthermore, this gene plays a role in the response to hydric stress. The tobacco-transgenic plants showed a larger tolerance to hydric stress as determined by the larger relative water content and proline amount as well as a larger activity of the superoxide dismutase enzyme (Guo et al., 2021). An *Arabidopsis thaliana* transgenic plant was created to express the transcription factor CkWRKY33 isolated from *Caragana korshinskii* which is native to the desert zones of China. The transgenic plant showed a larger resistance to hydric stress as demonstrated by the increased plant survival percentage, and larger amount of soluble sugars and proline (Li et al., 2021). A transcription factor NAC type, designated as SlNAC2, was isolated from tomatoes and introduced into tobacco plants. It was found that the tobacco-transgenic plants showed a better recovery from hydric stress without a reduction in plant biomass (van Beek et al., 2021).

Transgenic rice plants expressing a chimeric construct including a promoter with responsive elements to the hydric stress controlling the expression of a transcription factor OsTZF5, which belongs to the group of zinc finger, showed better resistance to the hydric stress and a larger yield (Selvaraj et al., 2020).

Transgenic plants of maize expressing the tobacco NPK1 gene under the transcription control of a constitutive promoter showed better tolerance to hydric stress and higher photosynthetic activity under hydric stress (Shou et al., 2004).

A transgenic plant of wheat expressing the transcription factor DREB isolated from soybean showed a higher yield and better photosynthetic efficiency under hydric stress compared with the isogenic lines (Zhou et al., 2020).

With the objective of improving resistance to hydric stress, a soybean transgenic plant was created expressing a construct in which the gene codA from soybean encoding the choline oxidase enzyme is under the control of the promoter of the gene rd29A, which is inducible in response to hydric stress. It was reported that the transgenic plants showed more biomass compared with the control plants (Ta et al., 2020).

1.5.3 TROPICAL FRUIT TRANSGENIC PLANTS

Transgenic plants of banana were developed by overexpressing the gene MaSIP2-1 which encodes a plant aquaporin gene of the small and basic intrinsic protein group (Ishikawa et al., 2005). It was recorded that the transgenic plant showed better resistance to hydric stress based on the amount of chlorophyll, proline, soluble sugars, and abscisic acid. Moreover, they showed a shorter recovery time from hydric stress compared with the isogenic lines (Xu et al., 2020).

A transgenic plant of the yellow passion fruit (*Passiflora edulis* f. flavicarpa) expressing a construct in which the gene encoding the enzyme 1-aminocyclopropane-1 carboxylic acid was located in antisense. The objective was to reduce ethylene synthesis and the ripening phenomena of the fruit. The presence of the construct was found in three out of nine plants. Due to the long-life cycle of the plant, the effects of the construct on fruit ripening phenomena were not reported (Quoirin et al., 2002).

In order to induce resistance in pineapple towards the development of the fungal disease known as black heart, transgenic pineapple plants were created expressing the gene encoding the enzyme polyphenol oxidase. It was found that 38% of the lines showed resistance to black heart disease, regardless of the methodology used to introduce the construct into the genome (Ko et al., 2013).

Pineapple transgenic plants with the goal of improving the resistance to *Phytophthora nicotianae* var. parasitica fungi infection were created by introducing the genes encoding for a chitinase from *Phaseolus vulgaris* and an anti-phytophthora gene from *Nicotiana tabacum*. After eight years of studying the agricultural characteristics of the transgenic plant, no differences in size or fruit composition were found as compared with the isogenic lines. However, the effects on the plant's resistance to the fungi attack are still underway (Yabor et al., 2021).

With the objective of creating plum plants resistant to hydric stress, transgenic plum plants expressing the gene encoding for a superoxide dismutase isoenzyme from *Spinacia oleracea* and the gene encoding for an ascorbate peroxidase isoenzyme from *Pisum sativum* were created. It was reported that the transgenic plants showed better resistance to hydric stress as measured by the photosynthetic capacity and the control of oxidative stress induced by the low availability of water (Diaz-Vivancos et al., 2016).

A transgenic papaya plant was created by expressing the gene encoding the coat protein of the papaya ringspot virus with the objective of increasing resistance to the mentioned virus. It was found that the transgenic plants showed better resistance to virus infection, good agronomic characteristics, and no negative organoleptic effects on the fruit (Meissner et al., 2021).

Several transcription factors from the WRKY family were isolated from papaya. It is known that WRKY transcription factors play a role in the plant's resistance to different abiotic stresses, including hydric stress. It is envisioned that papaya transgenic plants expressing these transcription factors will be created to improve the papaya plant's resistance to diverse abiotic stresses (Abu Bakar et al., 2020).

Transgenic lychee plants overexpressing the gene LcMYB1, which plays a role in the biosynthetic metabolic pathway of flavonoids, were created. It was reported that a red color developed in the plant roots. Based on these results, the authors suggest creating lychee plants with desirable characteristics (Qin et al., 2021).

Transgenic avocado plants expressing genes encoding the enzyme S-adenosyl L-methionine hydrolase and an antifungal protein were created. Obtaining whole transgenic avocado plants was difficult, and the results of experiments studying the effect of the transgenes on the fruit are not yet available (Litz et al., 2005).

1.6 CONCLUSIONS AND PERSPECTIVES

Based on the analysis included in this chapter, it is crystal clear the positive effect of DNA recombinant technology on the sustainability of agricultural systems. It is important to mention that the analysis was focused on hydric stress; however, this technology can be utilized to reduce the negative effects of other abiotic stresses such as high temperatures, salinity, low temperatures, high solar irradiation, and the presence of heavy metals such as mercury, among others.

On the other hand, it is important to mention that there is widespread fear about the utilization of this technology by the public, which has hindered its use. Bibliographic analysis carried out by the corresponding author of this chapter clearly showed that the possible negative effects of this technology can occur, but the probability is rather low.

With the development of genome editing protocols, which are still being improved, it will be possible in the future to design plants with better resistance to both biotic and abiotic stresses, as well as fruits and vegetables with a longer postharvest shelf life, better organoleptic characteristics, and higher nutritional qualities.

KEYWORDS

- **agricultural systems**
- **DNA recombinant technology**
- **food security**
- **nutritional qualities**
- **organoleptic characteristics**
- **postharvest shelf life**
- **stresses**
- **sustainability**

REFERENCES

Abu Bakar, F., Paramalingam, P., & Hasan, K. (2020). The construction of plant expression vector harboring *Carica papaya* L. WRKY gene in *Escherichia coli*. *Pertanika Journal of Science and Technology, 28*, 47–57. <Go to ISI>://WOS:000658825100005.

Ahmad, R., Anjum, M. A., Naz, S., & Balal, R. M. (2021). Applications of molecular markers in fruit crops for breeding programs: A review. *Phyton-International Journal of Experimental Botany, 90*(1), 17. https://doi.org/10.32604/phyton.2020.011680.

Balasaravanan, T., Pius, P., Kumar, R. R., Muraleedharan, N., & Shasany, A. (2003). Genetic diversity among South Indian tea germplasm (*Camellia sinensis, C. assamica,* and *C. assamica* spp. *lasiocalyx*) using AFLP markers. *Plant Science, 165*(2), 365<Go to ISI>:// WOS:000658825100005372.

Calhoun, D., Gebeyehu, G., Miranda, A., Rajaram, S. V., & Van Ginkel, M. (1994). Choosing evaluation environments to increase wheat grain yield under drought conditions. *Crop Science, 34*(3), 673–678.

Chakravarti, A. (1991). A graphical representation of genetic and physical maps: The Marey map. *Genomics, 11*(1), 219–222. https://doi.org/10.1016/0888-7543(91)90123-V.

Chen, L., & Yamaguchi, S. (2002). Genetic diversity and phylogeny of the tea plant (*Camellia sinensis*) and its related species and varieties in the section Thea genus Camellia determined by randomly amplified polymorphic DNA analysis. *The Journal of Horticultural Science and Biotechnology, 77*(6), 729–732.

Cohen, S. N., Chang, A. C., Boyer, H. W., & Helling, R. B. (1973). Construction of biologically functional bacterial plasmids in vitro. *Proceedings of the National Academy of Sciences of the United States of America, 70*(11), 3240–3244. https://doi.org/10.1073/pnas.70.11.3240.

Davey, M., Rech, E., & Mulligan, B. (1989). Direct DNA transfer to plant cells. *Plant Molecular Biology, 13*(3), 273–285.

Díaz-Caravantes, R. E., & Wilder, M. (2014). Water, cities and peri-urban communities: Geographies of power in the context of drought in northwest Mexico. *Water Alternatives, 7*(3), 499–517.

Díaz-Vivancos, P., Faize, L., Nicolás, E., Clemente-Moreno, M. J., Bru-Martínez, R., Burgos, L., & Hernández, J. A. (2016). Transformation of plum plants with a cytosolic ascorbate peroxidase transgene leads to enhanced water stress tolerance. *Annals of Botany, 117*(7), 1121–1131.

Dobler-Morales, C., & Bocco, G. (2021). Social and environmental dimensions of drought in Mexico: An integrative review. *International Journal of Disaster Risk Reduction, 55*, 102067.

Doebley, J. F., Gaut, B. S., & Smith, B. D. (2006). The molecular genetics of crop domestication. *Cell, 127*(7), 1309–1321. https://doi.org/10.1016/j.cell.2006.12.006.

FAO and UN Water. (2021). *Progress on Level of Water Stress. Global Status and Acceleration Needs for SDG Indicator 6.4.2, 2021*. Rome. https://www.unwater.org/publications/progress-level-water-stress-2021-update.

Gimenez, E., Salinas, M., & Manzano-Agugliaro, F. (2018). Worldwide research on plant defense against biotic stresses as an improvement for sustainable agriculture. *Sustainability, 10*(2), Article 391. https://doi.org/10.3390/su10020391.

Gong, Z. Z., Xiong, L. M., Shi, H. Z., Yang, S. H., Herrera-Estrella, L. R., Xu, G. H., Chao, D. Y., Li, J. R., Wang, P. Y., Qin, F., Li, J. J., Ding, Y. L., Shi, Y. T., Wang, Y., Yang, Y. Q., Guo, Y., & Zhu, J. K. (2020). Plant abiotic stress response and nutrient use efficiency. *Science China-Life Sciences, 63*(5), 635–674. https://doi.org/10.1007/s11427-020-1683-x.

Grover, A., & Sharma, P. (2016). Development and use of molecular markers: Past and present. *Critical Reviews in Biotechnology, 36*(2), 290–302.

Guo, J. J., Li, S., Li, H. Y., Li, W., & Li, D. H. (2021). Expression of the kale WRKY gene *BoWRKY10* in transgenic tobacco confers drought stress tolerance. *Russian Journal of Plant Physiology, 68*(1), 147–157. https://doi.org/10.1134/s1021443721010076.

Herrera-Estrella, L., Depicker, A., Van Montagu, M., & Schell, J. (1983). Expression of chimaeric genes transferred into plant cells using a Ti-plasmid-derived vector. *Nature, 303*(5914), 209–213.

Hittalmani, S., Shashidhar, H., Bagali, P. G., Huang, N., Sidhu, J., Singh, V., & Khush, G. (2002). Molecular mapping of quantitative trait loci for plant growth, yield, and yield-related traits across three diverse locations in a doubled haploid rice population. *Euphytica, 125*(2), 207–214.

Ishikawa, F., Suga, S., Uemura, T., Sato, M. H., & Maeshima, M. (2005). Novel-type aquaporin SIPs are mainly localized to the ER membrane and show cell-specific expression in *Arabidopsis thaliana*. *FEBS Letters, 579*(25), 5814–5820. https://doi.org/10.1016/j.febslet.2005.09.076.

Ko, L., Eccleston, K., O'Hare, T., Wong, L., Giles, J., & Smith, M. (2013). Field evaluation of transgenic pineapple (*Ananas comosus* (L.) Merr.) cv. '*Smooth Cayenne*' for resistance to blackheart under subtropical conditions. *Scientia Horticulturae, 159*, 103–108. https://doi.org/10.1016/j.scienta.2013.05.006.

Li, Z., Liang, F. P., Zhang, T. B., Fu, N., Pei, X. W., & Long, Y. (2021). Enhanced tolerance to drought stress resulting from *Caragana korshinskii* CkWRKY33 in transgenic *Arabidopsis thaliana*. *BMC Genomic Data, 22*(1), Article 11. https://doi.org/10.1186/s12863-021-00965-4.

Limera, C., Sabbadini, S., Sweet, J. B., & Mezzetti, B. (2017). New biotechnological tools for the genetic improvement of major woody fruit species. *Frontiers in Plant Science, 8*, 1418.

Litz, R. E., Raharjo, S. H. T., & Gomez-Lim, M. A. (2005). Plant recovery following transformation of avocado with anti-fungal protein and SAM hydrolase genes. *Acta*

Horticulturae [Proceedings of the international symposium on biotechnology of temperate fruit crops and tropical species]. International symposium on biotechnology of temperate fruit crops and tropical species, Daytona Beach, Florida, USA.

Lorence, A., & Verpoorte, R. (2004). Gene transfer and expression in plants. In *Recombinant Gene Expression* (pp. 329–350). https://pubmed.ncbi.nlm.nih.gov/15269435/.

Martínez-Nieto, M. I., Estrelles, E., Prieto-Mossi, J., Roselló, J., & Soriano, P. (2020). Resilience capacity assessment of the traditional *Lima bean* (*Phaseolus lunatus* L.) landraces facing climate change. *Agronomy, 10*(6), 758. https://www.mdpi.com/2073-4395/10/6/758 (accessed on 30 May 2024).

Meissner, P. E. M., Vilarinhos, A. D., De Oliveira, V. J. D., Da Silva, D. D. S., Santos, V. D., & Dantas, J. L. L. (2021). Resistance of transgenic papaya trees to papaya ringspot in Brazil. *Pesquisa Agropecuaria Brasileira, 56*, Article e01954. https://doi.org/10.1590/S1678-3921.pab2021.v56.01954.

Ortiz, R., Braun, H. J., Crossa, J., Crouch, J. H., Davenport, G., Dixon, J., Dreisigacker, S., Duveiller, E., He, Z., & Huerta, J. (2008). Wheat genetic resources enhancement by the International Maize and Wheat Improvement Center (CIMMYT). *Genetic Resources and Crop Evolution, 55*(7), 1095–1140.

Paajanen, P., Kettleborough, G., López-Girona, E., Giolai, M., Heavens, D., Baker, D., Lister, A., Cugliandolo, F., Wilde, G., & Hein, I. (2019). A critical comparison of technologies for a plant genome sequencing project. *Gigascience, 8*(3), giy163.

Pask, A., & Reynolds, M. (2013). Breeding for yield potential has increased deep soil water extraction capacity in irrigated wheat. *Crop Science, 53*(5), 2090–2104.

Prasanna, B., & Hoisington, D. (2003). Molecular breeding for maize improvement: An overview. *Indian Journal of Biotechnology, 2*(1), 85–98.

Qin, Y. Q., Wang, D., Fu, J. X., Zhang, Z. K., Qin, Y. H., Hu, G. B., & Zhao, J. T. (2021). *Agrobacterium rhizogenes*-mediated hairy root transformation as an efficient system for gene function analysis in *Litchi chinensis*. *Plant Methods, 17*(1), Article 103. https://doi.org/10.1186/s13007-021-00802-w.

Quoirin, M., Winkler, L. M., & Ayub, R. (2002). *Agrobacterium tumefaciens*-mediated transformation of yellow passion fruit (*Passiflora edulis* f. *flavicarpa*) with the CMe-Aco1 gene. *Acta Horticulturae* [Citrus and other subtropical and tropical fruit crops: Issues, advances and opportunities]. 26th International Horticultural Congress, Toronto, Canada.

Reynolds, M., Dreccer, F., & Trethowan, R. (2007). Drought-adaptive traits derived from wheat wild relatives and landraces. *Journal of Experimental Botany, 58*(2), 177–186.

Sands, G. R., & Podmore, T. H. (2000). A generalized environmental sustainability index for agricultural systems. *Agriculture, Ecosystems & Environment, 79*(1), 29–41. https://doi.org/10.1016/S0167-8809(99)00147-4.

Sanger, F., Air, G. M., Barrell, B. G., Brown, N. L., Coulson, A. R., Fiddes, J. C., Hutchison, C., Slocombe, P. M., & Smith, M. (1977). Nucleotide sequence of bacteriophage φX174 DNA. *Nature, 265*(5596), 687–695.

Sanger, F., Nicklen, S., & Coulson, A. R. (1977). DNA sequencing with chain-terminating inhibitors. *Proceedings of the National Academy of Sciences, 74*(12), 5463–5467. https://doi.org/10.1073/pnas.74.12.5463.

Sayre, K., Acevedo, E., & Austin, R. (1995). Carbon isotope discrimination and grain yield for three bread wheat germplasm groups grown at different levels of water stress. *Field Crops Research, 41*(1), 45–54.

Schmidt, B., & Hildebrandt, A. (2017). Next-generation sequencing: Big data meets high-performance computing. *Drug Discovery Today, 22*(4), 712–717. https://doi.org/10.1016/j.drudis.2017.01.014.

Selvaraj, M. G., Jan, A., Ishizaki, T., Valencia, M., Dedicova, B., Maruyama, K., Ogata, T., Todaka, D., Yamaguchi-Shinozaki, K., Nakashima, K., & Ishitani, M. (2020). Expression of the CCCH-tandem zinc finger protein gene OsTZF5 under a stress-inducible promoter mitigates the effect of drought stress on rice grain yield under field conditions. *Plant Biotechnology Journal, 18*(8), 1711–1721. https://doi.org/10.1111/pbi.13334.

Shou, H. X., Bordallo, P., & Wang, K. (2004). Expression of the *Nicotiana* protein kinase (NPK1) enhanced drought tolerance in transgenic maize. *Journal of Experimental Botany, 55*(399), 1013–1019. https://doi.org/10.1093/jxb/erh129.

Sussex, I. M. (2008). The scientific roots of modern plant biotechnology. *The Plant Cell, 20*(5), 1189–1198. https://doi.org/10.1105/tpc.108.058735.

Ta, D. T., Ngo, D. M., Nguyen, N. H., Pham, N. B., Do, P. T., & Chu, H. H. (2020). Production of drought tolerant transgenic soybean expressing *coda* gene under the regulation of a water stress-inducible promoter. *Pakistan Journal of Botany, 52*(3), 793–799. https://doi.org/10.30848/pjb2020-3(32).

van Beek, C. R., Guzha, T., Kopana, N., van der Westhuizen, C. S., Panda, S. K., & van der Vyver, C. (2021). The *SlNAC2* transcription factor from tomatoes confers tolerance to drought stress in transgenic tobacco plants. *Physiology and Molecular Biology of Plants, 27*(5), 907–921. https://doi.org/10.1007/s12298-021-00996-2.

Wachira, F. N., Waugh, R., Powell, W., & Hackett, C. (1995). Detection of genetic diversity in tea (*Camellia sinensis*) using RAPD markers. *Genome, 38*(2), 201–210.

Wang, H. L., Wang, R. F., Liu, B., Yang, Y. B., Qin, L., Chen, E. Y., Zhang, H. W., & Guan, Y. N. (2020). QTL analysis of salt tolerance in *Sorghum bicolor* during whole-plant growth stages. *Plant Breeding, 139*(3), 455–465. https://doi.org/10.1111/pbr.12805.

Xu, Y., Li, J., Song, S., Liu, J., Hou, X., Li, Y., Wang, A., Huang, D., Wang, J., & Xu, Z. (2020). A novel aquaporin gene *MaSIP2-1* confers tolerance to drought and cold stresses in transgenic bananas. *Molecular Breeding, 40*(7), 1–13.

Yabor, L., Gomez, D., Perez-Bonachea, L., Martinez, J., Escalante, D., Garro, G., Hajari, E., Sershen, & Lorenzo, J. C. (2021). Pineapple fruits from transgenic plants have limited differences in mesocarp biochemical component contents. *Acta Physiologiae Plantarum, 43*(1), Article 6. https://doi.org/10.1007/s11738-020-03181-2.

Yao, M., Huang, H., Yu, J., & Chen, L. (2005). Analysis of applicability of ISSR in molecular identification and relationship investigation of tea cultivars. *Journal of Tea Science, 25*(2), 153–157.

Zhang, K., Raboanatahiry, N., Zhu, B., & Li, M. (2017). Progress in genome editing technology and its application in plants. *Frontiers in Plant Science, 8*, 177.

Zhou, Y. B., Chen, M., Guo, J. K., Wang, Y. X., Min, D. H., Jiang, Q. Y., Ji, H. T., Huang, C. Y., Wei, W., Xu, H. J., Chen, X., Li, L. C., Xu, Z. S., Cheng, X. G., Wang, C. X., Wang, C. S., & Ma, Y. Z. (2020). Overexpression of soybean DREB1 enhances drought stress tolerance of transgenic wheat in the field. *Journal of Experimental Botany, 71*(6), 1842–1857. https://doi.org/10.1093/jxb/erz569.

Zupan, J., Muth, T. R., Draper, O., & Zambryski, P. (2000). The transfer of DNA from *Agrobacterium tumefaciens* into plants: A feast of fundamental insights. *The Plant Journal, 23*(1), 11–28.

CHAPTER 2

Methodological Proposal to Measure Environmental Sustainability in the Agricultural System of Table Grape in Sonora

D. FÉLIX-GURROLA,[1] K. TERÁN-SAMANIEGO,[2] J. L. PAZ-LUNA,[3] and J. M. ROBLES-PARRA[2]

[1]*Programa de Doctorado en Desarrollo Regional, Centro de Investigación en Alimentación y Desarrollo Regional, A. C. Carretera Gustavo Enrique Astiazarán Rosas, C. P., Hermosillo, Sonora, Mexico*

[2]*Coordinación de Desarrollo Regional, Centro de Investigación en Alimentación y Desarrollo Regional, A. C. Carretera Gustavo Enrique Astiazarán Rosas, C. P., Hermosillo, Sonora, Mexico*

[3]*Estudios Sociales, Revista de Alimentación Contemporánea y Desarrollo Regional, Centro de Investigación en Alimentación y Desarrollo Regional, A. C. Carretera Gustavo Enrique Astiazarán Rosas, C. P., Hermosillo, Sonora, Mexico*

ABSTRACT

Sustainability of productive activities is a commitment assumed by nations and establishes an ethical responsibility that must be fulfilled. One of the productive activities that has met this commitment is export agriculture, especially table grapes. This commitment is derived from the strong pressure exerted by consumers who demand the fruit, requiring attributes that are not visible in the product, but are directly related to the business organization's practices and the three areas of sustainable development. This market pressure is expressed in

Sustainable Agricultural and Food Systems Engineering: New Research, Practices, and Developments. Jose Juan Buenrostro-Figueroa, Julio César Tafolla-Arellano, Leonardo Sepúlveda Torre, Porteen Kannan, A. K. Haghi (Eds.)

certifications, seals, labels, and audits that companies of the table grape system in Sonora must comply with. Although the antecedents show intentions to develop procedures that characterize the productive activities, no procedures have been found that methodologically measure the environmental sustainability this system has achieved. It is in this sense that the present research work seeks to make its contribution: to develop a methodological procedure that allows measuring the environmental sustainability of agricultural systems, especially table grapes in the state of Sonora. The methodological procedure consisted of several stages: 14 managers from an equal number of companies were interviewed, chosen from a representative sample with 95% confidence and 7.5% error. The number of certifications held by each company was identified through a guided interview. An analysis was made of the categories and attributes of each of the areas of sustainable development (environmental, social, and economic), particularly the review points incorporated in the environmental area of each certification. The percentage achieved by each firm in the system in each of the categories of the environmental component of sustainable development was determined, and based on this, an average progress in environmental sustainability was obtained for each business organization. The results indicate that the system of table grape exporting companies in the state of Sonora has an environmental sustainability of 70%.

2.1 INTRODUCTION

The growing concern for environmental protection and people's healthcare has been reflected in the need to promote sustainable practices in agricultural production. This has led business organizations to seek and adopt strategies that support their efforts in this area, that is, they have acquired certifications, seals, badges, and audits (Contreras, 2019).

The exporting companies of the table grape system in the State of Sonora are submerged in this scenario, where agricultural certifications reflect their commitment and performance in terms of environmental, social, and economic sustainability. Compliance with these dimensions allows companies in this system to move towards an increasingly sustainable organizational structure (Solís et al., 2017).

This research project focuses on the analysis of environmental sustainability based on agricultural certifications. The methodological procedure consisted of applying semi-structured interviews to 14 managers from the same number of table grape exporting companies, which were taken from a universe of 41.

Relevant information for the analysis was compiled, and matrices were elaborated. This allowed the visualization of the following categories of the environmental component of sustainability: fertilizers; flora and fauna care; greenhouse effect gases; garbage management; conventional and hazardous waste; land use and management; plague; water use, energy use; use and management of pesticides; and use and management of chemicals (Terán, 2022).

The objective of this research is to provide a methodological procedure that allows dimensioning environmental sustainability in the Sonoran table grape system. It looks for the incorporation of most of the elements contained in said dimension, which emanate from the demands of a specialized market.

2.2 LITERATURE REVIEW

Research references have mainly focused on agronomic aspects of the agricultural system. In the review of the state of the art, it was found that the following have been studied: factors affecting labor productivity in production (Montaño, 2017), transportation logistics (Hurtado et al., 2018), social and environmental responsibility (Hurtado et al., 2018), agricultural certifications (Contreras, 2019), work methods, and time and motion studies (Montaño & Preciado, 2018), leadership styles and corporate social responsibility (A la Torre, 2018), the role of the market in building sustainable business organizations (Solis et al., 2017), the market structure of the Sonoran table grape system (Velderrain, 2019) in addition to the profitability of grape production systems in Aguascalientes (Borja-Bravo, 2016); however, no research with a methodological procedure that allows sizing environmental sustainability in the Sonoran table grape system has been conducted.

2.2.1 SUSTAINABILITY

The concept of "sustainability" was devised in the Brundtland report by the United Nations, where it was defined as: "meeting the needs of present generations without compromising the ability of future generations to meet their own needs" (WCED, 1987).

Guillén et al. (2020) argue that this concept applied in development promotes a condition of balance, resilience, and interconnection that allows human society to meet its needs, without exceeding the capacity of its ecosystems to renew themselves and to avoid biological diversity decreases.

For Ávila (2018), sustainability is a wake-up call for current generations. The author comments that this concept should represent a warning regarding the current use of the planet's resources, which should be managed in such a way that future generations can develop a standard of living with the same or better possibilities than those managed by the generations of the present.

Zarta (2018) argues that sustainability must be related to a system of life that offers the basic environmental, social, and economic services that all members of a community should receive, but without endangering the viability of the natural and social systems on which this supply of services depends.

Thus, the concept of sustainability should be defined in terms of three basic and interrelated dimensions that constitute complementary aspects for the achievement of the same purpose: economic sustainability, social sustainability, and environmental sustainability (García-Orozco et al., 2020).

An investigation concerning the degree of maturity of sustainability in a company is the one conducted by Baumgartner & Ebner (2010), where they presented the different maturity profiles for some business organizations in Austria, so that they could incorporate sustainable strategies in their operations. The profiles were based on a classification according to the three dimensions of sustainability (social, ecological, and economic), divided in turn into different specifications.

The basic methodology consisted of – (i) defining the concept of sustainable business organizations; (ii) establishing sustainability strategies; (iii) developing sustainability strategies that adhere to the different sustainability profiles; (iv) determining the degree of maturity of the firm in terms of sustainability; and (v) concluding the degree of maturity of the company and designating the corresponding sustainability strategies (Baumgartner & Ebner, 2010).

In this way, they obtained strategies that they called introverted, extroverted, transformative-extroverted, conservative, and visionary. The study constitutes support to guide the process of assessing the sustainable efforts of companies and support the estimation of the degree of sustainable maturity (Baumgartner & Ebner, 2010).

It is evident the effort of researchers to obtain a degree that corresponds to the progress each company has been able to achieve in the different points reviewed, in the three dimensions of sustainability; however, it is important to clarify that these review points are raised in a generalized manner. In this type of evaluation, a checklist with specific review points that leads to a deeper analysis of the company is required to determine the degree of maturity of the business organizations in their transition towards sustainability (Terán, 2022).

On the other hand, Jacobo et al. (2015) reviewed hotels in Ciudad Obregón, Sonora, and determined their performance in sustainability based on economic, social, and environmental dimensions. For its realization, the authors followed the steps mentioned below: (i) literature review; (ii) identification of indicators through a process of operationalization of variables using a matrix associated with sustainable performance and its dimensions; (iii) design of a questionnaire to collect the information; and (iv) concluding whether the establishments are totally sustainable.

Their analysis allowed them to conclude that these hotels are not entirely sustainable. Additionally, it was possible to identify areas of opportunity that hotels should consider improving their performance (Jacobo et al., 2015). Notwithstanding the above, the analysis was mainly based on the following: (a) economic aspect (sales results and profits); (b) social element (access to incentive programs and carrying out charitable activities); and (c) environmental component (separation of solid waste into organic and inorganic).

The above indicates that the study considered only a small part of the review points that should be taken into account in each sustainability area. It also lacks the basis to determine a degree, level, or percentage of sustainability in the hotels reviewed. The research is pragmatic and responds to the needs of the companies reviewed, but it lacks representativeness considering the numerous review points that each sustainability area should encompass (Terán, Robles, & Paz, 2023).

Rueda-Mijangos & Mercado-Salgado (2020) conducted research aiming to identify the link between sustainable performance and the results of excellent management in the context of Mexican stone mines. This study was based on the application of a questionnaire following the standards of the Global Reporting Initiative and the performance indicators of the European Foundation for Quality Management Model.

The procedure of the aforementioned research consisted of the following steps: (i) literature review; (ii) presentation of the results of sustainable performance in stone mines in the State of Mexico under the Global Reporting Initiative (GRI) guidelines in the economic, environmental, and social areas of the company; (iii) collection of primary information on the degree of administrative excellence (or performance) of the mines, through the European Foundation for Quality Management (EFQM) model; (iv) application of the questionnaire to owners, managers, or representatives of stone mines in the State of Mexico; and (v) testing the internal consistency of the sustainable performance scales and the EFQM results block, through Cronbach's alpha (Rueda-Mijangos & Mercado-Salgado, 2020; ANDI, 2022).

In the research project, a score was assigned corresponding to environmentally sustainable performance, and another score was obtained corresponding to economic sustainable performance, but no score related to social sustainable performance was achieved (Rueda-Mijangos & Mercado-Salgado, 2020).

For the research of Rueda-Mijangos & Mercado-Salgado (2020) to be a reference in the sustainability measurement of a company, it is necessary to include the level, degree, percentage, or score of the three major components of sustainability. In this case, the one that was omitted was the social part (Pinedo-Taco et al., 2021).

Up to this level of review, the existing research on the measurement of sustainability and/or sustainable performance of companies is scarce or lacking in representativeness. Some review elements that should be included within the three areas of sustainability have been omitted. It is even possible that one of these areas (economic, social, and environmental) is not taken into account in the review process (Pinedo-Taco et al., 2021).

2.2.1.1 ENVIRONMENTAL SUSTAINABILITY

The term environmental sustainability was introduced because of the Brundtland report, presented by the World Commission on Environment and Development of the United Nations (UN). It was evidenced that human actions generated environmental destruction, increased poverty, and vulnerability of communities (Rodríguez, 2019).

For Schönrock (2020), environmental sustainability revolves around the concepts of management of natural resources and environmental quality. It encompasses aspects related to the balanced management of ecosystems; the preservation of biological diversity; the maintenance of clean environments; the efficient use of natural resources; the importance of creating greater environmental awareness among consumers; and the prevention of factors that lead to climate change.

Oláh (2020) comments that the environment contemplates the operation of natural systems and social systems. The author clarifies that without the support offered by nature, there can be no life in human beings, nor development. In other words, human activities must be carried out within the limits of the assimilation capacity of the environment to maintain the integrity of the processes required for the functioning of an entity.

In this line, Rajala et al. (2016) presented a study in the manufacturing industry in Finland and referred that environmental sustainability should be incorporated in companies to obtain a better development of their products and business operations, increase sales, discover new markets, improve corporate image, achieve product differentiation, and achieve a higher return on investment.

In this study, the principles of reduce, reuse, recover, recycle, and redesign were evidenced, improving the materials recycling process, eliminating dependence on fossil fuels, and trying to find new technologies for production. Because of these new values and behaviors, the collective identity of the company was changing, entering a new culture of business sustainability (Rajala et al., 2016).

Notwithstanding the above, this study lacks an evaluation of the initial level of sustainability of the companies, and the level acquired by the companies once they incorporated the principles.

Franco (2021) elaborated his research based on the quantitative evaluation of the environmental sustainability of agricultural activities on a local scale. To compare the supply and demand of natural resources in a local agricultural system, a methodology based on the ecological footprint approach was adopted.

The methodological process was based on three main phases: (i) assessment of total biocapacity (BCTOT), based on the different soil types within the agricultural area of the region: Cropland under agricultural production (BCC) and other bio-productive areas: Forest land, grazing land, built-up land, other arable land (BCA); (ii) evaluation of the total ecological footprint (EFTOT), caused by the different production processes (crops and livestock) and other management work directly linked to agricultural activity; and (iii) calculation of the ecological balance (EB = BCTOT – EFTOT), which represents the value of the synthetic indicator capable of establishing the sustainability of the local agricultural system (Franco, 2021).

The author assures that the results of the study offer starting points for future research on the possibility of defining whether an agricultural production system is "sustainable," and likewise, he says the investigation allows the introduction of a regional brand of sustainability as a key to local development (Franco, 2021).

Hashemi et al. (2022) explored in their research the long-term effects of agricultural production on the sustainability of blue water consumption in northern Iran; the study was conducted with reference to two periods: before (1984–1997) and during (1998–2018) the operation of the Tajan irrigation and drainage network (TIDN).

The authors applied a combination of different methods to estimate hydrological EFRs of rivers (Environmental Flow Requirements), traditional water reservoirs, and groundwater resources. To evaluate water quality, they focused on three main contaminants in the region's water resources, including nitrogen, phosphorus, and salinity (Hashemi et al., 2022).

The researchers calculated monthly agricultural water footprints through the application of the AquaCrop model and then compared them to the region's BWA (Blue Water Availability), which was calculated by subtracting the monthly EFR from the monthly natural flow. The logical sense of the decision taken by the authors was as follows: if the water flow exceeds the blue water availability, the production system should be considered as having unsustainable water consumption (Hashemi et al., 2022).

The methodological process used by the researchers was as follows: (i) definition of the study area; (ii) quantification of water flow; (iii) calculation of EFRs and violations related to the limits previously established in the law; and (iv) determination of groundwater EFRs and violations related to the limits indicated by the law (Hashemi et al., 2022).

What these researchers also sought to demonstrate in their research was that sustainable food security depends primarily on a redefinition of agricultural systems in hydrological units, and that this must consider the sustainability of components that are entirely natural, such as water resources (Hashemi et al., 2022).

The literature reviewed so far indicates that existing research on the calculation of a level or degree of environmental sustainability is limited, and those that exist consider methodologies that are commonly based on only one aspect of the environment, such as the ecological footprint (Franco, 2021). Other authors rely on considering minimal aspects, such as the loss of fertile soil, deforestation, and resource depletion (Vera et al., 2019).

From the above, it is evident the need to include in the calculation of a level or degree of environmental sustainability, attributes such as: care of flora and fauna (Meza & Rodríguez, 2021); conventional waste and hazardous waste management (Castro-Garzón, Contreras, & Rodríguez, 2020); water use (Hashemi et al., 2022); energy use (Fernandes et al., 2020); pesticide use and management; and chemical use and management, among others (Jáquez-Matas et al., 2022).

2.2.2 AGRICULTURAL SUSTAINABILITY

Agriculture remains the backbone of global development, providing humans with food and other goods. However, it is also one of the main reasons why the use of natural resources has been extended beyond its limits, confirming the pressure exerted by modern agriculture on sustainability (Awan et al., 2021).

Agricultural activity processes have been oriented to supply the food needs of the population; generate an important economic flow; and attend consumers who have changed their consumption patterns. Thus, these production systems seek the production of agricultural products that have the nutritional characteristics demanded by consumers, but also, that have practices oriented to the protection of the environment and the care of people's health (Arias, Rodríguez, & Beduschi, 2021).

Therefore, any route that is chosen to improve the global scenario must drastically reduce the environmental footprint of agriculture. Although stakeholders across the globe, from farmers to international decision-makers, unanimously agree on the broad goals of sustainability, the reality is that a controversy prevails, and that is that there is no orderly approach to ideas on how these goals should be achieved (Acuña & Ccasani, 2022).

Although there is no unified and agreed definition of agricultural sustainability, it can be said that the idea starts from the ethical principle of sustainable development. The concept was devised in the Brundtland Report in 1987. That is, agricultural sustainability must be able to meet the needs of current generations, without compromising the ability of future generations to meet their own needs. It must consider the three pillars of sustainability: environmental, social, and economic (Rizo-Mustelier, Vuelta-Lorenzo, & Lorenzo-García, 2017).

Considering the above, the Food and Agriculture Organization of the United Nations (FAO) defines agricultural sustainability as the management and conservation of natural resources and the orientation of technological and institutional change to ensure the achievement and continuity of the satisfaction of human needs, for the present and the future. Therefore, land, water, plant, and animal genetic resources must be conserved, and their activities must be directed to be environmentally friendly, technically appropriate, economically viable, and socially acceptable (Oberč & Arroyo-Schnell, 2020).

In this regard, Soulé et al. (2021) conducted an in-depth analysis of the conceptual and methodological framework behind environmental sustainability assessment methods in agriculture. They analyzed all the environmental sustainability assessment methods they found in their review and proposed a hierarchical classification of the methods according to their contribution. For example, whether they are for providing knowledge, for raising awareness, or for ranking and selecting options in problem-solving.

The methodological process they used consisted of the following stages: (i) systematic review in agricultural sciences, derived from PRISMA (Preferred Reporting Items for Systematic Reviews and Meta-Analyses); (ii) establishment of the use of the terminology of "dimension/theme/sub-theme/indicator," which is widely used and easy to understand; (iii) characterization of the main themes; and (iv) a multivariate analysis was performed to look for trends and clusters (Soulé et al., 2021).

The research results in 4 clusters where methods are classified according to the action they address and the issues they deal with. The researchers speak of emerging issues that should be included in a review method to assess environmental sustainability, for example: ecosystem services (mostly related to soil), organic carbon and nitrogen (mostly related to water), the relationship between indicators, and the scales established (Soulé et al., 2021).

On the other hand, Pin, Corzo, & Rezabala (2021) proposed as an objective in their research to determine, through social, economic, and environmental indicators, the agro-productive sustainability in the savanna association of the Florón sector of the Portoviejo canton in Ecuador. To achieve this, the following methodological procedure was used:

- A non-experimental, exploratory design with descriptive scope was used.
- The questionnaire was developed on the QuestionPro website to be applied virtually. The reliability and validity of the survey were determined by means of Cronbach's alpha coefficient in SPSS software, with a margin of error of 0.05% and a confidence level of 95%.
- The survey was sent electronically to the sample under study.
- Observation was carried out in person during the farmers' work schedules. Investigators observed the interaction between farmers, the type of production, their production levels, how they carried it out, and the activities they had in the environment.
- Subsequently, the analysis of the methodology was applied according to the criteria for the evaluation of management systems integrating sustainability indicators (MESMIS).

Researchers presented the most relevant conclusions of their study – the existence of gaps and inequities in the social, economic, and environmental spheres; insufficient financial support from authorities of local and regional agricultural institutions; lack of access to water for year-round cultivation;

and lack of farmers' training in environmental issues; among others (Pin, Corzo, & Rezabala, 2021).

The literature reviewed in this section presents research that analyzes the conceptual basis and methodological procedure of some models for evaluating environmental sustainability in agriculture. However, they do not propose holistic methodologies that contemplate a large part of the components of the environmental scope of sustainability (Lagunas-Vázquez et al., 2017).

2.2.3 ENVIRONMENTAL SUSTAINABILITY IN THE SONORAN TABLE GRAPE SYSTEM

In recent years, agricultural companies have been pressured by consumers because of the principles of environmental and social sustainability that must exist in their practices. With this, a new stream of business activities arises. In this context, companies and their owners pursue, in addition to economic prosperity, social and ecological objectives. This is the case of the export business organizations that belong to the table grape system in Sonora (Terán, 2022).

One of the studies in this line is presented by Aranda (2016), who sought to identify how leadership with a gender perspective contributes to the construction of sustainable companies in the Sonoran table grape system. The methodological basis used to meet this objective and verify this assumption was based on direct interviews with managers of table grape-producing companies. Companies were selected through a statistical sample with 95% confidence and 7.5% error, resulting in a group of 13 subjects. It was concluded that leaders show knowledge of the markets, understand their demands, and internalize them in their companies.

The study shows the main requirements of consumers are those related to quality and safety, such as the size, color, and sugar level of the grape. At the same time, they demand the fruit to be free of pesticide and agrochemical residues that may affect the health of consumers. The degree of perception of leaders has led to adjustments in business towards more systemic and sustainable structures. They also allow their employees to participate in decision-making, explain to team members what is expected from them, and place their trust by delegating responsibilities to them (Aranda, 2016).

Similarly, A la Torre et al. (2018) carried out a study that aimed to identify the leadership styles that enable Sonoran table grape producing companies to incorporate corporate social responsibility (CSR) criteria. The research adopted a descriptive cross-sectional design, using a questionnaire to survey managers of grape-producing companies.

The results indicated that transformational and transactional leadership styles lead to higher levels of CSR, a finding that aligns with the literature reviewed by the researchers. This, in turn, confirms that these styles help companies comply with the environmental and social principles required by new market demands (A la Torre et al., 2018).

In the review of the state of the art, particularly in research related to the Sonoran table grape system, it was found that the following has been studied: the factors that affect labor productivity in agricultural production (Montaño et al., 2018), transport logistics (Hurtado, Preciado, & Bañuelos, 2018), social and environmental responsibility (Terán, 2022), agricultural certifications (Contreras, 2019), work methods and leadership styles (A la Torre et al., 2018), and the role of the market in building sustainable companies (Solís et al., 2017). However, no research oriented towards the formulation of a methodology that contributes to the knowledge of the degree of environmental sustainability of the Sonoran table grape system has been carried out.

The present investigation seeks to propose a methodological procedure to measure environmental sustainability in the companies that belong to the Sonoran table grape system. This starts from a conceptual framework related to environmental sustainability in that system, founded on compliance with certifications, seals, badges, and audits. The foregoing, because the procedure is intended to incorporate most of the elements contained in the environmental dimension of sustainability, which are based on the demands of a specialized market.

2.3 METHODOLOGICAL PROCEDURE

The present work consisted of the application of a semi-structured interview script to a sample of 14 managers from the same number of companies. These were obtained from a universe of 41 business organizations distributed in the regions of Caborca, Pitiquito, Altar, Hermosillo, Carbó, San Miguel de Horcasitas, Guaymas, and Empalme. The sample size (n) was estimated using the following formula for finite populations:

$$n = \frac{K^2 pqN}{(E^2(N-1)+K^2 pq} \tag{1}$$

where; 'n' is the sample size; 'N' is the size of the population; 'K' is the desired level of reliability; 'E' is the estimation error; 'P' is the probability of being chosen; 'q' is the probability of not being chosen.

Substituting it results in:

$$n=\frac{1.96^2(41)(0.97)(0.03)}{[0.075^2(41-1)]+[1.96^2(0.97)(0.03)]}$$

An n = 13.60 is obtained and rounded to 14 business organizations with a reliability of 95% and 7.5 error. Companies' managers were asked about certifications to complement the interviews. The options were:

PRIMUS GFS, USDA ORGANIC, GLOBAL GAP, FAIR TRADE, C-TPAT, MEXICO CALIDAD SUPREMA, SENASICA, SMETA DEALTI, DEAR, KROGER, and RAINFOREST.

The matrix containing this information was prepared (Table 2.1). A binary assignment of 1 and 0 was carried out (0 was assigned to companies that did not comply with the corresponding certification, and 1 to those that did).

The review mechanism for each company was based on the implemented certifications, and it was carried out as follows:

- A list of attributes (verification points) was drawn up in Excel, considering the certifications in the three main areas of sustainable development.
- 353 environmental attributes were grouped as follows: pests, external matter, and crop diseases; use and management of chemicals; use and management of pesticides; use and management of soils; use and management of fertilizers; use of water; use of energy; air pollution; care of flora and fauna; management of garbage, conventional and hazardous waste; and greenhouse gases.
- 584 attributes of the social part were also grouped into the following sets: care for workers' rights (social justice); safety and occupational health and hygiene; training and education; hazard analysis and critical control points (HACCP); food safety and sanitation; organic practices; safe and clean conditions in facilities; protectors and preservatives; and audits, inspections, and monitoring.
- 12 economic attributes were clustered into the following groups:
    - Generalities (e.g., level of established salary, frequency of payroll, withholdings, deductions, etc.); and
    - Particularities (e.g., economic capacity to pay for children's education, to acquire a house or the facilities to build it, and to make the payment of electricity, water, and telephone services).
- The matrix of categories and attributes of the three dimensions of sustainability was prepared (Table 2.2) and it was based on the three

dimensions of sustainable development, their categories, and attributes. The result was 21 categories and 949 attributes. The number 0 was given to those certifications that did not maintain any attribute, and 1 to those that did.

- The matrix of categories and attributes of environmental sustainability (Table 2.3) was made. It is important to comment that this research focuses only on the environmental dimension of sustainability.

  These categories and attributes were identified and distributed as follows:

  - Water use: 68 attributes.
  - Fertilizers: 63 attributes.
  - Conventional and hazardous waste management: 50 attributes.
  - Soil management and use: 51 attributes.
  - Pesticides use and management: 39 attributes.
  - Pests, external matter, and crop diseases: 35 attributes.
  - Chemicals use and management: 17 attributes.
  - Care of flora and fauna: 23 attributes.
  - Energy use: 6 attributes.
  - Greenhouse gases: 1 attribute.

  Number 0 was assigned to certifications that did not maintain any attribute, and 1 to those that did.

- Finally, the average of the percentages obtained by each company in the categories is calculated according to the environmental sustainability attributes addressed, which allows determining the progress of each business organization towards environmental sustainability, as well as the progress of the entire table grape system.

This section focuses on the primary research proposal, which lists the phases to be followed to measure the level of environmental sustainability of the companies in the Sonoran table grape system. It provides a perspective of the commitment and performance in terms of environmental sustainability practices that these companies maintain.

2.4 RESULTS

The percentages achieved by the business organizations of the Sonoran table grape system (hereafter ORG 1, ORG 2, ORG 3, ORG 4, ORG 5, ORG 6, ORG 7, ORG 8, ORG 9, ORG 10, ORG 11, ORG 12, ORG 13, and ORG

TABLE 2.1 Matrix of Certifications

ORGANIZATION	CERTIFICATIONS										
	PRIMUS GFS	CCOF	GLOBAL GAP	FAIR TRADE	MEXICO C.S.	SENA SICA	SMETA	DEALTI	DEAR	KROGER	RAIN FOREST
ORG1	1	1	1	1	1	1	0	1	0	0	0
ORG2	0	0	1	0	0	1	1	0	0	0	0
ORG3	0	0	1	0	0	0	1	0	0	0	0
ORG4	1	1	1	1	1	1	1	0	1	1	1
ORG5	1	1	1	1	0	0	1	0	0	0	0
ORG6	1	0	0	0	0	1	1	0	0	0	0
ORG7	1	1	1	1	0	1	1	1	0	0	0
ORG8	0	0	1	0	1	1	0	1	0	0	1
ORG9	1	1	1	1	1	1	1	1	0	1	1
ORG10	1	0	1	1	0	1	1	1	0	0	0
ORG11	1	0	0	0	0	0	0	0	0	1	0
ORG12	0	0	1	0	0	0	1	0	0	1	0
ORG13	1	1	1	0	0	0	1	0	0	0	0
ORG14	1	1	1	0	0	0	0	0	0	0	0

TABLE 2.2 Matrix of Categories and Attributes of the Three Dimensions of Sustainability

DIMENSIONS	CATEGORIES	ATTRIBUTES	CERTIFICATIONS										
			PRIMUS	CCOF/ USDA	GLOBAL GAP	FAIR TRADE	MÉXICO C.S.	SENASICA	SMETA	DEALTI	DEAR	KROGER	RAIN FOREST
Environmental	USE AND MANAGEMENT OF PESTICIDES	A RECORD OF EXPIRED PESTICIDES SHOULD BE KEPT	0	0	0	0	1	1	1	0	0	0	0
Environmental	LAND USE AND MANAGEMENT	RISK ASSESSMENT (IF THERE WAS ANIMAL FARMING)	1	1	1	0	0	0	1	0	0	0	0
Social	CARE FOR THE RIGHTS OF WORKERS (SOCIAL JUSTICE)	WORKERS MUST RECEIVE WRITTEN INFORMATION ABOUT THEIR RIGHTS	0	0	0	1	0	0	0	0	0	0	0
Social	SAFETY, OCCUPATIONAL HEALTH AND HYGIENE	ELECTRICAL SYSTEMS MUST INCLUDE CIRCUIT BREAKERS OR FUSE BOXES	0	0	0	0	0	0	0	0	0	1	0
Economic	GENERALITIES	WORKERS MUST BE PAID AT LEAST ONCE A MONTH	0	0	0	1	0	1	1	0	0	0	1
Economic	GENERALITIES	WAGE DEDUCTIONS ARE NOT TO BE USED AS A DISCIPLINARY MEASURE.	0	0	0	1	0	1	1	0	0	1	0
Environmental	C..	A...	...	...	...	...	...	...	...	...	...	...	...
Social	C..	A...	...	...	...	...	...	...	...	...	...	...	...
Economic	C..	A...	...	...	...	...	...	...	...	...	...	...	...
Total	**Total Categories (21)**	**Total Attributes (949)**	**155**	**263**	**217**	**89**	**223**	**249**	**237**	**44**	**47**	**165**	**163**

TABLE 2.3 Matrix of Categories and Attributes of the Environmental Dimension of the Certifications

			CERTIFICATIONS										
DIMENSIONS	CATEGORIES	ATTRIBUTES	PRIMUS	CCOF/ USDA	GLOBAL GAP	FAIR TRADE	MÉXICO C.S.	SENASICA	SMETA	DEALTI	DEAR	KROGER	RAIN FOREST
Environmental	USE AND MANAGEMENT OF PESTICIDES	A RECORD OF EXPIRED PESTICIDES SHOULD BE KEPT	0	0	0	0	1	1	1	0	0	0	0
Environmental	LAND USE AND MANAGEMENT	RISK ASSESSMENT (IF THERE WAS ANIMAL FARMING)	1	1	1	0	0	0	1	0	0	0	0
Social	CARE FOR THE RIGHTS OF WORKERS (SOCIAL JUSTICE)	WORKERS MUST RECEIVE WRITTEN INFORMATION ABOUT THEIR RIGHTS	0	0	0	1	0	0	0	0	0	0	0
Social	SAFETY, OCCUPATIONAL HEALTH AND HYGIENE	ELECTRICAL SYSTEMS MUST INCLUDE CIRCUIT BREAKERS OR FUSE BOXES	0	0	0	0	0	0	0	0	0	1	0
Economic	GENERALITIES	WORKERS MUST BE PAID AT LEAST ONCE A MONTH	0	0	0	1	0	1	1	0	0	0	1
Economic	GENERALITIES	WAGE DEDUCTIONS ARE NOT TO BE USED AS A DISCIPLINARY MEASURE.	0	0	0	1	0	1	1	0	0	1	0
Environmental	C..	A...	...	...	...	...	...	...	...	...	...	...	...
Social	C..	A...	...	...	...	...	...	...	...	...	...	...	...
Economic	C..	A...	...	...	...	...	...	...	...	...	...	...	...
Total	**Total Categories (21)**	**Total Attributes (949)**	**155**	**263**	**217**	**89**	**223**	**249**	**237**	**44**	**47**	**165**	**163**

14) are presented in this section. This result is supported by the analysis of Tables 2.1–2.3, as well as the compliance of the companies with each of the environmental sustainability categories.

2.4.1 *ENVIRONMENTAL SUSTAINABILITY OF TABLE GRAPE SYSTEM BUSINESS ORGANIZATIONS*

Figure 2.1 shows that business organizations 4 and 9 meet all 68 attributes concerning water use, according to the certifications they have. This means they have made progress in environmental sustainability, based on the certification's attributes. On the other hand, ORG 11 obtained 29% of the attributes, so it is necessary to improve water management and reduce its impact on the environment.

ORG 7 and ORG 10 are also found with 94% of the attributes, followed closely by ORG 1 with 88%. ORG 5, ORG 6, ORG 2, and ORG 8 obtained percentages between 71% and 79%. Among the companies with moderate scores were ORG 3, ORG 12, and ORG 13 with 54% of the attributes, while ORG 14 obtained 47%.

The appropriate and responsible use of water in agriculture is essential to ensure the sustainability of water resources and the quality of the environment. When using recycled water, it is crucial that the processes are under the supervision of a competent authority; it is also important to comply with local and national regulations (USDA, 2021; Mexico Calidad Suprema, 2021; SENASICA, 2021).

It is essential to carry out and document risk assessments that consider, among other aspects, the identification of water sources and their historical results, application methods, timing of water use, water contact with the crop and growth stage, as well as the purity of the water used in applications of phytosanitary products (USDA, 2021; Mexico Calidad Suprema, 2021; SENASICA, 2021).

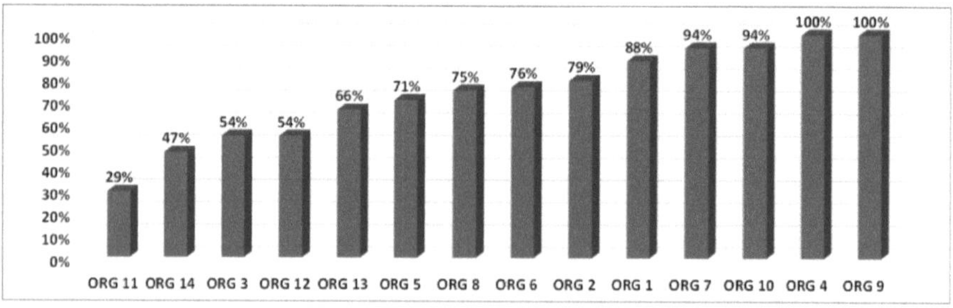

FIGURE 2.1 Water use.

On the other hand, Figure 2.2 shows a wide variety of percentages in the use of fertilizers for different companies. Business organizations 1, 4, 7, and 9 meet all 63 attributes denoted by the certifications, demonstrating compliance with environmental sustainability, which in this category is 100%. ORG 11 only complies with 10 of these attributes, representing 16% compliance.

ORG 6, 2, 8, 14, 5, 13, and 10 achieved percentages between 60% and 81%. It is assumed they are trying to continue implementing strategies to make their agricultural practices more sustainable. Finally, there are business organizations 3 and 12, which have the lowest levels, so it is necessary for them to improve practices related to the use of fertilizers.

The proper use and management of fertilizers in agriculture is essential to ensure environmental sustainability and production efficiency. It is necessary to keep detailed records of all fertilizer applications, including geographic location, trade name, type and concentrations, amount of product applied, method and/or equipment used, and the name of the operator who made the application (Primus GFS, 2021; USDA, 2021; Mexico Calidad Suprema, 2021; SENASICA, 2021).

Fertilizer storage should be in properly ventilated areas free of rainwater or condensation. All fertilizers should be stored in a manner that minimizes the risk of contamination to water sources (GlobalGAP, 2014, 2017; Primus GFS, 2021; USDA, 2021; Mexico Calidad Suprema, 2021; SENASICA, 2021).

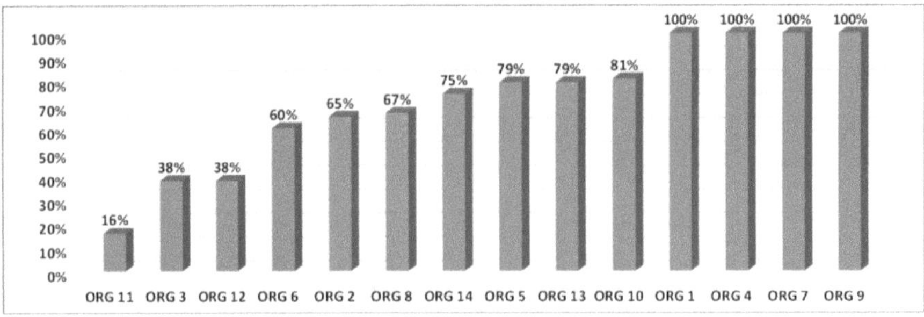

FIGURE 2.2 Fertilizers.

Figure 2.3 shows ORG 4 and ORG 9 comply with all 50 attributes related to conventional and hazardous waste management, which means their progress towards sustainability in terms of these attributes and certifications is 100%. On the other hand, ORG 11 only satisfies 7 of the attributes, which means it only reaches 14% in this category.

ORG 14 has 40%, which shows that its efforts to adopt environmental protection practices are just beginning. Business organizations 3, 12, 8, 13, 2, 6, 1, 5, 7, and 10, acquired percentages between 72% and 94%, that is because they already have worked on implementing more sustainable agricultural practices.

Proper management of garbage, conventional, and hazardous waste requires a comprehensive, current, and documented plan that addresses waste and pollution reduction, as well as the implementation of recycling. Organic waste should be composted to condition the soil, while contaminated machinery wash water should be collected and disposed of with responsibility (RainForest Alliance, 2020; Primus GFS, 2021; USDA, 2021; Mexico Calidad Suprema, 2021; SENASICA, 2021).

Empty pesticide containers and hazardous materials should be washed, punctured, and not reused. Pesticide residues must not exceed established maximum limits and must be analyzed at least once a year in official or recognized laboratories (GlobalGAP, 2014, 2017; Primus GFS, 2021; USDA, 2021; Mexico Calidad Suprema, 2021; SENASICA, 2021).

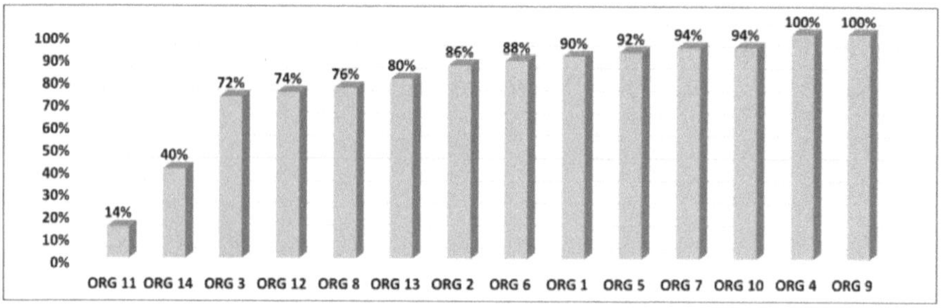

FIGURE 2.3 Conventional and hazardous waste management.

Figure 2.4 shows ORG 4 and ORG 9 meet all the 51 criteria related to land management and use that are included in the certifications; in other words, they obtained a 100% compliance rate in this aspect. In contrast, ORG 11 only complies with 14 of these standards, resulting in a compliance rate of 27% in this category. On the other hand, ORG 3, ORG 12, ORG 14, and ORG 13 obtained percentages between 27% and 51%. They are currently initiating the implementation of environmental sustainability strategies. ORG 6, ORG 2, ORG 8, and ORG 5 are between 59% and 71%; they have progressed in achieving environmental sustainability. Finally, ORG 10, ORG 1, and ORG 7 reached percentages between 92% and 96%, which means efforts to implement practices that protect the environment are yielding results.

Management and use of soils in agriculture require a risk management plan that addresses aspects such as proximity to animal operations, composting, water intrusion, and dust. An environmental impact analysis should include the history of the area, water, and soil management (GlobalGAP, 2014, 2017; FairTrade, 2020; Primus GFS, 2021; USDA, 2021; Mexico Calidad Suprema, 2021; SENASICA, 2021).

It is essential to analyze productive zones considering factors such as altitude, climate, water availability, luminosity, winds, soil quality, and topography. A soil analysis should include texture, physicochemical, phytopathological, and microbiological characteristics, as well as topography and contour lines (GlobalGAP, 2014, 2017; Rainforest Alliance, 2020; USDA, 2021; Mexico Calidad Suprema, 2021; SENASICA, 2021).

There must be written justification for the use of soil fumigants, and the use of methyl bromide is not allowed. Finally, programs must be established to prevent soil erosion through practices such as mulching, drainage systems, and minimum tillage (GlobalGAP, 2014, 2017; USDA, 2021).

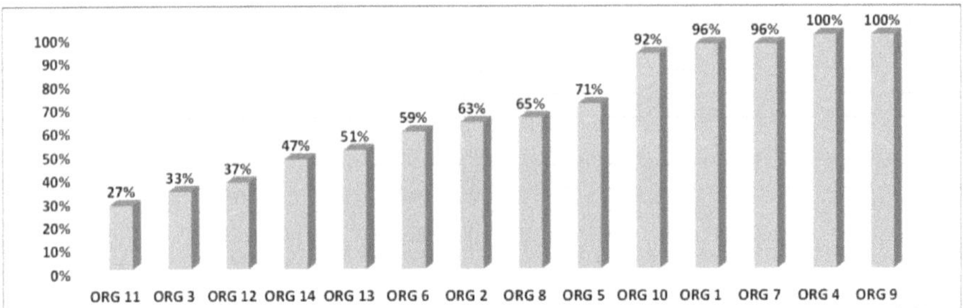

FIGURE 2.4 Soil management and use.

Figure 2.5 shows business organizations 10, 1, 7, 4, and 9 meet all 39 attributes concerning pesticide use and management, which denotes a level of 100% compliance in this category of certifications.

ORG 11 and ORG 14 have no attributes in this category. ORG 3, ORG 12, and ORG 13 have a percentage of 56%, indicating they are still implementing pesticide management measures. Regarding business organizations 5, 8, 6, and 2, they reached a percentage between 67% and 90%, which means they retain an approach that aims to be more sustainable in the use and management of pesticides.

The use and management of pesticides in agriculture must be rational and justified, focusing on selective products with low impact on wildlife and human health. It is necessary to comply with national regulations and follow labeling recommendations. Likewise, application equipment should be kept in good condition and calibrated annually (Mexico Calidad Suprema, 2021; SENASICA, 2021; RainForest Alliance, 2020).

Protective clothing and equipment should be stored separately from pesticides, and those responsible for checking and adjusting equipment should be competent. Detailed records of pesticide applications must be kept, and products that are not legally approved in the country must not be used (Mexico Calidad Suprema, 2021; SENASICA, 2021).

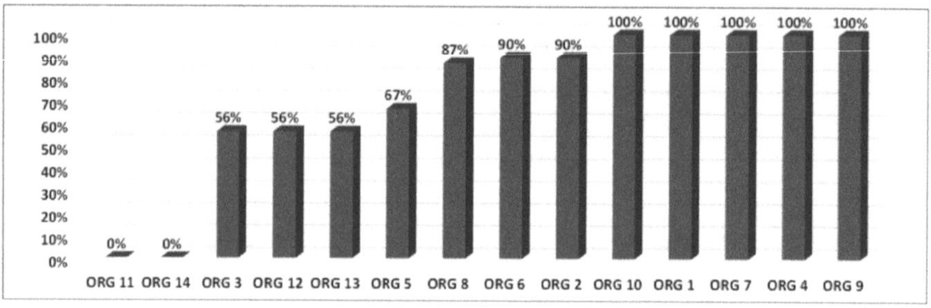

FIGURE 2.5 Pesticides use and management.

Figure 2.6 shows ORG 4 and ORG 9 comply with all 35 attributes related to pests, external matter, and crop diseases, according to the certifications they have, i.e., their progress in environmental sustainability is the highest since they obtained 100% of the attributes.

In contrast, ORG 11, ORG 3, and ORG 12 only comply with 9 of the mentioned attributes, which translates into 26% of environmental sustainability. On the other hand, ORG 6 and ORG 2 obtained a percentage of 49% and 51%, respectively, which indicates they are implementing measures to control the mentioned attributes.

ORG 8, ORG 13, and ORG 14 obtained percentages between 63% and 71%, which means they have implemented effective measures to reduce the environmental impact associated with the management of pests, external matter, and crop diseases. Finally, ORG 5, ORG 10, ORG 1, and ORG 7 with percentages of 74%, 77%, 94%, and 94%, respectively, demonstrated they are making progress in environmental sustainability.

An effective pest control program should be implemented in production, storage, and packaging areas to avoid contact between control devices and food. The producer must demonstrate the implementation of production practices that reduce pest incidence and the need for intervention (GlobalGAP, 2014, 2017; USDA, 2021; Rain Forest Alliance, 2021).

It is advisable to prioritize non-chemical pest control and follow anti-resistance strategy recommendations. If biological agents are used for pest control, they should be used in accordance with current legislation (Mexico Calidad Suprema, 2021; SENASICA, 2021).

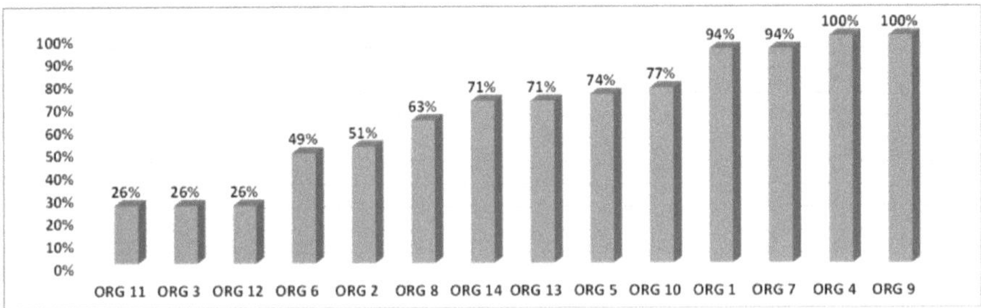

FIGURE 2.6 Pests, external matter, and crop diseases.

On the other hand, Figure 2.7 shows that organizations 4 and 9 fully comply with the 17 criteria associated with the use and management of chemicals, according to the certifications they hold, which translates into an advanced level of environmental sustainability, since they obtained 100% of the attributes.

Companies with the lowest percentage in this aspect are ORG 6 and ORG 8 with 59%, followed by ORG 11 with 65%, and ORG 2 and ORG 3 with 71%. ORG 1, ORG 5, ORG 7, ORG 10, ORG 13, and ORG 14 achieved 76%, and ORG 12 achieved 94%. Data suggest most of the companies have adequate chemical use and management, which contributes positively to their transition towards environmental sustainability.

Chemicals use and management in agricultural production require the availability of technical safety data sheets and a clear index for consultation. Producers should have an accident procedure that includes detailed information and emergency contact numbers. Chemical storage areas should have eyewash facilities, clean water sources, and first aid kits (GlobalGAP, 2014, 2017; USDA, 2021; Kroger, 2021).

Chemicals should be stored securely in designated and labeled areas, with copies of substance labels available. In addition, a chemical inventory and usage logbook must be kept. To prevent chemical contamination of products, cleaning agents and lubricants should be stored in designated secure areas away from the product (GlobalGAP, 2014, 2017; USDA, 2021; Kroger, 2021).

Figure 2.8 shows ORG 4 and ORG 9 satisfy all 23 attributes associated with the care of flora and fauna, according to their certifications. This means their journey towards environmental sustainability in this category is the most advanced because they obtained 100% compliance in the attributes. In contrast,

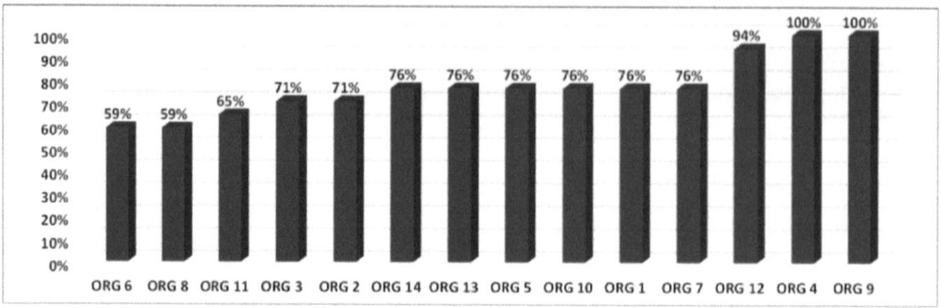

FIGURE 2.7 Chemicals use and management.

ORG 11 does not comply with any environmental sustainability attribute in this category.

On the other hand, ORG 6, ORG 3, ORG 12, ORG 13, ORG 14, and ORG 2 achieved percentages between 17% and 30%, followed by ORG 5 with a percentage of 52%. It is also evident that business organizations 10, 1, 7, 8, 4, and 9 show a percentage higher than 60%, which indicates significant work in the implementation of practices aimed at addressing these attributes.

It is essential to have a written action plan that aims to improve habitats and maintain biodiversity in agricultural fields. It is important to implement reforestation and natural regeneration plans in unproductive or special areas. Producers should analyze how to improve the environment for the benefit of the local community (GlobalGAP, 2014, 2017; USDA, 2021; SENASICA, 2021; Rainforest Alliance, 2021).

An environmental impact analysis focused on the protection of fauna and flora should be carried out. Local zoning plans and existing regulations should be consulted during crop planning. Similarly, the conversion of unproductive sites into ecologically focused areas to encourage natural flora and fauna should be considered (GlobalGAP, 2014, 2017; USDA, 2021; SENASICA, 2021; Rainforest Alliance, 2021).

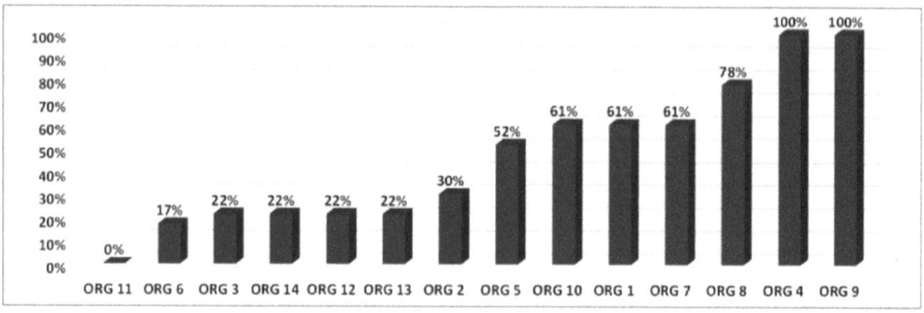

FIGURE 2.8 Care of flora, and fauna.

Figure 2.9 shows ORG 3, ORG 12, ORG 13, ORG 2, ORG 5, ORG 10, ORG 7, ORG 4, and ORG 9 meet all six attributes associated with energy use, which means a great advance towards environmental sustainability since they met 100% of the attributes.

In contrast, ORG 11 does not have any environmental sustainability attribute achieved in this category. On the other hand, ORG 1, ORG 8, and ORG 14 have a percentage of 50% and ORG 6 has 67%, which indicates they are implementing improvements to achieve energy efficiency.

It is essential to have records of energy consumption (invoices) in the fields and agricultural practices, so the producer knows how and where this consumption is carried out. An energy efficiency improvement plan should include minimizing the use of non-renewable energy (GlobalGAP, 2014, 2017; USDA, 2021).

It is important to investigate renewable energy opportunities, such as solar, wind, geothermal, and biomass. It is recommended to map energy use over time as a proportion of production and set targets for reduction (GlobalGAP, 2014, 2017; USDA, 2021).

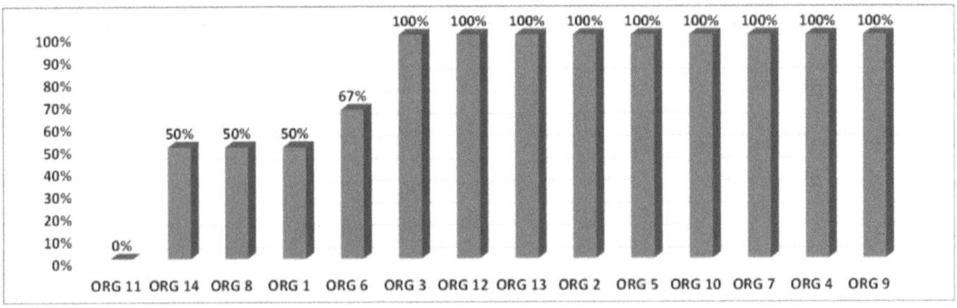

FIGURE 2.9 Energy use.

Figure 2.10 indicates that only ORG 8, ORG 4, and ORG 9 comply with the single category related to greenhouse gas emission control, so they have managed to move towards environmental sustainability by obtaining 100% compliance in this category. The rest of the companies did not meet this category.

Producers are responsible for documenting net greenhouse gas (GHG) emissions from the main sources in their production and processing operations. These emissions come from different activities, including fossil fuel and electricity use, fertilizer application, waste management, wastewater treatment, and land use change (Rainforest Alliance, 2021).

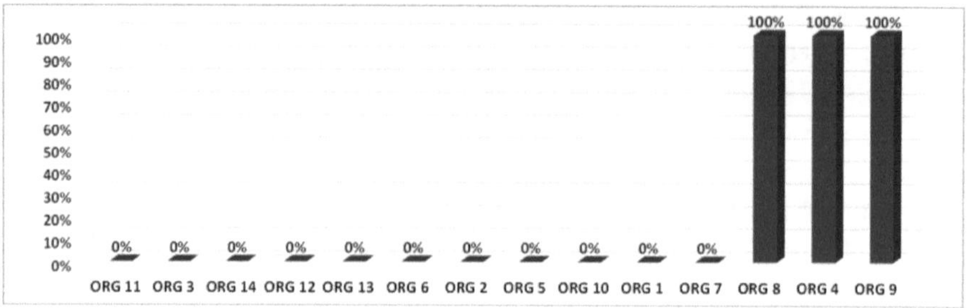

FIGURE 2.10 Greenhouse gases.

2.4.2 ENVIRONMENTAL SUSTAINABILITY OF THE TABLE GRAPE SYSTEM

The percentage of environmental sustainability based on the attributes denoting agricultural certifications is a key indicator to evaluate the commitment and compliance of companies in caring for the environment and implementing sustainable agricultural practices.

The following is an analysis of the environmental sustainability levels of the companies and the systems to which they belong, based on the agricultural certifications they hold. ORG 11 presents a 20% compliance in the 353 attributes, which indicates there is a need to increase efforts to acquire certifications that support their commitment to the environment.

ORG 3 and ORG 14 show a proportion of 48%, which suggests they have achieved moderate progress in the implementation of sustainable agricultural practices; however, they could still improve and obtain additional certifications that support their progress in becoming more environmentally sustainable.

With a value of 50%, ORG 12 has achieved considerable progress in terms of compliance with agricultural certifications. This indicates there is already a solid foundation that motivates the company to improve its practices and achieve higher levels of environmental sustainability.

ORG 13 and ORG 6 have achieved a grade of 66%, demonstrating a strong commitment to sustainable agricultural practices and the possibility of growth in terms of agricultural certifications. ORG 2 and ORG 8 present values of 70% and 71% respectively, that is, they maintain a responsible approach oriented towards compliance with environmental sustainability through their agricultural practices.

With results of 75% and 87%, ORG 5 and ORG 10 prove the commitment they have to environmental sustainability and the implementation of responsible agricultural practices, through the certifications they manage. ORG 1 and ORG 7 have reached 90% and 93% respectively and stand out for their good performance in the field of environmental sustainability, due to the timely implementation of agricultural certifications that are aimed at protecting the environment and human health.

Finally, ORG 4 and ORG 9 have reached a level of compliance of 100% of the attributes, i.e., they have a degree of excellence in environmental sustainability. These companies have been pioneers in the adoption of sustainable agricultural practices and the implementation of the most relevant agricultural certifications for decades.

Both business organizations represent the ideal model of a company that is supported by practices aimed at protecting the environment and caring for people's health. They constantly seek to acquire the certifications that are necessary to confirm the commitment they have acquired before society, which is to proceed responsibly with natural resources (Figure 2.11).

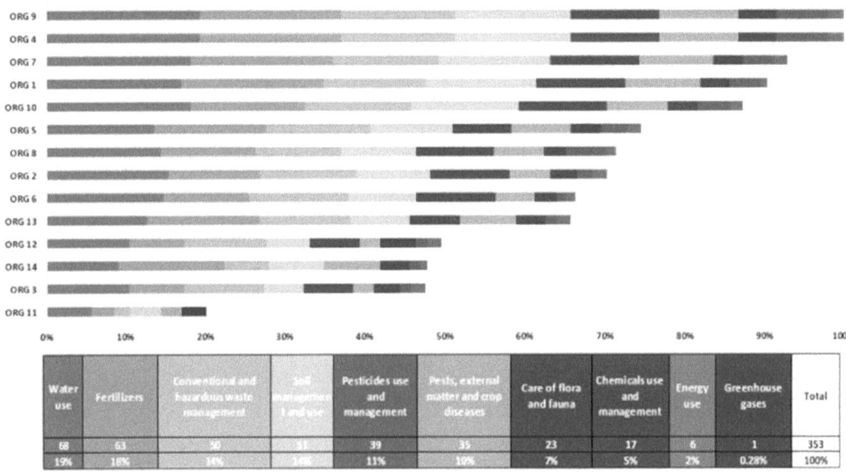

FIGURE 2.11 Environmental sustainability of table grape system.

The percentage reached by each company for the fulfillment of the attributes of certifications they have illustrates a general perspective of the scenario in which the companies of the Sonoran table grape system are found in the field of environmental sustainability.

Thus, the methodological procedure proposed in the present investigation has been useful to measure the environmental sustainability of the Sonoran table grape system. It is possible to assert this system has reached 70% in environmental sustainability.

This denotes certifications have played a crucial role in the adoption of more sustainable practices to protect the environment and people's health. However, it is important to clarify there are still opportunity areas to promote improvements in this area of sustainability.

2.5 DISCUSSION

Business organizations provide information related to the certifications and audits that each one manages. They recognize that complying with such processes is part of what the market demands from them, which is consistent with what was stated by Robles & Garza (2011); Contreras (2019); and Terán (2022). In addition, they confirm that compliance is directly related to progress, in this case, in the environmental field of sustainability, which allows table grape system companies to acquire the tools to move towards sustainable development (Rajala, 2016; Franco, 2021).

Progress in environmental sustainability can be achieved through the methodological procedure proposed in this research project, which is made up of the following steps: (i) definition of the object of study; (ii) obtaining the sample; (iii) preparation and application of the instrument; (iv) preparation of the matrix of certifications that business organizations of table grape system in Sonora have; (v) elaboration of the matrix of categories and attributes of the three dimensions of sustainability, based on the certifications; (vi) preparation of the matrix of categories and attributes of the environmental dimension of sustainability, based on the certifications; and (vii) calculation of the average achieved by the compliance of each company in the categories that encompass the different attributes of the environmental part of sustainability.

This methodological proposal differs from the approach presented by Baumgartner & Ebner (2010), which is a much more pragmatic method in which the review points belonging to the environmental field of sustainability are not specified; only the categories are mentioned, which are the following: air emissions, water emissions, soil emissions, conventional and hazardous waste, biodiversity, and environmental aspects of the product's life cycle.

The methodological plan presented here does not agree with what Rajala (2016) proposes in his research, since this author contemplates only some of the principles of the environmental field of sustainability, such as: Reduce, reuse, recover, recycle, and redesign. That is, it is mostly focused on the product and not on the practices of the company; it does not consider categories such as Fertilizers; flora and fauna care; greenhouse gases; management of garbage, conventional and hazardous waste management and use; land use and management, etc.

On the other hand, Franco (2021) presents in his research a quantitative evaluation of the environmental sustainability of agricultural activities in a system of local companies, with the purpose of comparing the supply and demand of natural resources, for which he adopted a methodology based on the ecological footprint approach. Nevertheless, this investigation has no representativeness because it only considers one category of environmental sustainability.

Thus, the methodological plans shown by the authors represent an attempt to measure environmental sustainability in agricultural systems. However, it is the present methodological procedure, which in a forceful and specific way, offers the possibility of carrying out a deep measurement in the matter, in the case of the companies of the table grape system in Sonora. The premise of this research is the proposed methodological procedure includes 353 attributes of the environmental part of sustainability. That is, it proposes the review of 353 particularities that are included in 10 categories belonging to the environmental component of sustainable development, according to the certifications they have.

2.6 CONCLUSION

The theoretical, referential, and methodological analysis obtained from the literature review allows us to confirm that the proposed methodological procedure is effective in measuring the environmental sustainability of the system of table grape companies in the state of Sonora. It also shows that the referential evidence was relatively limited with respect to studies discussing the measurement of environmental sustainability in the agricultural industry (primary sector), and even more limited considering the concepts and methodological procedures that support this research work. This study addressed in depth the analysis of environmental sustainability in 14 table grape-producing business organizations in the municipalities of Caborca, Pitiquito, Altar, Hermosillo, Carbó, San Miguel de Horcasitas, Guaymas, and Empalme. Based on the above, the result was that the system of these companies has reached 70% progress in environmental sustainability.

The study identified areas of opportunity to improve the commitment companies must maintain with environmental sustainability, through the adoption of responsible agricultural practices, that is, through certifications. Companies 4 and 9 stood out, as they have achieved considerable progress towards environmental sustainability. This research emphasizes the importance of addressing the issue of environmental sustainability in agricultural systems. It shows that it is essential for companies in this area to continue working to obtain certifications that support their efforts to guarantee a more sustainable and responsible future.

KEYWORDS

- **agricultural sustainability**
- **agricultural systems**
- **environmental protection**
- **environmental sustainability**
- **sustainability**
- **table grape system**

REFERENCES

A la Torre, M., Robles, J., Rodríguez, M., Gómez, B., & Flores, B. (2018). Estilos de liderazgo en exportadoras de uvas de mesa sonorense. *Revista Mexicana de Agronegocios, 42*(1345-2018-3664), 943–954.

Acuña, L., & Ccasani, J. (2022). Estimación de la huella ecológica en la producción de café pergamino. Caso de los caficultores de la cooperativa agraria cafetalera la prosperidad de chirinos. *Revista Campus, 27*(33), 1–22. Retrieved from: https://www.aulavirtualusmp.pe/ojs/index.php/rc/article/viewFile/2208/2697 (accessed on 30 May 2024).

ANDI. (2020). Global reporting iniciative. marco de la sostenibilidad corporativa. Retrieved from: https://www.andi.com.co/uploads/GRI-sostenibilidad-2022.pdf (accessed on 30 May 2024).

Aranda, A. N. (2016). *Liderazgo y Organizaciones Sustentables en el Sistema de Vid de Mesa Sonorense* (Master's thesis, CIAD, Hermosillo, Sonora).

Arias, J., Rodríguez, A., & Beduschi, L. C. (2021). Perspectivas de la agricultura y del desarrollo rural en las Américas Una mirada hacia América Latina y el Caribe. *Blog IICA*. Retrieved from: http://52.165.25.198/bitstream/handle/11324/1850/CDCR22068589e.pdf?sequence=2&isAllowed=y (accessed on 30 May 2024).

Ávila, Z. (2018). La sustentabilidad o sostenibilidad: un concepto poderoso para la humanidad. *Tabula Rasa, 28*, 409–423.

Awan, S., Ahmed, S., Ullah, F., Nawaz, A., Khan, A., Uddin, M. I., & Alyami, H. (2021). IoT with blockchain: A futuristic approach in agriculture and food supply chain. *Wireless Communications and Mobile Computing, 2021*, 1–14.

Baumgartner, R. J., & Ebner, D. (2010). Corporate sustainability strategies: sustainability profiles and maturity levels. *Sustainable Development, 18*(2), 76–89.

Borja-Bravo, M., García-Salazar, A., Reyes-Muro, L., & Arellano-Arciniega, S. (2016). Rentabilidad de los sistemas de producción de uva (*Vitis vinífera*) para mesa e industria en Aguascalientes, México. *Agricultura, Sociedad y Desarrollo, 13*(1), 151–168.

Campbell, B. M., Beare, D. J., Bennett, E. M., Hall-Spencer, J. M., Ingram, J. S., Jaramillo, F., & Shindell, D. (2017). Agriculture production as a major driver of the Earth system exceeding planetary boundaries. *Ecology and Society, 22*(4).

Castro-Garzón, H., Contreras, E. J., & Rodríguez, J. P. (2020). Análisis ambiental: impactos generados por los residuos agrícolas en el municipio de El Dorado (Meta, Colombia). *Revista ESPACIOS*, ISSN, *798*, 1015.

Cedeño, J. R. P., Bacallao, J. A. C., & Encalada, Y. A. R. (2021). Agricultura local sustentable. Caso de estudio asociación La Sabana del sector El Florón del Cantón Portoviejo. *Polo del Conocimiento: Revista Científico-Profesional, 6*(8), 37–61.

Contreras, A. C., Preciado, J. M., Báez, R., & Velderrain, R. A. (2019). Impacto de la implementación de la certificación comercio justo en una empresa del sistema VID de mesa sonorense. *Horizontes Empresariales, 18*(1), 4–15.

Contreras, C., Preciado, M., Báez, R., Robles, J., Taddei, C., & Benítez, V. (2018). Certificaciones agrícolas como conducta estratégica del sistema vid de mesa sonorense. *Revista Iberoamericana de Tecnología Postcosecha, 19*(1).

Fernandes, R., dos Reis, M., Fernandes, L., Rodrigues, B. J., Pozzi, C., Michael, R., & Urquiaga, S. (2020). Energy use efficiency in soybean crops in different regions of Brazil eficiência energética na cultura da soja cultivada em diferentes regiões brasileiras. *Semina: Ciências Agrárias, Londrina, 41*(6 suplemento 2), 2991–3010.

Franco, S. (2021). Assessing the environmental sustainability of local agricultural systems: How and why. *Current Research in Environmental Sustainability, 3*, 100028.

García-Orozco, D., Moreno, I. C. E., Alfaro-García, V. G., & Merigó, J. M. (2020). Sustentabilidad en México: un análisis bibliométrico de la investigación científica presentada en los últimos 28 años. *Inquietud Empresarial, 20*(2), 101–120.

GLOBALGAP. (2014). Buenas Prácticas Agrícolas. Melissa Ganoza. Lima, Perú. Retrieved from: http://repositorio.promperu.gob.pe/handle/123456789/745?show=full (accessed on 30 May 2024).

GLOBALGAP. (2017). Implementation and Certification Process. Ignacio Antequera. Technical KAM. Tashkent, Uzbekistán. Retrieved from: https://www.globalgap.org/uk_en/Profiles/Antequera/ (accessed on 30 May 2024).

Guillén, J., Calle, J., Gavidia, A. M., & Vélez, A. G. (2020). Desarrollo sostenible: Desde la mirada de preservación del medio ambiente colombiano. *Revista de Ciencias Sociales, 26*(4), 293–307.

Hashemi, S. Z., Darzi-Naftchali, A., Karandish, F., Ritzema, H., & Solaimani, K. (2022). Assessing agro-environmental sustainability of intensive agricultural systems. *Science of the Total Environment, 831*, 154994.

Hermida, M., & Manté, M. (2019). Los indicadores ambientales para la medición del desarrollo sostenible de Tierra del Fuego. *Informes Científicos Técnicos-UNPA, 11*(1), 1–27.

Hurtado, B., Robles, J., & Preciado, M. (2018). Responsabilidad social y ambiental en empresas exportadoras de uva de mesa sonorense. Enviado y aceptado para su publicación en la *Revista UPIICSA Investigación Interdisciplinaria.*

Hurtado, B., Robles, J., Preciado, M., & Bañuelos, N. (2018). Logística de transporte y desarrollo local en organizaciones exportadoras de uva de mesa sonorense. *Estudios Sociales, 28*(51), 0–0.

Jacobo, C. A., Ochoa, S., & Leyva, B. A. (2015). Desempeño sustentable en establecimientos de hospedaje de Ciudad Obregón, Sonora, México. *Revista Interamericana de Ambiente y Turismo-RIAT, 10*(1), 55–66.

Jáquez-Matas, S. V., Pérez-Santiago, G., Márquez-Linares, M. A., & Pérez-Verdín, G. (2022). Impactos económicos y ambientales de los plaguicidas en cultivos de maíz, alfalfa y nogal en Durango, México. *Revista Internacional de Contaminación Ambiental, 38*.

Kroger. (2018). The Kroger family of companies 2018. sustainability report setting, The table for a sustainable future. Retrieved from: https://www.thekrogerco.com/wp-content/uploads/2021/07/Kroger-2018-ESG-Report.pdf (accessed on 30 May 2024).

Kroger. (2021). Tablas de desempeño de sostenibilidad. Retrieved from. https://www.thekrogerco.com/sustainability/2021-performance-tables/ (accessed on 30 May 2024).

Lagunas-Vázquez, M., Almendárez-Hernández, M. A., Beltrán-Morales, L. F., & Ortega-Rubio, A. (2017). Propuesta metodológica para medir la sostenibilidad costera local en zonas áridas: su aplicación en la Reserva de la Biósfera El Vizcaíno. *Estudios Sociales, 27*(50), 0–0.

México Calidad Suprema. (2021). Trabajando por el campo mexicano. Retrieved from: https://www.mexicocalidadsuprema.org.mx/quienes-somos/ (accessed on 30 May 2024).

Meza, L. E., & Rodríguez, A. G. (2021). Soluciones basadas en la naturaleza para la sostenibilidad de la agricultura: ruta para la sinergia entre las convenciones de Río y la recuperación pos-Covid-19. Retrieved from: https://repositorio.cepal.org/bitstream/handle/11362/47574/1/S2100775_es.pdf (accessed on 30 May 2024).

Montaño, K., & Rodríguez, M. (2017). La productividad del trabajo en la producción de uva de mesa sonorense. *Transitare, 3*(2), 58–82.

Montaño, K., Preciado, M., Robles, J., & Chávez, L. (2018). Métodos de trabajo para mejorar la competitividad del sistema de uva de mesa sonorense. *Estudios Sociales. Revista de Alimentación Contemporánea y Desarrollo Regional, 28*(52), 0–0.

Oberč, B. P., & Arroyo-Schnell, A. (2020). Approaches to sustainable agriculture. Exploring the pathways. *486*.

Oláh, J., Aburumman, N., Popp, J., Khan, M. A., Haddad, H., & Kitukutha, N. (2020). Impact of industry 4.0 on environmental sustainability. *Sustainability, 12*(11), 4674.

Pinedo-Taco, R., Borjas-Ventura, R., Alvarado-Huamán, L., Castro-Cepero, V., & Julca-Otiniano, A. M. (2021). Sustentabilidad de los sistemas de producción agrícola: una revisión sistemática de las metodologías empleadas para su evaluación. *Tropical and Subtropical Agroecosystems, 24*(1), 1–16.

Primus GFS. (2015). Estándar PrimusGFS. Retrieved from: http://primusgfs.com/wp-content/uploads/2019/07/PrimusGFS_Standardv2.1-2_SP.pdf (accessed on 30 May 2024).

Rainforest Alliance. (2016). Requirements and guidelines for use of the Rainforest Alliance trademarks. Retrieved from: https://www.rainforest-alliance.org/wpcontent/uploads/2018/07/rainforest-alliance-marks-guide.pdf (accessed on 30 May 2024).

Rainforest Alliance. (2020). RainForest Alliance sustainable agriculture standard farm requirements. Retrieved from: https://www.rainforest-alliance.org/wpcontent/uploads/2020/06/2020-Sustainable-Agriculture-Standard_FarmRequirements_Rainforest-Alliance.pdf (accessed on 30 May 2024).

Rainforest Alliance. (2021). What does it mean "Rainforest Alliance Certified"? Retrieved from: https://www.rainforest-alliance.org/insights/what-does-rainforest-alliance-certified-mean/ (accessed on 30 May 2024).

Rajala, R., Westerlund, M., & Lampikoski, T. (2016). Environmental sustainability in industrial manufacturing: re-examining the greening of Interface's business model. *Journal of Cleaner Production, 115*, 52–61.

Rizo-Mustelier, M., Vuelta-Lorenzo, D. R., & Lorenzo-García, A. M. (2017). Agricultura, desarrollo sostenible, medioambiente, saber campesino y universidad. *Ciencia en su PC* (2), 106–120.

Robles, J., & Garza, C. (2018). Nuevas condicionantes en las organizaciones de los sistemas alimentarios: el caso del sistema vid de mesa de Sonora. En L. Huesca (Coord.), *Bienestar y desarrollo en el siglo XXI* (pp. xx-xx). México, D. F.: CIAD-Plaza y Valdez.

Rueda-Mijangos, J. F., & Mercado-Salgado, P. (2020). Desempeño sustentable y resultados de excelencia administrativa en minas pétreas en el Estado de México 2019: un estudio exploratorio. *Minería y Geología, 36*(3), 268–283.

Schönrock, P. (2020). Enfoque integrado y dimensión ambiental del desarrollo sostenible en América Latina y el Caribe: análisis de los informes nacionales voluntarios presentados ante el primer ciclo del Foro Político de Alto Nivel (2016–2020). Retrieved from: https://bvearmb.do/handle/123456789/1075 (accessed on 30 May 2024).

SENASICA. Gobierno de México. Servicio Nacional de Sanidad, Inocuidad y Calidad Agroalimentaria. (2012). El SENASICA simplifica trámites certificados. Para salvaguardar la sanidad del campo mexicano, el SENASICA, expide los certificados de importación y exportación, el cual hace constar que los productos cumplen con los requisitos sanitarios correspondientes. Retrieved from: https://www.gob.mx/senasica/prensa/el-senasicasimplifica-tramites-certificados (accessed on 30 May 2024).

SENASICA. Gobierno de México. Servicio Nacional de Sanidad, Inocuidad y Calidad Agroalimentaria. (2018). Agricultura. Héroes de la alimentación. ¿Qué hay detrás de una agricultura sana? #3 Sistema de Reducción de Riesgos de Contaminación. Retrieved from: https://www.gob.mx/senasica (accessed on 30 May 2024).

Solís, D., Robles, J., Preciado, M., & Hurtado, B. (2017). El papel del mercado en la construcción de organizaciones sustentables. *Estudios Sociales, 27*(49), 274–294.

Soule, E., Michonneau, P., Michel, N., & Bockstaller, C. (2021). Environmental sustainability assessment in agricultural systems: A conceptual and methodological review. *Journal of Cleaner Production, 325*, 129291.

USDA. Departamento de Agricultura de E.U. Servicio de Comercialización Agrícola. (2021). Retrieved from: https://www.ams.usda.gov/rules-regulations/organic (accessed on 30 May 2024).

Velderrain, R., Preciado, M., Báez, R., Taddei, C., León, J., & Contreras, A. (2019). Estructura de mercado de sistema vid de mesa sonorense. *Revista Iberoamericana de Tecnología Postcosecha, 20*(2), https://www.redalyc.org/journal/813/81361553002/81361553002.pdf.

Vera, J., Conejero, W., Conesa, M. R., & Ruiz-Sánchez, M. C. (2019). Irrigation factor approach based on soil water content: A nectarine orchard case study. *Water, 11*(3), 589.

WCED. World Commission on Environment and Development, United Nations. (1987). *Our Common Future*. Oxford, England: Oxford University Press.

Zarta, P. (2018). La sustentabilidad o sostenibilidad: un concepto poderoso para la humanidad. *Tabula Rasa,* (28), 409–423.

CHAPTER 3

Short Consumer Circuits for Sustainable Agriculture: The Case of Maize in Alamos, Sonora

J. A. GARCÍA-FIGUEROA,[1] K. TERÁN-SAMANIEGO,[2]
P. C. ESPINOZA-LÓPEZ,[3] and J. M. ROBLES-PARRA[2]

[1]*Programa de Doctorado en Desarrollo Regional, Centro de Investigación en Alimentación y Desarrollo, Calle Gustavo Enrique Astiazarán Rosas, Apartado, Hermosillo, Sonora, Mexico*

[2]*Coordinación de Desarrollo Regional, Centro de Investigación en Alimentación y Desarrollo, Carretera Enrique Astiazarán Rosas, C.P., Hermosillo, Sonora, Mexico*

[3]*Coordinación de Desarrollo Regional, Centro de Investigación en Alimentación y Desarrollo, Calle Gustavo Enrique Astiazarán Rosas, Apartado, Hermosillo, Sonora, Mexico*

ABSTRACT

Mexico currently faces a problem of food sovereignty and food security. According to data from the agri-food and fisheries information system (SIAP, by its acronym in Spanish), about 19.8 million tons of yellow maize were consumed in Mexico in 2021, and 16.7 million tons were imported, which represents 84% of imports. If we add white maize to this scenario, of which 24.4 million tons were demanded and only 0.6 tons were imported, the level of imports is reduced to 40% in total. This is still alarming since the import of tons of maize is so large that it makes Mexico highly dependent on the production of other countries. In addition, it has been shown that the sanitary conditions in which maize is imported are inadequate because almost all of

Sustainable Agricultural and Food Systems Engineering: New Research, Practices, and Developments.
Jose Juan Buenrostro-Figueroa, Julio César Tafolla-Arellano, Leonardo Sepúlveda Torre, Porteen Kannan, A. K. Haghi (Eds.)

it corresponds to transgenic maize and/or with a high content of chemicals that have a high degree of harm to health and the environment. Agroecology is a technological procedure that allows maize production on a commercial scale to be carried out without harmful agrotoxics for human health. For this reason, this chapter seeks to analyze how short consumer circuits contribute to sustainable agriculture through the concept of agroecology. The methodological procedure that supports this research consists of the application of two types of instruments to obtain data. The first instrument was a survey directed to maize agricultural producers to define the productive status. The second one corresponded to interviews with final and intermediate consumers of the short circuits of maize. The results of the interviews carried out with producers and consumers of maize indicate a high degree of interest in agroecological maize. Therefore, we conclude there is an adequate level of demand capable of stimulating the production and commercialization of agroecological maize located no more than 25 km away from the consumer.

3.1 INTRODUCTION

Native maize is an essential component of agricultural practices and identity in much of Mexico. For its inhabitants, maize is linked to their existence, their ancestors, their culture, and their customs. The maize productive system represents a significant part of the livelihood of Indigenous people, who reflect on their knowledge and better-preserved agricultural practices. Mexico has a wide variety of native maize that plays a crucial role in the daily diet of the peasants who cultivate it (Santillán, 2022).

This great variety of maize is an expression of the large geographical and cultural diversity of the country. The multiple shapes, colors, flavors, and expressions that maize presents are a reflection of the domain of people and communities. The choice of processes to improve some of these characteristics has had an influence on the evolution of native varieties, which today reach a total of 59 unique Mexican breeds. For the communities of Mexico, the cultivation of white and yellow maize is essential for their economy and for the country (Hernández, 2022).

In recent years, maize has presented an increase in demand. This has caused a problem that applies direct pressure on natural resources and leads to an increment in the number of territories dedicated to agricultural activity. Planting and harvesting activities have been carried out without considering the instability and nutritional needs of the soil, which eventually results in low production efficiency and a high risk of environmental degradation (Moraga, 2021).

Favila & Reyes (2022) present in their research an approach to the problem of maize. Between the years 2011 and 2020, maize production in Mexico increased from 17.6 million tons to 27.4 million tons, representing a 55% increase. The harvested area also increased from 6 million hectares to 7.1 million hectares. Similarly, the yield per hectare grew during that time but dropped to 3.8 tons per hectare in 2020. Mexico is self-sufficient in the production of white maize but highly dependent on the production of yellow maize (*see* Figures 3.1 and 3.2) (SIAP, 2022).

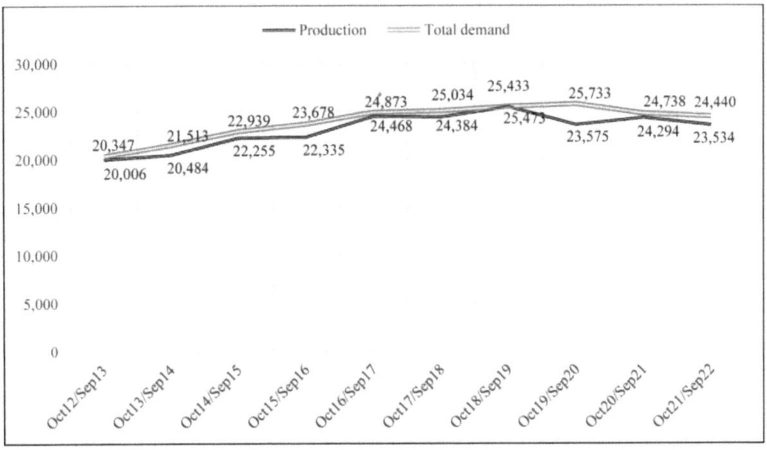

FIGURE 3.1 Production and total demand for white maize in Mexico.

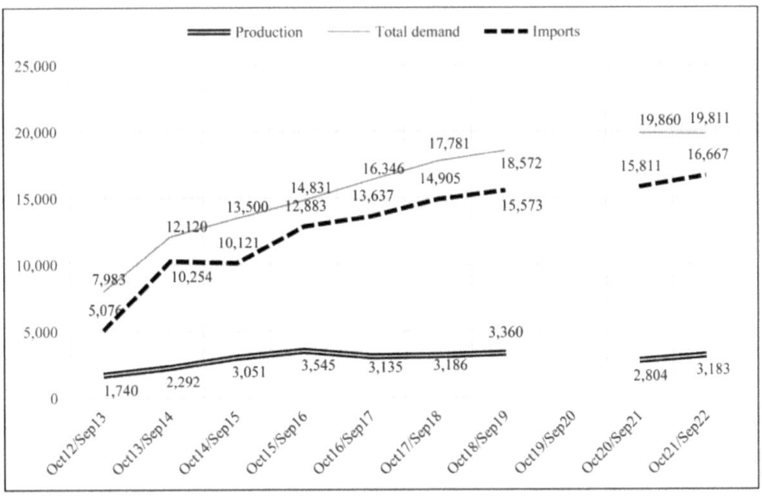

FIGURE 3.2 Total production and demand of yellow maize in Mexico.

There is no doubt Mexico has a deficit in its maize trade balance, which makes it difficult to achieve the goal of food sovereignty and food security. To guarantee a sufficient supply of maize to meet domestic demand, Mexico needs to reach a competitive position in the market and implement some strategies to move towards sustainability. This must involve paying special attention to issues such as profitability, yield in tons per hectare, and the productivity of the production process (Government of Mexico, 2021; Bastida, 2021; Cuadras-Berrelleza et al., 2021; Favila & Reyes, 2022).

Short circuits of consumers are the answer for this scenario since they have the capability to strengthen family farming systems, foster social ties, look for equity in commercial exchanges, establish autonomy among participants, and reduce the use of transport and packaging. It is for these very reasons that short circuits of consumers can also achieve sustainability in a local context (CEPAL, 2016; Ranaboldo, Arosio, & Díaz, 2016).

The aforementioned will allow people to have physical, social, and economic access to safe and sufficient food that meets their nutritional requirements and food preferences, enabling them to lead an active and healthy life. Likewise, it enables people to define and control their food and production systems in a way that is respectful to the environment and careful of people's health (López-Gonzálcz ct al., 2018).

3.2 LITERATURE REVIEW

3.2.1 SHORT CIRCUITS OF MAIZE CONSUMERS

The concept of short consumer circuits has its beginnings in 1965, in Japan. It was devised by a group of housewives who were concerned about industrial food production and the irrational use of chemicals and pesticides. The movement was called "teikei" and was based on the development of a market where the confidence in producing healthy food in exchange for the total purchase of the product could dominate (Villaseñor, 2017).

These circuits are a commercialization mechanism that makes it possible to reduce the gap between the consumer and the producer, establish a more balanced economic relationship between them, develop an educational-transforming dimension, eliminate the figure of the middleman as much as possible, favor local production, reduce the carbon footprint of marketing, and value what the territory has historically produced as biogenetic heritage (Saravia, 2020).

In the case of maize commercialization, short consumer circuits have individual farmers informally organized to produce a differentiated product without certification. In it, a maximum of participants between the producer and the consumer is one intermediary. In most cases, no prior agreements are defined, and close or proximity relationships are developed (Maceín & Barba, 2018).

Within the short circuits of maize consumers, there are producers with a certain degree of association that is verbally understood. In this kind of commercialization, it is easy to find an offered product with attributes related to the local and cultural aspects of the community where producers are established. A clear example of that is family farming (Maceín & Barba, 2018).

Although all this product differentiation is not supported and explained through a brand, seal, or certification, this is precisely the reason that motivates consumers to acquire the product. In this manner, they seek much more than a good price, they look for direct contact with the producer, the generation of trust relationships, product quality, food safety, and the generation of social, economic, and environmental changes (Maceín & Barba, 2018).

Regarding producers, they seek to obtain greater value from their production and save on other components of the chain, such as brands, transport, and packaging, among others. Farmers also look to create value from non-material assets, for example, product differentiation, territorial anchorage, authenticity, social bond, etc. (CEPAL, 2014).

This type of circuit facilitates the possibility of generating empathy between the producer and the consumer, as well as knowing in depth the preferences and expectations of the market. However, short consumer circuits at the same time present challenges, and one of them is that quality requirements are higher than in a long circuit of consumers (Marty & Ibáñez, 2016).

Short circuits of maize consumers maintain relationships that are restricted to local markets. This is characterized by the production, in this case, of fresh or artisan maize, in quantities that might seem insignificant for food supply, but globally favor agricultural production systems (Lizarazu, 2014). The success of short circuits is due to a growing demand from consumers, who are looking for local, authentic, healthy, and seasonal products, which incorporate processes that are directed towards caring for the environment and people's health (CEPAL, 2014).

Devisscher & Elías (2014); CEPAL (2016); Maceín & Barba (2018); Saravia (2020); and Paz & Infante (2020) mention in their research the most important characteristics of short consumer circuits (*see* Figure 3.3).

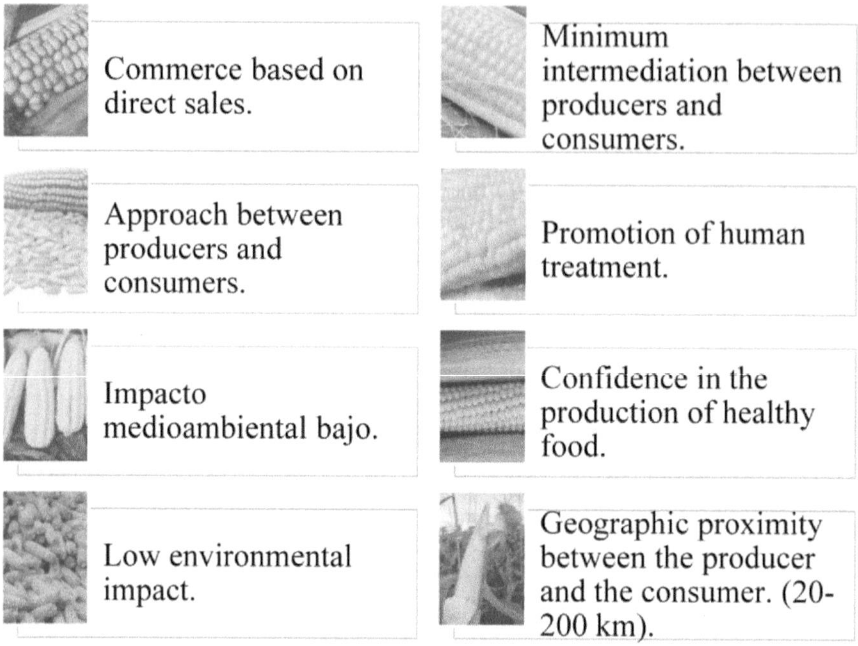

FIGURE 3.3 Characteristics of short consumer circuits.

The characteristics coincide with the three dimensions of sustainable development: economic, which pursues the economic well-being of the parties involved; environmental, which is aimed at protecting the environment; and social, which promotes the care of consumers and people who participate in the production process (Flores et al., 2020).

According to Flores et al. (2020), to achieve the attributes related to the environmental part of sustainability, the application of techniques that promote environmentally friendly production processes is required. And, to comply with those particularities that are linked to the social element of sustainability, it is necessary for these agricultural groups to formalize their structure and social denomination, and to promote equitable social participation.

One response to this approach is agroecology, which has a holistic vision in which man and his environment are considered within the same system. The mode of production and commercialization of short consumer circuits rescues the criteria of agroecology, which is based on the ethical principle of sustainable development. That is, it uses few inputs, has high biological diversity, manages a concept of respect for nature, promotes the fair participation of farmers, and rescues ancestral knowledge (Díaz et al., 2020).

3.2.2 AGROECOLOGY AND AGRICULTURAL SUSTAINABILITY

Agroecology, an integral part of agricultural sustainability, is an essential component in the global response to the environmental and social problems that maintain agricultural systems. It is focused on solving issues such as massive deforestation, water scarcity, biodiversity loss, soil depletion, social inequity, hunger, and extreme poverty (FAO, 2018).

The term "agroecology" has been used since the end of the 20th century by scientists who referred to the application of ecological principles to agriculture. This concept has developed more strongly in Europe than in other parts of the world over the last three decades. For agroecology, the transition towards more sustainable food systems must include environmental, economic, social, cultural, and political issues (Migliorini et al., 2020).

Agroecology is based on territorial and bottom-up processes that help provide contextualized solutions to local problems. Agroecology innovation strategies are founded on co-creating knowledge by combining science with the traditional, practical, and local knowledge of farmers. Agroecology empowers producers and communities as key agents of change, rather than adjusting to unsustainable farming system practices; it tries to transform food and farming systems by addressing the causes of problems in an integrated way and providing holistic and long-term solutions (FAO, 2018).

In order to provide guidance to countries to transform their agricultural and food systems, integrate sustainable agriculture on a large scale, achieve the Zero Hunger Challenge and reach many other Sustainable Development Goals (SDGs), 10 elements were established in the regional agroecology workshops: (i) diversity; (ii) synergies; (iii) efficiency; (iv) resilience; (v) recycling; (vi) co-creation and knowledge sharing (describing the common characteristics of agroecological systems, basic practices, and innovation criteria); (vii) human and social values; (viii) food culture and traditions (highlighting contextual aspects); (ix) circular and solidarity economy; and (x) responsible governance (addressing the enabling environment) (FAO, 2018).

3.2.3 SYSTEMS APPROACH: MAIZE AGRICULTURAL PRODUCTION SYSTEMS IN ALAMOS, SONORA

Agricultural production systems are undergoing marked changes due to rapid shifts in consumer demands, input costs, concerns for food safety, and environmental and social impacts. They include all activities related to the production and distribution of food products and maintain the commitment to fulfill the function of supplying the need for human nutrition in a given society (Walters et al., 2016).

For Mexico, the challenge of guaranteeing food for a rapidly growing population, estimated to reach 138.1 million inhabitants in 2030, results in food production requirements up to 14% more. The foregoing requires redirecting the policies under which these systems are governed, where a vision and commitment to building agricultural production systems that are productive, fair, healthy, inclusive, and sustainable, are expected (SADER, 2021).

Agricultural production systems have been oriented toward supplying the population's food needs. They generate a significant economic spillover and learn to meet consumers who have changed their consumption patterns. Thus, these systems look up the production of products that have the nutritional characteristics that consumers demand and are oriented to the specialization of their offer, which has shown significant growth in recent years (INFOAGRO, 2018).

Maize productive systems are made up of production and service organizations, which can be agricultural units, agri-food companies, commercial companies, and restaurants, among others, which seek to associate according to certain characteristics and operations, in a specific territory. It is important to mention the participation of small producers does not exclude them (Jaramillo et al., 2018).

Considering the above, it is necessary to note how the structure of the short consumer circuits is inserted in the general system of maize production on a commercial scale. Likewise, it is crucial to show how the institutional environment and research and development activities are also fundamental parts of this system to achieve the objectives: sustainability, equity, and competitiveness (*see* Figure 3.4).

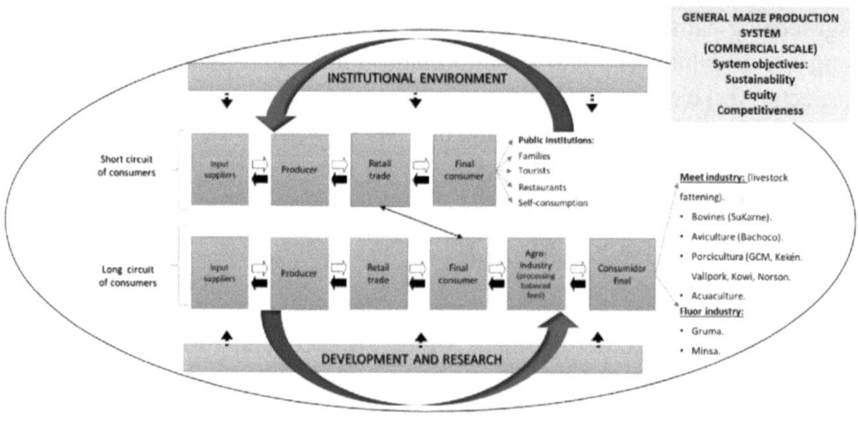

FIGURE 3.4 General maize production system.

In Mexico, Sonora stands out as one of those large agricultural areas of high productivity, with towns and municipalities that are joining efforts to promote their path. The promotion is towards sustainability and to consolidate themselves as one of the agricultural pillars in the face of population changes and climate variability. Among its main products are wheat, soybeans, cotton, sesame, and maize. An outstanding municipality in maize production is Alamos, Sonora (PIEAES, 2021).

Alamos is in the southeastern part of the state of Sonora. It is bordered to the north by Yécora, to the south by the state of Sinaloa, to the east by the state of Chihuahua, and the west by Navojoa. Its main economic activity is agriculture, and one of its main crops is maize, in which the social and private sectors participate (Government of the State of Sonora, 2016).

Alamos has an incipient mixture of agricultural knowledge that has been built through a system of traditional indigenous ecological knowledge and new Mesoamerican and European technologies; mainly derived from the recolonization process that took place in the locality in the 17th century (Colín, Gándara, & Saldaña, 2015).

In Alamos, agriculture has two modalities: (i) rainfed (dependent on weather conditions); and (ii) irrigated (assisted by water supply through various artificial methods). Its main resource is the family labor force to sustain its self-sufficiency economy (Colín, Gándara, & Saldaña, 2015). These maize-producing families are grouped in different ways and are mainly characterized by a reduced (or even non-existent) number of intermediaries between consumers and producers, and by a limited geographic distance between them (Sosa-Cabrera et al., 2022).

Thus, the system of short consumer circuits has become an option for the producers of Alamos, Sonora, because it is presented as a consumption strategy that provides greater benefits to producers and consumers in terms of variety, food safety, proximity, stability of supply and demand, cost-benefit, social equity, competitiveness, and sustainability (Sosa-Cabrera et al., 2022).

3.3 METHODOLOGICAL PROCESS

The methodology of this research is based on the following details:

1. Visual identification of the basic points where some maize derivatives are sold, such as grain buyers, processors, and direct sellers to the consumer.
2. Corroboration of the points of sale detected visually with those provided by the offices of the Secretariat of Tourism.

3. Sample selection. For producers and points of sale (restaurants), it was considered to take a sample for convenience due to the accessibility to collect information in terms of space and time (Hernández, Fernández, & Baptista, 2014). The foregoing resulted in the application of 15 interviews with producers, from a population of 28, which represents 53.57%; and for restaurants, 11 interviews were derived from a population of 23, which represents 47.82%.
4. Design of the semi-structured interview.

    i. For entrepreneurs at points of sale, the script is based on the following required information: General data of the point of sale; general characteristics of the business; What is their gastronomic offer?; Who is their maize supplier?; What is the basis for their purchase decision?; What are the characteristics of the consumer market?; What kind of demand do they have for the dishes they offer?; What kind of marketing processes do they have?; What are their commercialization strategies?; What are the quality and sanitation control criteria?; Do they already buy agroecological maize? Or are they interested in it? finally, do they have maize certifications? and, what are those certifications? among others.
    ii. For the producers, the script is based on the following required information: Whom do they buy the maize from?; On what basis do they make the decision about who to buy the maize from?; In which geographic areas do they buy maize?; What is the approximate distance from the maize purchase area to the production or transformation area?; What is the distance from the geographic area where the maize is purchased to the area where the purchased product is stored?; What is the distance from the production area to the storage area?; Who are your customers?; In which geographic areas do you sell your products?; among others.

5. The strategy followed to contact the person in charge of each of these points was the "snowball" strategy. The production conditions were reviewed to evaluate their impact on the producer.

Through the researchers' participant observation, photographic documentation, and conversations, the quantitative methods were complemented to identify the characteristics of maize food marketing that exist in each of the points reviewed, and thus determine whether it is possible to establish relationships between the maize producer and the short circuits of consumers.

Finally, with the support of the information gathered, it is possible to characterize how short consumer circuits contribute to sustainable agriculture through the concept of agroecology.

3.4 RESULTS

This section presents a characterization of the maize production system. This was considering the different agroecological lighthouses. An agroecological lighthouse is an expression to indicate different spots where technical knowledge and agroecological processes are shared to guide local producers. The guide is towards the design of more sustainable agricultural systems (Ranaboldo & Venegas, 2007; Damian & Toledo, 2016).

It is also based on data obtained through the instrument applied by the coordinators of the Field Schools (ECA by its acronym in Spanish) in the lighthouses (Figure 3.6). It is important to mention that ECAs are spaces to promote and strengthen the development of learning communities that exchange technical and organizational knowledge in favor of the agroecological transition (Romero et al., 2022).

In Mexico, some lighthouses and ECAs have been distributed to implement sustainable strategies in production processes, and then to transition in favor of the agroecological transition. Of the above, 1 lighthouse and 3 ECAs correspond to Alamos, Sonora.

3.4.1 TAXONOMY OF MAIZE PRODUCTION SYSTEM IN ALAMOS, SONORA

Producers involved with the agroecological lighthouses in Sonora are in the southern municipality of Alamos, mainly in the towns of El Tezal, Minas Nuevas, Los Tanques, and La Aduana. There are 28 maize producers identified who have a total of 112 hectares of available land for the production of agroecological maize (with an average area of 4 hectares). There are also 23 recognized points of sale, such as restaurants.

This chapter is founded on an approach where agricultural production systems are based on three main components: agricultural production, the transforming process, and consumers. With this starting point, it is possible to determine the tons that the short circuits contribute, to address aspects related to food security and sovereignty (*see* Table 3.1).

TABLE 3.1 Maize Producers in Alamos, Classified by Agriculture Modality (Surfaces of Land are Presented in Hectares)

Agriculture Modality	Maize Producers	Irrigated Surface	Rainfed Surface	Avg.
Irrigated	0	0	0	–
Rainfed	28	0	112	4

Producers who are in rainfed production systems have an average age of 52.5 years; the oldest one is 83 years old, and the youngest is 29 years old. Of the producers who are in this modality, 80% are men. Of all of them, only one producer identifies with an ethnic group (6%); 33% of the producers had the opportunity to study elementary school, 26% high school, 21% studied for a technical degree, and only 20% got a university degree (*see* Table 3.2).

TABLE 3.2 Sociodemographic Data About the Interviewees

Agriculture Modality	Average Age (Years)	Man pct. (%)		Women pct. (%)
Rainfed	52.5	80%		20%
Agriculture modality	Percentage of Producers Engaged in Another Activity			
	Cattle Raising	Other Agricultural Activity		Another Activity
Rainfed	6%	6%		60%
	Percentage of Educational Level Reached by Producers			
	Elementary School	High School	Technical Degree	University
Rainfed	33%	26%	21%	20%

Alamos does not have an irrigation system for maize production; therefore, all maize is produced with rainfed systems. This manner of production has been known as rudimentary because it doesn't include any kind of technology. Each activity in this production modality must coincide with the fact that work is manual and must align with the traditions and culture of the place to maintain the crop. In terms of handling, harvesting, and grain storage, the rainfed production system also has many particularities because it is a self-consumption system with many deficiencies in terms of technology.

Around 100% of producers said they don't utilize technology for production activities and don't receive technical assistance. Around 93% of them use a tractor in their tillage activities; however, only one of them could borrow a tractor, 27% use leased tractors, and 53% have their tractor.

Moreover, 67% of the producers stated they own threshers, motorized cranes, seeders, or some other machinery or equipment, but there are still

50% of producers using a yoke or plow. In reference to weed growth on farmland, 13% mentioned they use tools to control it.

Regarding harvesting activity, 100% of the producers specified that they practice it manually. Once maize is harvested, almost all producers store the grain in raffia sacks (60%), 13.3% keep it in storage tanks, 6.7% keep it in barns, and another 6.7% maintain the grain in their yards. The above shows there is a lack of adequate facilities for grain storage, such as silos, or properly installed warehouses (*see* Figure 3.5).

None of these producers use chemical fertilizers in their production; only one of them mentioned the use of compost as a natural fertilizer. Around 20% of producers use chemical insecticides to control pests; in particular, there is a chemical containing the abamectin compound from the Allister brand. This chemical is sold under the commercial name of Antex and is specially used to combat ants.

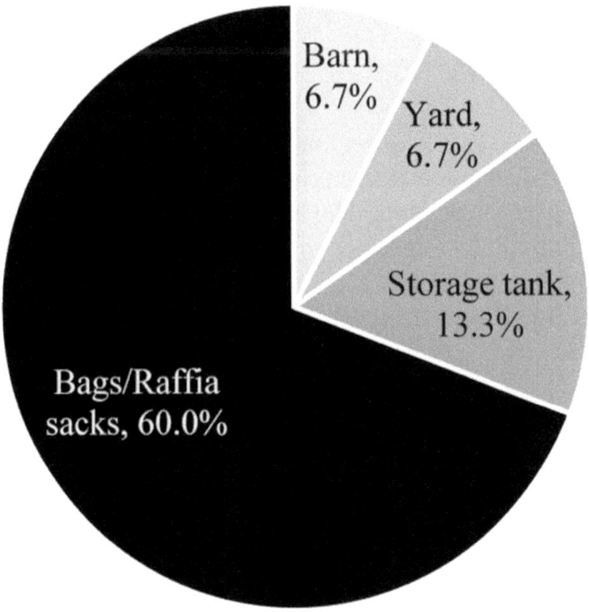

FIGURE 3.5 Grain storage technology.

On the other hand, the most widely used grain conservation technique is applying hydrated lime, used by 80% of the producers. Around 26.7% also apply a grain drying method, and only 13.3% select the grain to preserve the quality of genetic material (*see* Figure 3.6).

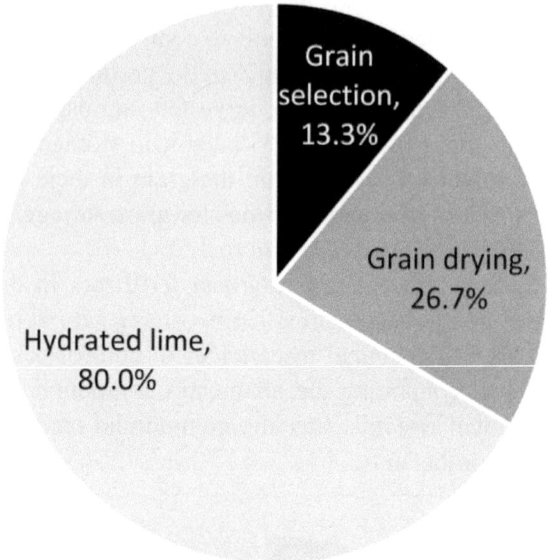

FIGURE 3.6 Grain preserves technology.

To demonstrate to producers that there are better production alternatives that can be applied for maize cultivation, it was proposed to carry out 5 experiments in different plots. The experimental planting design was made considering 5 divided plots, where different densities were tested in order to compare experimental planting with conventional methods.

To carry them out, different recommendations and adaptations were taken into account:

- Usually, producers perform production activities with animals and plows due to the lack of financing for agricultural machinery. For the experimental plot, the land was prepared with agricultural technology. These laborers had a regional unit cost of 900 pesos for the use of agricultural machinery, and 1,000 pesos for the use of animals.
- In the case of sowing, it was shown that the San Juan breed has great adaptability to the soil and climate conditions of the region, and a high germination potential compared to its commercial counterparts. This is because it registered 97% viable germination for every 100 seeds sown. For the experiments, a classification of the seed was made, establishing as criteria the damage to its endosperm, pericarp, germ, and pedicel.

- To control pests and diseases, producers have applied insecticides. However, it is important to mention that the use of entomopathogenic fungi, beneficial insects, and mating disruption pheromones was suggested for the experiments.
- In the case of fertilization, the construction of a reservoir for the production of leachate, and the use of fermented bat guano as a nitrogen source, were suggested for the experiments.

Below are the five treatment specifications:

- The first treatment (T1) called "the control" was based on the traditional way producers in the region sow, 80 cm apart from the plant, throwing 3 to 5 grains per step.
- Treatment T2 was sown conventionally, but with grains previously inoculated with mycorrhizae.
- Treatment T3 was planted with a density of 5 inoculated grains per meter every 20 cm approximately.
- Treatment T4 was planted with a density of 4 inoculated grains every 30 cm approximately; and
- Treatment T5 was planted with a density of 2 inoculated grains per meter every 40 cm approximately.

The results of the increase in performance are shown in Figures 3.7 and 3.8.

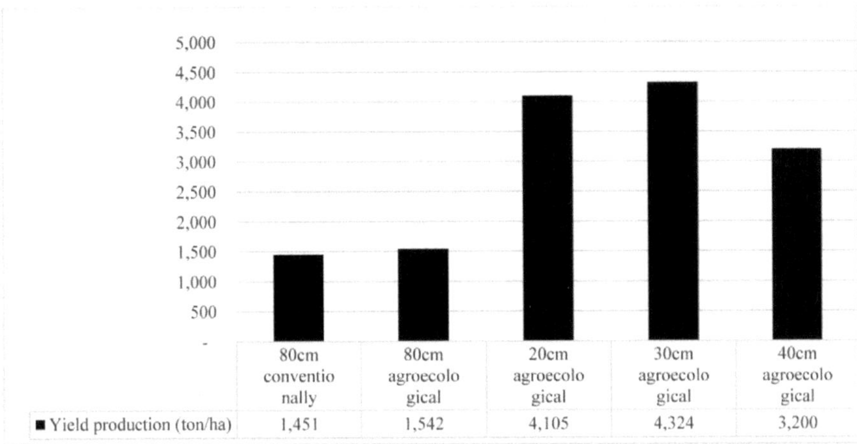

FIGURE 3.7 Results of agroecological treatment, yield of maize (grain) production in Alamos.

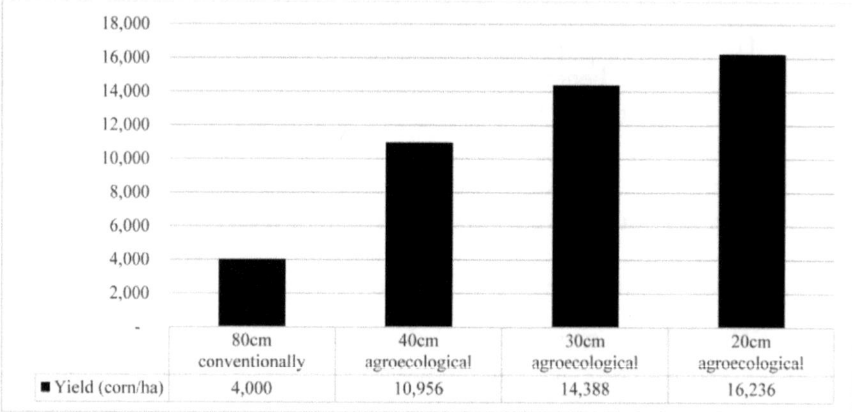

FIGURE 3.8 Results of agroecological treatment, yield of maize (corn) production in Alamos.

3.4.2 MAIZE PRODUCTION AND SALES OF SHORT CONSUMER CIRCUITS IN ALAMOS, SONORA

All producers from Alamos's lighthouse use native and Creole varieties of maize seeds[1]; specifically races called San Juan and mayobachi. All these producers produce in rainfed systems, and their decisions are based on the information obtained on fieldwork.

Native maize continues to be the livelihood of thousands of rural Mexican families because it represents the base of main meals, but mostly because it is native maize and not genetically improved or hybrid. The value of maize in Sonora remains sovereign because it is a popular source of food which is included in a great diversity of recipes that can be prepared based on it.

Results of the experiments in the ECA installed in the Minas Nuevas community showed maize yield production can be increased considerably. It was not the same for the historical yields from 2017 to 2021, which were very low. They are shown in Figure 3.9 to make a reference to this situation.

It is necessary to mention that the analysis of this chapter was conducted under a systematized market approach regarding the short circuit of consumers, in which maize is typically sold in two ways: fresh as corn, and in grain for nixtamalized use, such as flour, tortilla, or derivatives.

According to data from the last production cycle, out of 112 hectares available for maize production at the Alamos lighthouse, maize was only

[1] Creole maize varieties, are varieties that have been cultivated and selected by farmers for generations, retaining a distinct identity and lacking formal crop improvements (Cowan, 2019).

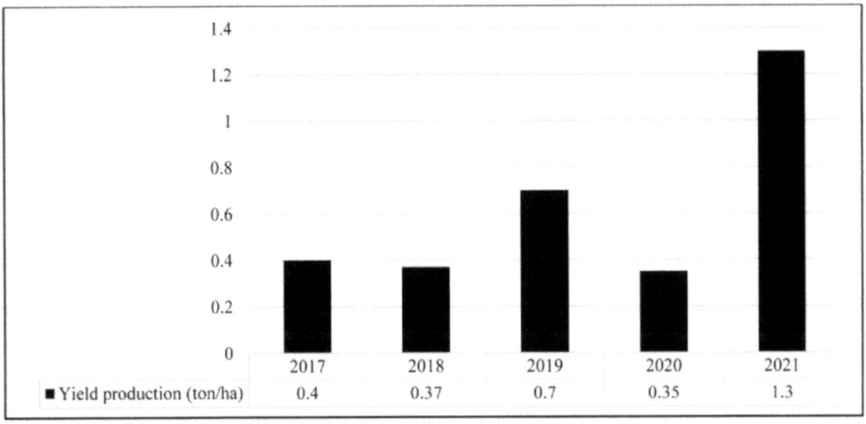

FIGURE 3.9 Yield of maize production in Alamos, from 2017 to 2021.

produced in 84 hectares, and 80% of that production was destined for self-consumption.

In reality, only a small portion of maize was sold as fresh corn; in the last five years, only 60% of the producers sold fresh corn, with 27% of the amount sold to local stores and 7% to intermediaries. The participation of small producers in Alamos, Sonora, in the commercialization of fresh corn, is low, historically aimed only at self-consumption. Based on the above information, the potential market size of Alamos can be predicted for maize in grain or fresh corn, as shown in Tables 3.3 and 3.4.

TABLE 3.3 Maize Production (Corn) in Alamos – Agroecological Treatment vs. Traditional One

Production System	Yield (#corn/ha)	Market Price ($/corn)[2]	Production Value (Price* yield)	Production Cost	Profit per Hectare (Production Value – Production Cost)	Available Surface in Lighthouse (Hectares)	Income Estimates per Year in Alamos Lighthouse ($ pesos)
Traditional (2021)	4,000	$3.75	$15,000.00	$3,700.00	$11,300.00	112	$1,265,600.00
Agroecological	12,600[3]	$3.75	$47,250.00	$4,722.00	$42,528.00	112	$4,763,136.00

[2]Based in the minimum price per dozens of corns registered for the last three years in supply center Francisco I. Madero in Hermosillo, Sonora ($45/dozen) (SNIIM, 2023).

[3]For the production of corn, the density of the exercise T4 and T5 is suggested, in which the increase in the number of ears amounted to 10,956 ears and 14,388 corn, respectively. Therefore, an average is considered for the table (12,600).

TABLE 3.4 Maize Production (Grain) in Alamos – Agroecological Treatment vs. Traditional One

Production System	Yield (ton/ha)	Market Price ($/ton)[4]	Production Value (Price*yield)	Production Cost	Profit per Hectare (Production Value – Production Cost)	Available Surface in Lighthouse (Hectares)	Income Estimates in Alamos Lighthouse ($ pesos)
Traditional (2021)	1.30	$6,805.00	$8,846.50	$3,700.00	$5,146.50	112	$576,408.00
T1	1.45	$6,805.00	$9,867.25	$3,700.00	$6,167.25	112	$690,732.00
T2	1.54	$6,805.00	$10,479.70	$4,722.00	$5,757.70	112	$644,862.40
T3	4.11	$6,805.00	$27,968.55	$4,722.00	$23,246.55	112	$2,603,613.60
T4	4.32	$6,805.00	$29,397.60	$4,722.00	$24,675.60	112	$2,763,667.20
T5	3.20	$6,805.00	$21,776.00	$4,722.00	$17,054.00	112	$1,910,048.00

3.4.3 GENERAL CHARACTERISTICS OF SHORT CONSUMER CIRCUITS IN ALAMOS, SONORA

This section provides a general description of the short consumer circuits at the Alamos lighthouse. It shows how actors are directly or indirectly inter-related: producers (agro-producers in agroecological transition), corn and grain traders (intermediaries), product processors (tortillerías), points of sale (supermarkets, local marketers, local establishments with food preparation service and restaurants), and final consumers.

The gastronomic route, as a relevant component of the short consumer circuits of Alamos, is integrated by 25,000 inhabitants. It is one of the main tourist destinations in Sonora in terms of history and colonial architecture. This municipality belongs to the "pueblos mágicos" program[5] since 2005 and in recent years has experienced significant growth in terms of tourism infrastructure.

There is a diversified gastronomic offer in restaurants with food preparation services, regular food, Mexican "antojitos," fish and seafood, pizzas, international cuisine, and coffee shops. According to information provided by tourism authorities, its gastronomic industry is made up of 27 establishments, of which, 23, dishes prepared with some derivatives of maize are offered.

The product with high demand is the "tortilla" because it is used to accompany and prepare a wide variety of recipes. On the other hand, the

[4]Market price based on guaranteed prices imposed in 2023 by SEGALMEX (Mexican food safety by its acronym in Spanish).

[5]"Pueblos Mágicos" is a Program of the Ministry of Tourism that acts directly on the towns as a distinctive brand of tourism in Mexico. The program seeks to optimize the rational use of natural and cultural resources and attractions, strengthen infrastructure, quality of services, innovation and development of tourism products, marketing, and modernization, to trigger the growth of the tourism market in the localities that support the program (SECTUR, 2014).

sale of fresh corn represents the only way in which local producers sell their maize. They do not add any value due to a lack of capital and inconvenience in production costs and sale prices.

A few selling points offer other products derived from maize. It can be mentioned that nixtamalized corn and corn flour are for food processing, although the suppliers of these products come from other municipalities within the region. Basically, the maize trade is carried out on a local scale; therefore, the geographical distance between supply and final consumption is very short, a characteristic of short consumer circuits.

Consumers of different products derived from maize have certain quality requirements. Specifically for tortillas, they require above all a good flavor and to be easily fried on the stove without losing their composition. For fresh corn, the consumer looks for a good size, which is complete with the absence of impurities (without fungus). In the case of seeds, consumers require good germination and hard grain.

The proposed modalities of short circuits of consumption are local markets and festivals, organic street markets, sales in local restaurants, and gastronomic tourist routes as a meeting space between the demand for local agroecological products and the producers.

The main consumers of restaurants and different selling points are national tourists (especially from Sonora and Sinaloa), followed by foreign tourists, and in third place, the local population. These consumers are the ones who stimulate the consumption of sustainable maize because they ensure that maize is clean and free of chemicals.

These consumers have high incomes, a high level of academic preparation, and a high awareness of sustainability aspects, i.e., it is a specialized market that can ask for maize free from the use of chemicals. They want producers to promote environmental protection and people's health care. They want minimum intermediation between producer and consumer, seek humane treatment, but also support the family farmer and small producers.

There are no official data on the number of annual visiting tourists who come to the municipality. Still, only during the period of Easter holidays, in 2022, 49,323 people arrived at the place, leaving an economic spill of about 6 million pesos (Palomares, 2022).

Likewise, the temporality of the flow of visitors and tourists to the area is marked by the high season and the low season. The high season takes place mainly in the holiday periods of Easter, summer, and New Year's Eve. Similarly, during holidays and weekends, there are a greater number of visitors.

It is necessary to highlight that every year at the end of January, the international cultural festival Alfonso Ortiz Tirado is held in the town; this event favors the participation of the highest number of visitors to the municipality in the year.

The level of tortilla volume that selling points need is determined by the seasonality of the flow of visitors. In high season, each restaurant has an average daily consumption that goes from 20 to 30 kilos of tortillas. In low seasons, the average drops from 5 to 15 kilos.

The following information (Table 3.5) describes the behavior and characteristics of customers and food supply in 2023 for Alamos, Sonora.

TABLE 3.5 Characteristics of the Gastronomic Offer and Consumption of Maize and its Derivatives, in Alamos

Most consumed maize derivatives	• Tortillas • Chilaquiles • Enchiladas • Tortilla soup • Tamal • Meat toast • Gorditas • Pozole • Totopos
Maize products used in selling points and restaurants	• Tortillas • Nixtamal • Corn • Corn flour
Corn input suppliers	• Local tortilla shops • Small supermarkets in Álamos and Navojoa
Main consumers	• National tourism (Sonora and Sinaloa mostly) • Foreign tourism • Local population
Visitor flow timing	• High season; weekends, holidays, and periods of vacation. Months: January, April, June, July, September, November, and December. Low season: Months: February, April, May, August, and October
Average volume (kg) of daily consumption of tortillas by selling point or restaurant	High season: 20 30 kg of maize tortillas Low season: 5 15 kg of maize tortillas

In this sense, the present analysis focuses on gastronomic tourist routes and sales in local restaurants as a marketing strategy that allows for reducing the gap between the producer and the consumer in a balanced economic relationship. In addition, it seeks to stimulate local production through sustainable maize agriculture.

3.5 DISCUSSION

Producers and restaurant owners recognize that Alamos has the necessary elements to achieve sustainable maize agriculture through the agroecological model. On one hand, there is high interest from local producers in participating in the agroecological transition of maize, and on the other, there is awareness from restaurant owners and point-of-sale entrepreneurs in consuming clean maize. This information matches the investigation of Sosa-Cabrera et al. (2022), who indicate that the components of short consumer circuits should work in an integrated manner and be directed toward the implementation of practices related to sustainability.

The experiments carried out for producers in the field plots showed that it is possible to increase yield in production from 1.3 tons to 4 tons in the field. The results were endorsed by the comparison of traditional methods of maize sowing with those related to agroecology. This coincides with what Contreras et al. (2019) establish in their study, where they point out that agroecological practices increase soil fertility and, thus, productivity.

Short consumer circuits of agroecological maize represent a tool for the development of local sustainable agriculture. This is where producers can benefit from advice on production techniques, an increase in earnings, and low production costs; on the other hand, consumers can benefit from healthy food, proximity to producers, and humane treatment.

This last affirmation is supported by FAO's investigation (2018). In this investigation, the organization sustains that agroecology strategies empower producers and communities as key agents of transformation to renovate food and farming systems by addressing the causes of problems and providing holistic and long-term solutions.

Visitors and tourists in Alamos represent the most important consumers for the proposed gastronomic route. The consumers ask for clean maize and have the economic capacity to buy it. They want maize free from the use of chemicals and supported by practices that promote environmental protection and people's health care.

Given this, Palomares (2022) comments that Alamos is considered the tourist jewel of Southern Sonora, due to its history. Its colonial infrastructure and its "magic" are why it receives thousands of tourists every year. The author places the municipality as a model to follow due to the gastronomic offer it has prepared for the tourists who visit it.

The structure of short consumer circuits can be inserted into the general system of maize production on a commercial scale because it has agricultural units (restaurants, informal selling points, and the final consumer) that try to be associated because of certain characteristics and operations in a specific territory. This agrees with Jaramillo et al. (2018) who argue that small producers are not excluded in maize general productive systems.

3.6 CONCLUSION

Short consumer circuits have elements that promote the development of sustainable agriculture. These include trade based on direct sales, closer relationships between producers and consumers (with minimal intermediation), low environmental impact, support for family farming and small producers, humane treatment, trust in the production of healthy food, and geographical proximity between producers and consumers, typically within 20 to 200 km.

Sustainable agriculture, through the agroecological model, offers a wealth of production and storage techniques, protects the environment, and aims to expand natural resources while making the best use of nonrenewable ones. This strategy strengthens the transition from the agroecological model to a commercial one.

Producers in Alamos benefit from technical advice provided by agroecological specialists to implement sustainable agriculture in their production processes. This support enhances the possibility for producers to achieve production surpluses in a short period of time and to introduce agroecological maize into the local and regional markets.

Gastronomic tourism routes as a modality of short consumer circuits are a market opportunity for both local maize producers who practice sustainable agriculture through agroecology and for restaurant owners who look to offer safe food to tourists and visitors. It will be essential to create collaborative synergies between government institutions, scientific research centers, local maize producers, tourism entrepreneurs, and business associations related to tourism in the municipality of Alamos, Sonora. If this is not achieved, it will be quite difficult to establish the gastronomic route and to position short consumer circuits within maize general productive systems.

KEYWORDS

- **agroecology**
- **Alamos**
- **food security**
- **food sovereignty**
- **gastronomic route**
- **maize general productive systems**
- **short consumer circuits**
- **sustainable agriculture**

REFERENCES

Bastida, O. (2021). Estadísticas agrícolas de maíz grano en México (1980-2020). Base de datos. Retrieved from: https://blogagricultura.com/estados-produccion-maiz-grano/ (accessed on 30 May 2024).

Bienestar, Estrategia de Acompañamiento Técnico. (2022). *Retrieved from: https://* repositorio-alimentacion.conacyt.mx/jspui/bitstream/1000/160/1/Libro%20 Agroecolog%C3%ADa%20web.pdf (accessed on 30 May 2024).

CEPAL, N. (2014). Fomento de circuitos cortos como alternativa para la promoción de la agricultura familiar. Retrieved from: https://repositorio.cepal.org/bitstream/handle/11362/37152/1/S1420696_es.pdf (accessed on 30 May 2024).

CEPAL, N. (2016). Encadenamientos productivos y circuitos cortos: innovaciones en esquemas de producción y comercialización para la agricultura familiar. Análisis de la experiencia internacional y latinoamericana. Retrieved from: https://repositorio.cepal.org/bitstream/handle/11362/40688/S1600739_es.pdf (accessed on 30 May 2024).

Colín, J. S., Gándara, M. G. R. L., & Saldaña, T. M. (2015). Agricultura guarijía en la cuenca media del río Mayo, Álamos, Sonora. *Revista de Geografía Agrícola, 54*, 9–24.

Contreras-Cruz, A., Sánchez-Morales, P., Romero-Arenas, O., Rivera-Tapia, J. A., Ocampo-Fletes, I., & Parraguirre-Lezama, J. F. C. (2019). Prácticas agroecológicas y su influencia en la fertilidad del suelo en la región cafetalera de Xolotla, Puebla. *Acta Universitaria, 29.*

Cowan, C. (2019, May 22). Tras los pasos del maíz criollo, 50 años después. CIMMYT. Retrieved from: https://www.cimmyt.org/es/noticias/tras-los-pasos-del-maiz-criollo-50-anos-despues/#:~:text=Las%20variedades%20de%20ma%C3%ADz%20criollo,la%20base%20de%20esta%20diversidad (accessed on 30 May 2024).

Cuadras-Berrelleza, A. A., Peinado-Guevara, V. M., Peinado-Guevara, H. J., López-López, J. D. J., & Herrera-Barrientos, J. (2021). Agricultura intensiva y calidad de suelos: retos para el desarrollo sustentable en Sinaloa. *Revista Mexicana de Ciencias Agrícolas, 12*(8), 1401–1414.

Damián, M., & Toledo, V. (2016). *Utopística agroecológica. Innovaciones campesinas y seguridad alimentaria en maíz.* Puebla: Benemérita Universidad Autónoma de Puebla. Primera edición, ISBN: 978-607-487-999-5. Dirección de Fomento Editorial de Benemérita Universidad Autónoma de Puebla. Puebla, México.

Devisscher, M., & Elías-Argandoña, B. (2014). *Del Productor al Consumidor: Una Alternativa Comercial Para la Agricultura Familiar en Bolivia.* AVSF Proyecto Mercados Campesinos y Fundación Xavier Albó. La Paz, Bolivia.

Díaz, F. V., Arteaga, C. C., Mora, X. G., & Galarza, G. V. (2020). Alternativas agroecológicas para el control y manejo de arvenses en competencia específica con el cultivo de maíz. *Revista Caribeña de Ciencias Sociales, 6.*

FAO. (2018). *Los Diez Elementos de la Agroecología. Guía Para la Transición Hacia Sistemas Alimentarios y Agrícolas Sostenibles.* FAO. Retrieved from: https://www.fao.org/3/i9037es/I9037ES.pdf (accessed on 30 May 2024).

Favila, A., & Reyes, Á. D. (2022). Indicadores de competitividad del maíz mexicano en el mercado de Estados Unidos. *RECAI Revista de Estudios en Contaduría, Administración e Informática*, 51–68.

Flores, S. G., Hernández, L. G. G., Almeraya-Quintero, S. X., Pérez-Hernández, L. M., & Sangerman-Jarquín, D. M. (2020). Evaluación de la sustentabilidad del cultivo de maíz en Villaflores y La Trinitaria, Chiapas. *Revista Mexicana De Ciencias Agrícolas, 11*(7), 1565–1578.

Government of Mexico. Servicio de Información Agroalimentaria y Pesquera. (2021). Base de datos. Retrieved from: http://infosiap.siap.gob.mx:8080/agricola_siap_gobmx/ResumenProducto.do (accessed on 30 May 2024).

Government of the State of Sonora. (2016). *Programa de Fortalecimiento de Vialidad para Movilidad en el Estado de Sonora ÁLAMOS. Reporte Final.* Retrieved from: http://seiot.sonora.gob.mx/PDF/PFVMU/ALAMOS/DOCUMENTO_COMPLETO_ALA.pdf (accessed on 30 May 2024).

Hernández, E. (2022). *Biodiversidad mexicana. Comisión Nacional para el Conocimiento y el Uso de la Biodiversidad. Razas de Maíz de México.* Retrieved from: https://www.biodiversidad.gob.mx/diversidad/alimentos/maices/razas-de-maiz (accessed on 30 May 2024).

Hernández, R., Fernández, C., & Baptista, P. (2014). *Metodología de la Investigación* (5th ed.). México, D. F.: McGraw-Hill.

INFOAGRO. (2018). *Retos y Oportunidades del Sistema Agroalimentario.* Retrieved from: https://mexico.infoagro.com/retos-y-oportunidades-del-sistema-agroalimentario/ (accessed on 30 May 2024).

Jaramillo, J. G., Peña, B. V., Hernández, J. H., Díaz, R., & Espinosa, A. (2018). Caracterización de productores de maíz de temporal en Tierra Blanca, Veracruz. *Revista Mexicana de Ciencias Agrícolas, 9*(5), 911–923.

Lizarazu, R. (2014). Los circuitos cortos de comercialización: impacto en la generación de ingresos de pequeños productores rurales. In *Del Productor al Consumidor: Una Alternativa Comercial Para la Agricultura Familiar* (pp. 102–130). La Paz: Agrónomos y Veterinarios sin Fronteras.

López-González, J. L., Damián-Huato, M. A., Álvarez-Gaxiola, J. F., Méndez-Espinosa, J. A., Rappo-Miguez, S. E., & Paredes-Sánchez, J. A. (2018). Maíz (*Zea mays* L.) y seguridad alimentaria en el municipio de Calpan, Puebla-México. *AGROProductividad, 11*(1).

Maceín, J., & Barba, A. (2018). Canales cortos de comercialización. *Serie Estudios IMIDRA*. Madrid, España.

Marty, E. E., & Ibáñez, S. A. P. (2016). Identificación de circuitos comerciales como estrategia para la comercialización de productos agrícolas orgánicos en Ministerios Públicos del cantón Machala. *Cooperativismo y Desarrollo: COODES, 4*(2), 179–187.

Migliorini, P., Bàrberi, P., Bellon, S., Gaifami, T., Gkisakis, V. D., Peeters, A., & Wezel, A. (2020). Controversial topics in agroecology: A European perspective. *Ciencia e Investigación Agraria: Revista Latinoamericana de Ciencias de la Agricultura, 47*(3), 159–173.

Moraga, M. E. (2021). *Enmiendas Orgánicas y Sintéticas y su Efecto en la Producción de Maíz (Zea mays L.) y Frijol (Phaseolus vulgaris L.) y en la Fertilidad del Suelo* (Doctoral dissertation). Universidad Nacional Agraria.

Palomares, J. (2022). Turismo en Sonora. Álamos y sus bellezas conquistan al turismo. Por su infraestructura, riqueza cultural y sus festivales, el Pueblo Mágico atrae a miles de visitantes cada año. *El Imparcial*. Noticias de Sonora.

Paz, R. G., & Infante, C. M. (2020). Circuitos cortos de comercialización: el juego entre lo disponible y lo posible en la agricultura familiar. *Economía y Sociedad, 25*(58), 35–49.

PIEAES. (2021). *La Agricultura del sur de Sonora, Pilar de la Alimentación en el Futuro. La Presión que el Crecimiento Poblacional Ejerce a la Agricultura Familiar y a Pequeña Escala Está Obligando a la Agricultura de Alta Productividad a Transformarse Para Ser Sustentable*. Patronato para la Investigación y Experimentación Agrícola del Estado de Sonora y Centro Internacional de Maíz y Trigo. Retrieved from: https://idp.cimmyt.org/la-agricultura-del-sur-de-sonora-pilar-de-la-alimentacion-en-el-futuro/ (accessed on 30 May 2024).

Ranaboldo, C., & Venegas, C. (2007). *Escalonando la Agroecología. Procesos y Aprendizajes de Cuatro Experiencias en Chile, Cuba, Honduras y Perú*. México: Plaza y Valdés.

Ranaboldo, C., Arosio, M., & Díaz, P. (2016). Circuitos cortos de comercialización. El caso de los mercados públicos institucionales. Centro Latinoamericano para el Desarrollo Rural, RIMISP. Retrieved from: https://www.rimisp.org/documentos/circuitos-cortos-de-comercializacion-el-caso-de-los-mercados-publicos-institucionales/ (accessed on 30 May 2024).

Romero, M., Castañeda, W., Gutiérrez, H., & Torres, J. (2022). Las Escuelas de Campo. Espacio Estratégico para la construcción de autonomía de los pequeños y medianos productores. *Libro. Revoluciones Agrícolas de México*. Secretaría de Agricultura y Desarrollo Rural – Subsecretaría de Autosuficiencia Alimentaria – Instituto Nacional de Investigaciones Forestales, Agrícolas y Pecuarias.

SADER. (2021). *Sistema agroalimentario de México, un desafío de bienestar*. Secretaría de Agricultura y Desarrollo Rural, Gobierno de México. Retrieved from: https://www.gob.mx/agricultura/articulos/sistema-agroalimentario-de-mexico-un-desafio-de-bienestar (accessed on 30 May 2024).

Santillán, A. (2022). *La Identidad Cultural del Cultivo de Maíz en México: Importancia y Retos Ante la Apertura Comercial de Transgénicos* (Doctoral dissertation). Universidad de Santiago de Compostela.

Saravia, P. (2020). Circuitos Cortos de Comercialización alimentaria: análisis de experiencias de la región de Valparaíso, Chile. *Psicoperspectivas, 19*(2), 32–43.

SECTUR. (2014). *Guía de Incorporación y Permanencia Pueblos Mágicos*. Gobierno de México. Retrieved from: https://www.sectur.gob.mx/wp-content/uploads/2014/10/guia-pueblos-magicos.pdf (accessed on 30 May 2024).

SNIIM. (2023). *Sistema Nacional de Información e Integración de Mercados*. Retrieved from: http://www.economia-sniim.gob.mx (accessed on 30 May 2024).

Sosa-Cabrera, E., González-Amaro, R. M., Contreras-Hernández, H. A., & Flores-Romero, C. I. (2022). Factibilidad para comercializar agrodiversidad de la milpa en Circuitos Cortos Agroalimentarios del centro de Veracruz. *Estudios Sociales. Revista de Alimentación Contemporánea y Desarrollo Regional, 32*(60).

Villaseñor, I. (2017). *Agenda Estratégica Para los Productores de Tilapia en el Estado de Morelos y Proyecto Para Desarrollar Puntos de Venta* (Master's thesis). Chapingo, Estado de México. Retrieved from: https://chapingo.orex.es/cgi-bin/koha/opac-detail.pl?biblionumber=110839&shelfbrowse_itemnumber=209652 (accessed on 30 May 2024).

Walters, J. P., Archer, D. W., Sassenrath, G. F., Hendrickson, J. R., Hanson, J. D., Halloran, J. M., Vadas, P., & Alarcon, V. J. (2016). Exploring agricultural production systems and their fundamental components with system dynamics modeling. *Ecological Modelling, 333*, 51–65.

CHAPTER 4

Water Hyacinth as a Source of Lignocellulosic Biomass to Obtain Bioethanol

N. LÓPEZ-RAMÍREZ[1] and G. N. JUÁREZ-LUNA[2]

[1]*División de Energías Renovables, Tecnológico Nacional de Mexico/ TES Chicoloapan, Camino al Monte s/n, Esquina con Camino a Tlalmimilolpan, Loma de Guadalupe, Ejido de Chicoloapan, Estado de Mexico, C.P., Mexico*

[2]*Departamento de Ingeniería Química, Industrial y de Alimentos, Universidad Iberoamericana, Prolongación Paseo de Reforma 880, Lomas de Santa Fe, Ciudad de Mexico, C.P., Mexico*

ABSTRACT

This chapter highlights the importance of generating new alternatives to obtain energy that are less susceptible to exhaustion and, above all, less polluting, thus avoiding environmental deterioration. Bioethanol, for example, has acquired great value due to its use as fuel and above all because it is a renewable energy source since it can be obtained from lignocellulosic biomass, which is the most abundant on earth and comes from crops, agricultural residues, crops, and other industrial residues, thus allowing their recovery, since they are mainly composed of polysaccharides such as hemicellulose and cellulose, which can be subjected to different treatments such as physical, chemical, and biological to obtain fermentable sugars. An alternative could be the use of the water hyacinth, which is considered a weed for the bodies of water in which it develops, due to its rapid growth, obtaining on the one hand a solution to the environmental problems it causes

Sustainable Agricultural and Food Systems Engineering: New Research, Practices, and Developments. Jose Juan Buenrostro-Figueroa, Julio César Tafolla-Arellano, Leonardo Sepúlveda Torre, Porteen Kannan, A. K. Haghi (Eds.)

and on the other hand it would have an abundant source of lignocellulosic biomass to produce bioethanol, besides that it is a biomass that would not compete with the biomass obtained from food crops.

4.1 INTRODUCTION

The increase in global energy demand from fossil sources has created a series of important problems related to extraction sources, climate change, and a decrease in its supply. It is estimated that reserves of petroleum and gas will run out by 2042, and mineral coal will be the only fossil energy source until 2112. Renewable energy sources provide us with a great alternative to partially replace fossil fuels and offer a feasible way to utilize some abundantly available materials in nature (Kheybari, Kazemi, & Rezaei, 2019).

The second generation of biofuels, produced from non-food value biomass as raw material, is emerging as an option for the transportation sector. Biofuels could serve as an effective substitute for fossil fuels with the aim of reducing greenhouse gas emissions, improving air quality, and tapping into new energy resources (Rocha-Meneses, Raud, Orupõld, & Kikas, 2017). In particular, ethanol obtained from lignocellulosic biomass has garnered significant attention as an alternative to petroleum-based fuels due to its low cost, abundance, and compatibility with internal combustion engines and the gasoline supply chain (Biswas & Biswas, 2022; Martín & Grossmann, 2012).

Several countries (e.g., India, Mexico, and Brazil) have made a great investment in research focused on valorizing lignocellulosic biomass rich in sugars (45%–75% of polysaccharides) to produce liquid fuels, mostly ethanol from renewable and non-food value resources (Li, Jiang, Wu, Wu, Yang, Song, Ahmad, & Jin, 2022). Lignocellulosic biomass is the most abundant renewable resource on Earth and represents the main fraction of municipal, agro-industrial, and forest wastes (Limayem & Ricke, 2012). The use of lignocellulosic biomass to produce biofuels presents important advantages, among them, it has no food value, it is cheaper than other sources, and it is eco-friendly. Therefore, the biofuels obtained from lignocellulosic biomass produce fewer greenhouse gas emissions and reduce the environmental impact, specifically climate change (Ahmed et al., 2022; Soccol et al., 2019).

The bioconversion of lignocellulosic biomass into fermentable sugars is the most important stage of biorefining because it is the basis for ethanol production. Although many studies have been carried out to determine the future challenges in bioenergy generation, currently there is no single process or technology that is self-sustainable for the conversion of lignocellulosic biomass into ethanol. The global process mainly comprises the hydrolysis of polysaccharides of the lignocellulosic matrix into fermentable sugars, followed by fermentation and finally distillation to recover the ethanol produced (Binod, Janu, Sindhu, & Pandey, 2011).

The most important reservoir of lignocellulosic biomass is found in the cell walls of vegetables. It is a composite built of two kinds of polysaccharides: hemicellulose and cellulose, and one polyphenolic material: lignin. Both polysaccharides are a source of fermentable sugars, while lignin could be useful to isolate other chemical products of interest. Generally, the waste of cereals (e.g., rice straw, wheat straw, corn stubble, and cane bagasse) possesses a high fraction of lignocellulosic biomass and represents the most important raw material for biorefineries (Hassan, Williams, & Jaiswal, 2018). Weeds also contain lignocellulose and, therefore are a potential source for manufacturing a usable grade of bioethanol that can fulfill the fuel demand and other industrial needs. Water hyacinth biomass has no food value, is rich in polysaccharides, and poor in lignin, so it is a very interesting material to valorize as the feedstock in the biological conversion into ethanol process due to its easy availability in a tropical country (Biswas & Biswas, 2022).

As mentioned above, the most important criteria for choosing a biomass or lignocellulosic source are – (i) high content of structural carbohydrates; (ii) low content of lignin; and (iii) lack of food value. Some studies have demonstrated that Water hyacinth (*Eichhornia crassipes*) meets these three criteria as it contains 18–55% hemicellulose, 18–35% cellulose, and 3.5–10% lignin (in dry weight) (Nigam, 2002). Although there is high variability in the reported composition, in general terms, water hyacinth is considered rich in structural carbohydrates with very low lignin content and, due to its growth conditions (aquatic), it has no food value for humans. This composition is suitable for extracting fermentable sugars. Water hyacinth can therefore be used as lignocellulosic biomass and a good source of fermentable sugars for producing ethanol, organic acids, and other chemicals (Biswas & Biswas, 2022; Kumar, Singh, & Ghosh, 2009).

The conversion of lignocellulosic biomass into ethanol follows a four-step strategy. The first step is pretreatment to separate the complex lignin-hemicellulose-cellulose structure to make it more susceptible to hydrolysis (chemical or enzymatic). The second step is the hydrolysis of hemicellulose and cellulose to release monosaccharides. The third and fourth steps involve fermentation and distillation, respectively (Hu, Heitmann, & Rojas, 2008).

The hydrolysis process is the critical stage in ethanol production from lignocellulosic biomass due to its effects related to the substrate and enzymes. Chemical, thermal, and mechanical strategies have been studied to hydrolyze carbohydrate polymers. However, all these methods demand high energy consumption and generate toxic compounds. For the reasons mentioned above, the enzymatic hydrolysis of carbohydrate polymers is considered a great alternative to release monosaccharides. Despite enzymatic hydrolysis offering the potential to obtain higher yields and selectivity and lower energy consumption than chemical, thermal, or mechanical processes, the mechanisms of enzymatic hydrolysis and the relationship between substrate and enzyme (glycosyl hydrolase) are not yet completely understood (Yang, Dai, Ding, & Wyman, 2011).

The success of enzymatic hydrolysis of cellulose and hemicellulose depends on the synergy between several enzyme activities. For cellulose hydrolysis, endoglucanases (EG), cellobiohydrolase (CBH), and β-glucosidase (βG) are necessary. For hemicellulose hydrolysis, endoxylanases, β-xylosidases, α-L-arabinosidase, acetylxylan esterase, and the α-glucuronidase should be present (Zhang & Viikari, 2014).

It is also important to emphasize that thanks to its composition, the water hyacinth biomass has potential in several applications such as the synthesis of composite materials (Ajithram, Jappes, Chintra, & Daphne, 2023), biogas and biodiesel production (Gebehiwot, Dekebo, & Endale, 2022; Ali et al., 2022), organic water removal or phytoremediation, for obtaining high-value chemicals (e.g., cellulose nanofibers and nanocrystals, furfurals, etc.) (Juárez-Luna, Favela-Torres, Quevedo, & Batina, 2019; Ilo, Simatele, Nkomo, Mkhize, & Prabhu, 2020) for hydrolytic enzymes production (López-Ramírez et al., 2018), and animal feed and biofertilizer (Ilo, Simatele, Nkomo, Mkhize, & Prabhu, 2020).

In the global context, it is mandatory to look for energy resources that permit us to diminish our dependence on fossil fuels. Lignocellulosic biomass is a promising raw material for ethanol production. This work aims to present a critical review of several research studies focused on obtaining ethanol from water hyacinth.

4.2 WATER HYACINTH

The water hyacinth (*Eichhornia crassipes*) is a plant native to the Amazon River in South America. It is the major freshwater weed in most of the frost-free regions of the world and is generally considered the most troublesome aquatic plant in tropical regions. It was widely planted as a water ornamental around the world because of its striking flowers. It belongs to the *Pontederiaceae* family. Morphologically, it possesses leaves, stems, and roots. The leaves have petioles, both covered with a thin cuticle. The stem is covered with a thin cuticle and lignified sclerenchyma binds the central cylinder with scattered vascular bundles. The roots have endodermis with Casparian bands and an absence of cuticle and lignified sclerenchyma in the structure. All these morphological characteristics make it a free-floating plant that has encountered suitable environmental conditions in lakes, rivers, and rice paddy fields (Rojas-Sandoval & Acevedo-Rodríguez, 2013; Ganorkar, Jadeja, & Desai, 2022).

Water Hyacinth (WH) is considered nowadays as one of the worst weeds in the world, causing severe economic and environmental threats due to its adaptability to tropical climates, rapid growth rate, and robust nature (Thu Ha Tran, & Nguyen, 2022). Under optimal conditions, especially in eutrophic water basins rich in nutrients, particularly phosphorus (P) and nitrogen (N), WH is one of the most productive plants on earth capable of covering huge areas. According to Ali et al. (2022), WH can duplicate its covered area in just 6–28 days and double its weight in 4–58 days. The average growth rate is around 220 Kg per ha/day, which is much higher than other plants in a similar environment (Pal et al., 2023). The spread and proliferation of WH can have negative consequences such as interfering with irrigation, and navigation, depleting oxygen in water, and reducing the efficiency of electricity generation. All these negative characteristics have a negative impact on economic development, human health, and the environment (Ali et al., 2022). Fortunately, the great quantity of WH generated is usable biomass, which can be a raw material for energy and value-added products due to its chemical composition (Pal et al., 2023).

Different applications have found that WH biomass, composed mainly of hemicellulose, makes it a good raw material for lignocellulose-based energy production (pyrolysis, bio-methanation, biohydrogen, bioethanol, etc.) and for other industrial processes like composting, vermicomposting, and absorption of hydrocarbons (Suthar, Sharma, Kumar, Banu, & Tyagi, 2022). Table 4.1 shows the cellulose, hemicellulose, and lignin content determined by several researchers. It has been determined that WH biomass is rich in polysaccharides fraction and has a low lignin concentration.

TABLE 4.1 Cellulose, Hemicellulose, and Lignin Content in Water Hyacinth Biomass

Component (% Dry Matter)			References
Cellulose	**Hemicellulose**	**Lignin**	
18.2	48.7	3.5	Nigam (2002)
19.5	33.4	9.27	Gunnarsson & Petersen (2007)
19.7	27.1	–	Mishima, Kuniki, Sei, Soda, Ike, & Fujita (2008)
20	33	10	Bhattacharya & Kumar (2010)
31.44	44.68	19.99	Das, Ghosh, Paul, Ghosh, Pati, & Mondal (2016)
19.80	49.00	4.80	Zhang, Wei, Han, & Weng (2018)
24.88	45.12	5.12	Shankar et al. (2020)
24.6	30.5	10.9	Thu Ha Tran, & Nguyen (2022)
65	35	2.5	Ajithram, Jappes, Chintra, & Daphne (2023)
57	25.6	4.1	Tanpichai, Biswas, Witayakran, & Yano (2023)

The main components of water hyacinth biomass correspond to polysaccharides, which are susceptible to being hydrolyzed via chemical or enzymatic processes to release monosaccharides fermentable to produce ethanol. However, its composition permits the use of this biomass in several applications, for example, in paper production, as a fertilizer, animal feed, for water treatment, phytoremediation, isolation of high-value chemical products, and transformation into biogas or biodiesel.

Water hyacinth has been used as a bio-absorbent due to its chemical composition, which is mostly carbon and oxygen. Water hyacinth petiole contains approximately 40% holocellulose, making it a good absorbent material as it can be defibrillated after acid or alkaline treatment (Kabir et al., 2022). Carbon-based materials produced by water hyacinths have been discovered to eliminate a large volume of organic contaminants in water sources. These organic substances comprise phenol, sulfachloropyridazine, fluoroquinolones, and tetracycline. Fibers isolated from water hyacinth have been employed to absorb volatile alkane hydrocarbons (Amalina, Razak, Krishnan, & Zularisam, 2022). Kabier et al. reported that water hyacinth alkaline treatment was capable of removing 97.13% of azithromycin from an initial solution of 2,500 mg/L under alkaline conditions (pH 9). Furthermore, the isotherm, kinetic, and thermodynamic models suggest chemisorption as the main mechanism of azithromycin removal due to the great variety of functional groups present in the water hyacinth lignocellulosic biomass surface (Kabir et al., 2022).

Also, WH biomass fibers have been isolated by different methods (chemical or physical) to create polymer composites to enhance their mechanical properties, particularly tensile strength. Now, natural fibers are being utilized to produce a variety of products with added value, making them more valuable in many fields (Ajithram, Jappes, Chintra, & Daphne, 2023).

Recently, cellulose nanoparticles (e.g., cellulose nanofibers and cellulose nanocrystals) have received a great deal of attention in a wide variety of applications such as biomaterials, packaging, electronic devices, etc. These high-value components derived from WH lignocellulosic biomass present a series of advantages compared to synthetic polymers given that they are biocompatible, biodegradable, have a low density, low thermal expansion coefficient, superior mechanical properties, and are cheaper (Ilo, Simatele, Nkomo, Mkhize, & Prabhu, 2020). Juárez-Luna et al. (2019) isolated nanocrystals from cellulose thermochemically extracted from water hyacinth stems and partially hydrolyzed with a cellulase complex. On the other hand, Tanpichai et al. (2023) obtained a similar result by treating the WH lignocellulosic biomass with an alkali solution followed by a maceration process. The nanoparticles obtained in both works have at least one dimension under 100 nm, meeting the main criteria to consider this kind of material as "nano."

The degradation of lignocellulosic biomass into CO_2 and water in nature is a very slow and complex process that requires the presence of several metabolic pathways. The degradation of cellulose is the result of a combination of β-glucosidases, cellobiohydrolases, and β1-4-glucanases enzymatic activities, while hemicellulose degradation requires activities such as acetyl-xylan esterases, α-glucuronidases, β-D-xylosidases, and endo-1,4-β-xylanases (Kumar et al., 2022). Some authors have used water hyacinth as a substrate to produce hydrolytic enzymes, including cellulase and hemicellulase complex. López-Ramírez et al. (2018), in a stirred bench-scale solid-state bioreactor, produced cellulases and hemicellulases with a strain of *Trichoderma harzianum*. The results show high hemicellulose activity (107 ± 0.3 U/g dm measured as xylanase) when the bioreactor was operated at 30°C and a mixing rate of 3 rpm. In this way, a culture of *T. harzianum* or another microorganism capable of producing hydrolytic enzymes could help accelerate the process of saccharification of water hyacinth biomass for ethanol production.

WH biomass presents the advantage of containing low lignin due to the degradation process requiring high energy, and the hemicellulose and cellulose are more easily converted into monosaccharides, resulting in an enormous amount of utilizable biomass for the biofuel industry. The WH biomass can be used to produce biogas, hydrogen, butanol, and ethanol (Bhattacharya & Kumar, 2010).

4.2.1 BIOETHANOL

The economic development has driven the world toward an increase in energy demand. At present, approximately 82% of energy consumption in the world comes from fossil fuels. Therefore, it remains the main energy source. However, the production of biofuels to replace fossil fuels has increased in recent years. Biotechnological processes to obtain biofuels offer attractive ways due to being environmentally friendly, green, and reasonably profitable. Currently, the production of biofuels such as biodiesel, bioethanol, or biobutanol using residual lignocellulosic biomass like water hyacinth is a potential alternative to develop a sustainable culture and green energy independence (Osorio-González et al., 2023; Rocha-Meneses, Raud, Orupõld, & Kikas, 2019).

Industrial research efforts have become more focused on low-cost large-scale processes for lignocellulosic feedstock originating mainly from agricultural and forest residues along with herbaceous materials and municipal wastes (Taouda, Chabir, Aarab, Miyah, & Errachidi, 2017). Ethanol is considered a great alternative to partially substitute fossil fuels, whether as a single fuel in modified engines or as an additive in fuel mixes without engine modification when ethanol content reaches a maximum of 30% (Talebnia, Karakashev, & Angelidaki, 2010).

The design of profitable processes for ethanol production implies the selection of raw materials and the selection and definition of the configuration of the process that makes possible the conversion of said raw materials into the final product that meets the specifications (Cardona & Sánchez, 2007).

Due to the abundance of lignocellulosic biomass, it can be considered a suitable material for bioconversion into ethanol. There are several options to consider as raw materials for bioethanol production, including biomasses like rice straw, sugarcane bagasse, wheat straw, cotton stalk, bamboo, and water hyacinth (Rezania et al., 2017). Ethanol production from lignocellulosic biomass (second-generation biofuels) presents the highest potential among all alternatives without compromising food safety, although it is not yet a reality at an industrial scale (Wang, Zheng, & Xue, 2019; Dias et al., 2012).

Water hyacinth infestation will never be eradicated; therefore, the only sustainable strategy to control it is to use the biomass naturally produced in several fields, including energy production (Hirphaye, Mezgebe, Tura, & Fanta, 2022). Water hyacinth biomass is a potential raw material for ethanol production because it has a high content of hemicellulose and cellulose (45% and 20%, respectively, in dry weight) and a very low content of lignin (10% in dry weight). Furthermore, this plant is considered a noxious weed, has no food interest for human consumption, grows at a high rate, is available in tropical regions, and is inexpensive (Zhang, Wei, Han, & Weng, 2018).

Considering its technical feasibility and abundance, the WH could be used as the feedstock for second-generation bioethanol. Recent studies have focused on two relevant aspects of the transformation: (i) the effects of different pretreatments; and (ii) the effects of different strains, enzymes, and catalysts in the simultaneous saccharification and fermentation (Wang, Zheng, & Xue, 2019). For example, *Saccharomyces cerevisiae, Pichia stipitis,* and recombinant strains of *Escherichia coli,* among others, have been evaluated for the bioconversion of WH lignocellulosic biomass into ethanol. On the other hand, alkaline and acid pretreatments for reducing sugar release have been evaluated.

Ethanol production is a four-stage process. The first stage is pretreatment for lignin removal. The second stage corresponds to hydrolysis for the release of reducing sugars. The third stage is bioconversion into ethanol (fermentation), and finally, the fourth stage is distillation for the recovery of the ethanol produced (Asshokkumar et al., 2022).

For example, Mishima et al. (2008) produced ethanol from WH using *Saccharomyces cerevisiae* NRBC 2346 with a yield of 14 g/l of ethanol. When a recombinant strain of *Escherichia coli* KO11 was used, the yield increased to nearly 17% after simultaneous saccharification and fermentation.

On his part, Shankar et al. (2018) pretreated leaves of WH with diluted sulfuric acid (2.5%). After that, it was saccharified enzymatically with a crude enzyme secreted by *Sphingobacterium* sp. and finally fermented to ethanol with *S. cerevisiae*. The maximum yield was 6.18 g/l.

In the work of Nigam (2002), it is reported that using WH pretreated with 1% (v/v) sulfuric acid and heat as a culture medium (30°C, pH 6.0, and 0.02 aeration rate), the highest ethanol yield (Yp/s) was 0.35 g EtOH/gS with a strain of *Pichia stipites* NRRL Y-7124.

Lignocellulosic biomass is a recalcitrant structure due to the interactions of the three biopolymers mentioned above (cellulose, hemicellulose, and lignin). To achieve the highest yield, pretreatment is considered a key stage that should address several challenges including costs, environmental concerns, and sustainability (Asshokkumar et al., 2022).

4.2.2 LIGNOCELLULOSIC BIOMASS

Vegetal cells produce sugar blocks directly from photosynthesis, which occupies around 75% of the lignocellulosic biomass. Also known as lignocellulose, they are a sustainable organic source with net-zero carbon emissions and do not compromise global food supplies (Asshokkumar et al., 2022). Lignocellulose is the most colossal energy reservoir in the ecosystem. Apart from being eco-friendly, renewable, and economically viable, lignocellulosic biomass has been regarded as the most suitable raw material for biofuel production.

It is composed of three biopolymers tightly bonded: cellulose, hemicellulose, and lignin. However, the proportion between the three polymers depends on the origin of the biomass and the part of the plant analyzed (e.g., leaves, fruits, etc.). The lignocellulosic biomass also contains proteins, extractives, inorganic compounds, and pectin in small amounts (Li, Jiang, Wu, Wu, Yang, Song, Ahmad, & Jin, 2022; Ahmed et al., 2022). Its composition is largely diverse with a complex hierarchical structure consisting mainly of cellulose (40–60%), hemicellulose (20–40%), and lignin (10–25%), along with other minor components, such as minerals, acetyl, and phenolic groups (Asshokkumar et al., 2022). The lignocellulosic biomass sources available are usually grouped into three categories: primary, secondary, and tertiary sources. The primary source includes key products or crops such as short-rotation plantations and sugarcane. Residues such as bagasse, straw, and rice husks are secondary sources of biomass produced during some production processes. Finally, the tertiary sources of lignocellulosic biomass are the residue end products of various processes, such as sewage treatment, i.e., the organic portion of the solid waste. Tertiary biomass sources are cheap and include the industrial and domestic waste collected in a specific area (Ahmed et al., 2022).

In lignocellulosic biomass, a matrix of cellulose and lignin is surrounded by hemicellulose chains. Cellulose is a homopolymer of β-D-glucose, while hemicelluloses are mainly a heteropolymer composed of at least five different monomers, called D-xylose, D-arabinose, D-mannose, D-glucose, and D-galactose. Additionally, D-glucuronic and D-galacturonic acids can be present in branched chains (Rezania et al., 2017; Houfani, Anders, Spiess, Baldrian, & Benallaoua, 2020). Lignin is an amorphous heterogeneous polymer insoluble in water and composed of three aromatic alcohols also called monolignols corresponding to propyl phenol units: p-coumaryl, coniferyl, and sinapyl alcohol (Houfani, Anders, Spiess, Baldrian, & Benallaoua, 2020).

Particularly, lignocellulose from WH is rich in cellulose and hemicellulose (40–70% in dry weight) and has low lignin content (3.5–9.5%). Lignin removal is a difficult process that requires the use of some chemicals and enzymes; for this reason, lignocellulosic biomass with lesser lignin content can improve the economic potential of biofuel production compared with biomass rich in lignin (Rezania et al., 2017; Bhatt & Shilpa., 2014).

To achieve the correct transformation of lignocellulosic biomass, a strategy of four steps is commonly followed. The first step is the pretreatment of the lignocellulosic biomass to eliminate the lignin. After, the hydrolysis step is necessary to convert cellulose and hemicellulose (polysaccharides) into monosaccharides. Once monosaccharides have been released, the fermentation process is carried out with the correct microorganism. Finally, the ethanol produced is recovered by distillation (Hu, Heitmann, & Rojas, 2008).

Pretreatment is a crucial step to improve the bioavailability of the polysaccharides to be transformed into ethanol. Nevertheless, it is so important that the cost of the pretreatment never exceeds 40% of the total cost of the process (Akram et al., 2023). Pretreatment techniques can be physical, chemical, biological, or a mix of them, mainly physicochemical (Asshokkumar et al., 2022).

4.2.2.1 CELLULOSE

Cellulose is the most abundant polysaccharide in the world, synthesized by several organisms including bacteria, algae, tunicates, and higher plants. In plants, it is the principal component of lignocellulosic biomass and represents between 40 and 50% of dry weight (Li, Bashline, Lei, & Gu, 2014; Shahzadi et al., 2014).

Cellulose is a biological unbranched linear homopolymer constituted of β-D-anhydroglucopyranose units bonded covalently by β-(1–4) glycosidic bonds. Specifically, acetal (glycosidic) bonding is formed between carbon number one of a ring of glucopyranose and carbon number four of another ring of glucopyranose. The fundamental unit that repeats in cellulose is the disaccharide called cellobiose. At the end of every chain, there exists a hemiacetal group that acts as a reducing agent, so it is called the reducing end (RE). At the other extreme (beginning of the chain), the reducing behavior is lacking, so it is denominated the non-reducing end (NRE) (Holding, 2016). The length and molecular weight of cellulose depend on its origin. Both molecular weight and length are measured by the degree of polymerization

(DP) or the number of units of β-D-anhydroglucopyranose per chain. In lignocellulosic biomass, the DP average is around 10,000 (Holding, 2016; Sharma et al., 2023). Cellulose has abundant hydroxyl (-OH) active functional groups that are capable of forming intra- and intermolecular bonds between polymeric chains, resulting in strong hydrogen bond networks. Hydrogen bonds formed between hydroxyl groups and oxygen atoms between neighboring cellulose chains, together with van der Waals interactions, lead to the formation of different patterns called crystalline cellulose (Li, Bashline, Lei, & Gu, 2014). The crystalline structure results from the assembly of individual chains into elementary fibrils with a cross-section of 3–5 nm and a length of several micrometers. When tens of elementary fibrils auto-assemble in rectangular arrays surrounded by hemicellulose and lignin, they form the so-called microfibrils (10–30 nm). A collection of microfibril hierarchical packing together finally constitutes the macroscopic cell wall of the plant (Zhao, Zhu, Cheng, Chen, Wu, & Yu, 2021).

Cellulose exists in nature as a semicrystalline polysaccharide, with amorphous regions and crystalline regions. Amorphous cellulose is easier to access, so it can be easily hydrolyzed by cellulases. However, in the case of crystalline cellulose, it is harder to access the enzymes, and consequently, these regions cannot be completely broken down. The main products obtained from the hydrolysis of cellulose from lignocellulosic biomass include ethanol, biogas, succinic acid (SA), levulinic acid (LA), sorbitol, hydroxy acetaldehyde, and many others (Sharma et al., 2023).

4.2.2.2 HEMI CELLULOSE

Hemicellulose is one of the most abundant components of lignocellulosic biomass. It represents around 15–35% of it. Due to its amorphous molecular configuration and highly branched structure, it is easier to convert into monosaccharides than cellulose (Luo, Liu, Zheng, Li, Zhou, Huang, & Shuai, 2019).

It is a heteropolymer with a lower degree of polymerization (DP) and molecular weight compared to cellulose. It contains both branching and linear polysaccharides that are composed of a variety of anhydrous sugars (xylose, arabinose, D-glucuronic, L-rhamnose, and others). Since hemicellulose components have an amorphous nature and low DP, they do not significantly increase the biomass recalcitrance (Sharma et al., 2023). The degree of polymerization for hemicellulose is around 200 on average. The content and type of polysaccharides depend on the source of the biomass.

For example, hardwood has a hemicellulose content of 10–15%, softwood 18–23%, and herbaceous plants 20–25%. Hemicellulose is a polymer that breaks down more quickly than cellulose when exposed to heat or chemicals. It contains a significant amount of carboxyl groups, unlike cellulose which contains both carboxyl and carbonyl groups (Ahmed et al., 2022).

Hemicellulose may also be closely bound to lignin by aromatic esters and to cellulose via hydrogen bonds, thus forming a bridge between cellulose and lignin. There are several types of hemicelluloses depending on the main sugar constituting the polymer. Xylan is built from xylose units linked by β(1-4) glycosidic bonds to L-arabinose substituted xylopyranose unit, 4-O-Methyl-D-glucuronic acid, acetyl feruloyl, and/or units of p-coumaroyl. Ferulic acid can potentially interact with each other through diester-forming ester bonds, which link two xylan chains or allow the association of hemicelluloses with lignin. Xyloglucans consist of a chain of D-glucose linked by β-1,4-glycosidic bonds and substituted by D-xyloses. Galacto-(gluco)-mannans consisting of glucose and mannose units represent the main hemicelluloses of angiosperms and coniferous trees. Other lignocellulosic materials may also contain considerable amounts of arabinogalactans and galactans. In general terms, it contains a significant amount of carboxyl groups, unlike cellulose which contains both carboxyl and carbonyl groups. It is the neutral and/or acidic side group forming the branches of hemicelluloses that make them non-crystalline or weakly crystalline (Ahmed et al., 2022; Houfani, Anders, Spiess, Baldrian, & Benallaoua, 2020).

4.2.2.3 LIGNIN

The third major macromolecule that constitutes lignocellulose biomass is lignin. This macromolecule (molecular weight of more than 10,000 Daltons) plays an essential role in the plant given that it is responsible for the force and mechanical resistance of cell walls. Furthermore, it actively participates in water transport and forms the basis of plant protection against pathogens and insects. Lignin is the most complex and renewable aromatic three-dimensional biopolymer. It contains a variety of functional groups such as phenyl-propane, methoxy, and phenolic compounds and contributes between 10 and 25% of lignocellulosic biomass. The great variety of functional groups provides active centers for chemical and biological modification (Zhang & Viikari, 2014; Asshokkumar et al., 2022; Sameni, Krigstin, & Sain, 2016).

Lignin is an irregular non-crystalline structure lacking a defined primary structure formed by free-radical polymerization of alcohol precursors. According to Vanholme et al. (2019), it is synthesized by several biosynthetic pathways including shikimate, phenylpropanoid, and the synthesis of monolignols by oxidation of hydroxycinnamyl alcohols. The three primary units in lignin are the p-hydroxyphenyl (H), guaiacyl (G), and syringyl (S) groups, derived from the oxidative coupling of p-coumaryl, coniferyl, and sinapyl alcohols (Asshokkumar et al., 2022; Vanholme, Meester, Ralph, & Boerjan, 2019).

Despite the fact that WH biomass contains a low concentration of lignin, it is necessary to apply a physical, chemical, or biological pretreatment (or a mix of them) since the presence of lignin, even in low concentration, prevents the accessibility of cellulase enzymes for saccharification (Umai, Jacob, & Kumar, 2022). Due to its abundance and chemical composition, lignin represents a potential source to produce adsorbents, chemical products, and polymer processing materials. Nowadays, lignin is considered a roadblock in lignocellulose biomass valorization due to its unproductive binding with cellulolytic enzymes (Sharma et al., 2023). Pretreatments aim to break the hydrogen bonds and cross-linked hydrophobic interactions between cellulose, hemicellulose, and lignin, thus increasing the accessibility of biomass to enzyme hydrolysis. An effective pretreatment may improve the accessibility of cellulases, pore size, and surface area while lowering its degree of polymerization (DP) and crystallinity (Sharma et al., 2023).

4.3 PRETREATMENTS

The pretreatment methods used to improve the conversion of polysaccharides into ethanol are considered vital tools to achieve a profitable process. It is known that pretreatments represent the most expensive stage in this kind of technology. The ideal cost of pretreatment should be between 20 and 40% of the production cost. In general, the objective of pretreatment is to disrupt the lignocellulosic structure by disintegrating lignin and hemicellulose and breaking cellulose into small fragments that are exposed to hydrolysis, either enzymatic or chemical (Ahmed et al., 2022).

Pretreatment methods for lignocellulosic materials can be classified as physical, chemical, biological, or a combination of them. It is always necessary to apply a physical pretreatment to reduce the particle size and, depending on the specific characteristics of the lignocellulosic biomass, choose the most suitable pretreatment. A desirable pretreatment is capable of improving porosity, removing lignin, preventing the formation of inhibitors, avoiding sugar degradation, reducing crystallinity, and requiring less energy input (Sharma et al., 2023).

For WH lignocellulosic biomass, acid or alkaline, diluted, or concentrated pretreatments are commonly preferred before the fermentation process (Bayrakci & Koçar, 2014). However, different existing biomass pretreatment processes have advantages and limitations in terms of technical, economic, and environmental aspects such as energy consumption, waste generation, chemical usage, and investment cost (Areepak et al., 2022).

4.3.1 PHYSICAL PRETREATMENT

The principal aim of physical pretreatment is to reduce the particle size of lignocellulosic biomass and increase the surface accessibility of polysaccharides. In this sense, commonly mechanical devices are utilized, for example, mills, chipping, grinding, and freezing (Sharma et al., 2023). The biomass surface area increases with the reduction of the size and crystallinity of particles through mechanical pretreatments. Typical comminution mechanical methods are conducted before chemical or biological processing (Sheng et al., 2021).

Besides, to eliminate and destroy lignin and hemicellulose fractions from lignocellulosic biomass, several cutting-edge techniques have been employed. For instance, irradiation with gamma rays and electric energy pulses, hydrothermal treatment, pyrolysis, microwave, ultrasound, and extrusion (Ahmed et al., 2022).

Until now, pretreatment with microwave and ultrasound has been demonstrated to be feasible to get a profitable pretreatment of the lignocellulosic biomass process. However, the efficiency of microwaves depend on the dielectric properties of the feedstock, to be able to convert electromagnetic energy into heat. High moisture and high inorganic content are necessary to achieve the hydrothermal conditions required for the hydrolysis of lignocellulosic biomass (Kumar et al., 2022). Most of the methods mentioned above are high energy consumers, high-cost processing, and require specialized instrumentation and equipment. Therefore, they are still unattractive for industrial applications (Sharma et al., 2023).

4.3.2 CHEMICAL PRETREATMENT

The chemical pretreatment has the objective of destroying lignin, swelling the hemicellulose and cellulose fibers, increasing the porosity, and exposing the polysaccharide fraction to enzymes, acid, or alkali attack. For this purpose, several chemical products have been evaluated. This includes solvent-based, acid-alkali, ionic liquids, steam explosion, inorganic salts, and many others (Ahmed et al., 2022).

Acid or alkali pretreatments are the conventional methods to fractionate WH lignocellulosic biomass. It has been demonstrated that the pretreatment with sulfuric acid (5% v/v) improves the yield of ethanol produced. However, there is a risk of the formation of growth inhibitor compounds like furfural, hydroxymethylfurfural, and formic and levulinic acids after 45 minutes of the process. Nevertheless, if sulfuric acid (4% v/v) pretreatment is carried out combined with hydrothermal pretreatment, the yield of ethanol increases to 5.1 g/l due to the alteration of the lignocellulosic matrix without the formation of growth inhibitor compounds (Rezania et al., 2017).

However, the conventional pretreatment methods generate hazardous wastes, difficult to handle and dispose of. The migration toward green chemistry requires the development of new alternatives such as Ionic Liquids (IL) and Deep Eutectic Solvents (DES) given that they are affordable, biodegradable, simple to prepare, recyclable, and produce fewer toxic byproducts (Kumar et al., 2022; Umai, Jacob, & Kumar, 2022).

The ethanol production from WH lignocellulosic biomass seems to be sustainable management of this weed considered noxious. It has been reported that acid-diluted pretreatment combined with a biological pretreatment is a good strategy to hydrolyze WH biomass and should possibly achieve a concentration of 430 mg g^{-1} of reducing sugars while the yield reaches 1.4 g L^{-1} of ethanol without adding enzymes (Zhang, Wei, Han, & Weng, 2018).

4.3.3 BIOLOGICAL PRETREATMENT

Biological pretreatment is the use of microorganisms or enzymes to degrade hemicellulose and lignin. Several strategies exist to carry out this kind of pretreatment. For example, the use of purified enzymes tailored to specific lignocellulosic biomass composition. Alternatively, treatment with fungi or bacteria has been evaluated. In the case of bacteria, axenic strains or consortia of bacteria have been employed. Biological pretreatments offer several advantages over physical and chemical pretreatments, including moderate operational conditions such as temperature, acidity (pH), and pressure. They are also cost-effective, do not form growth inhibitor compounds, and do not generate hazardous byproducts. However, they do have some limitations. For instance, there is less available surface area due to the structural complexity of biomass, a slow digestion rate, and the potential for contamination with fungi or unwanted microorganisms (Ahmed et al., 2022; Sharma et al., 2023).

Acid pretreatment followed by enzymatic hydrolysis has been considered a feasible method for obtaining reducing sugars from lignocellulosic biomass. Once lignin has been removed, the high selectivity of cellulolytic enzymes allows for the production of monosaccharides such as glucose and xylose (Luo, Liu, Zheng, Li, Zhou, Huang, & Shuai, 2019).

4.4 ENZYMATIC HYDROLYSIS

Hydrolysis of cellulose and hemicellulose is mainly carried out by inorganic acids (e.g., H_2SO_4, HCl) or by hydrolytic enzymes through cellulases and hemicellulases. Enzymatic hydrolysis is the most important stage during the transformation of lignocellulosic biomass into ethanol. Cellulases are a complex consisting of at least three different enzymatic activities that work simultaneously and synergistically as a mixture, thus increasing their catalytic activity. Hemicelluloses is a complex of at least six different enzymes. The objective of enzymatic hydrolysis is the release of monosaccharides (Sharma et al., 2023; Amit, Nakachew, Yilkal, & Mukesh, 2018).

4.4.1 CELLULASES

Cellulases are a group of three different enzyme activities (endoglucanases, exoglucanases, and β-glucosidases) required for the complete degradation of the cellulose fraction into glucose. Endoglucanases hydrolyze random interchain β1-4-glycosidic bonds. As reaction products, they release oligosaccharides, disaccharides (cellobiose), and monosaccharides (glucose). Additionally, exoglucanases hydrolyze the β1-4-glycosidic bonds of the reducing (RE) or non-reducing (NRE) ends of the cellulose chain. The reaction products of these enzymes release cellobiose. Finally, β-glucosidases convert the oligosaccharides and cellobiose into glucose, completing the saccharification process. They are classified into several glycosyl hydrolase (GH) families, including GH5, GH7, GH12, and GH45 (Devi, Bajar, Kour, Kothari, Pant, & Singh, 2022).

Cellulases have different applications in the paper industry, food, pharmaceuticals, and chemicals. However, their principal application is in the degradation of lignocellulosic biomass for biofuel production. The cellulases classification of the International Union of Biochemistry and Molecular Biology (IUBMB) is as follows: EC 3.2.1.4 Endoglucanases, EC 3.2.1.91 Cellobiohydrolases, and EC 3.2.1.21 β-glucosidases (Panchapakesan & Shankar, 2016).

4.4.2 *HEMICELLULASES*

Hemicellulases are a group of hydrolytic enzymes of industrial relevance, mainly in biotechnology applications like the saccharification of lignocellulosic biomasses (Xin & Geng, 2010). They are an enzymatic complex in which there is the participation of at least six different enzyme activities. In Table 4.2, the principal components of the complex, its substrate, and the products of the reaction are shown.

TABLE 4.2 Enzymatic Activities in the Hemicellulases Complexes

Activity	EC Classification	Substrate	Product
Endo-*β*-1,4-xylanase	EC 3.2.1.8	Xylan	Xylooligosaccharydes
β-xilosidase	EC 3.2.1.37	Xylooligosaccharides	Xylose
α-L-arabinofuranosidase	EC 3.2.1.55	Arabinan, arabinoxylan and arabinogalactan	L-arabinose
α-D-glucuronidase	EC 3.2.1.139	α-D-glucuronic	An alcohol and D-glucuronate
α-D-galactosidase	EC 3.2.1.22	Galactosides	D-galactose
Acetil xylan esterase	EC 3.1.1.72	Xylano and xylooligosaccharides	Acetate residues
Ferulic acid esterase	3.1.1.73	Xylan, and lignin	Ferulic acid (hydroxycinnamic acid)

The endo-β-1,4 xylanase (EC 3.2.1.8) acts randomly on the internal glycosidic bonds of xylan to release xylooligosaccharides. The β-xylosidase (EC 3.2.1.37) hydrolyzes the non-reducing ends (NRE) of xylan to release xylose. Also, depending on the specific hemicellulose composition, the participation of other enzymes shown in Table 4.2 could be possible (Madadi, Tu, & Abbas, 2017).

The second most abundant sugar found in lignocellulosic biomass is xylose, which comprises the backbone of xylan in hemicellulose. The enzymatic activities that are part of hemicellulases are hydrolytic enzymes of families GH5, GH8, GH10, and GH11 (Stickel, Elander, McMillan, & Brunecky, 2014).

4.5 CONCLUSION

The water hyacinth is one of the worst weeds that cause problems for the aquatic ecosystem. Although measures to control its population, including the use of herbicides and mechanical removal, have been widely applied, in most cases it remains inefficient. This review highlights the use of water lily as an alternative for the production of bioethanol, being a plant lignocellulosic biomass that is characterized by having a high content of hemicellulose and cellulose and a low amount of lignin. In addition to being cheap and easily available biomass due to its rapid development. Furthermore, it does not compete with food crops. Likewise, the conversion of lignocellulosic biomass into bioethanol requires an efficient pretreatment process that improves the digestibility of cellulose and hemicellulose, thus facilitating its subsequent conversion. The review analyzed the physical and chemical treatments that are commonly used and that are most efficient to increase the solubility of water lily polysaccharides. In the same way, the enzymes that are highly functional in the hydrolysis process are indicated to carry out the conversion of oligosaccharides into sugar monomers that are used for the production of bioethanol. Therefore, the water hyacinth is biomass with great potential for obtaining high-value-added compounds, including bioethanol.

KEYWORDS

- **bioethanol**
- **hemicellulose**
- **lignocellulosic biomass**
- **oligosaccharides**
- **polysaccharides**
- **sugar monomers**
- **water hyacinth**

REFERENCES

Ahmed, S. F., Mofijur, M., Chowdhury, S. N., Nahrin, M., Rafa, N., Chowdhury, A. T., Nuzhat, S., & Ong, H. C. (2022). Pathways of lignocellulosic biomass deconstruction for biofuel and value-added products production. *Fuel, 318*, 123618. https://doi.org/10.1016/j.fuel.2022.123618.

Ajithram, A., Jappes, J. T. W., Chintra, G. K., & Daphne, R. (2023). Serious environmental threat water hyacinth (*Eichhornia crassipes*) plant natural fibers: Different extraction methods and morphological properties for polymer composite applications. *Materials Today: Proceedings*. https://doi.org/10.1016/j.matpr.2023.03.431.

Akram, H. A., Imran, M., Javaid, A., Latif, S., Rizvi, N. B., Jesionowski, T., & Bilal, M. (2023). Pretreatment and catalytic conversion of lignocellulosic and algal into biofuels by metal-organic frameworks. *Molecular Catalysis, 539*, 112893. https://doi.org/10.1016/j.mcat.2022.112893.

Ali, S. S., Elsamahy, T., Abdelfattah, A., Mustafa, A. M., Khalil, M. A., Mastropetros, S. G., Kornaros, M., & Sun, J. (2022). Exploring the potential of anaerobic co-digestion of water hyacinth and cattle dung for enhanced biomethanation and techno-economic feasibility. *Fuel, 329*, 125397. https://doi.org/10.1016/j.fuel.2022.125397.

Amalina, F., Razak, A. S., Krishnan, S., & Zularisam, A. W. (2022). Water hyacinth (*Eichhornia crassipes*) for organic contaminants removal in water – A review. *Journal of Hazardous Materials Advances, 7*, 100092. https://doi.org/10.1016/j.hazadv.2022.100092.

Amit, K., Nakachew, M., Yilkal, B., & Mukesh, Y. (2018). A review of factors affecting enzymatic hydrolysis of pretreated lignocellulosic biomass. *Research Journal of Chemistry and Environment, 22*(7).

Areepak, C., Jiradechakorn, T., Chuetor, S., Phalakornkule, C., Sriariyanun, M., Raita, M., Champreda, V., & Laosiripojana, N. (2022). Improvement of lignocellulosic pretreatment efficiency by combined chemo-mechanical pretreatment for energy consumption reduction and biofuel production. *Renewable Energy, 182*, 1094–1102. https://doi.org/10.1016/j.renene.2021.11.002.

Asshokkumar, V., Venkatkarthick, R., Jayashree, S., Chuetor, S., Dharmaraj, S., Kumar, G., Chen, W. H., & Ngamcharussrivichai. (2022). Recent advances in lignocellulosic biomass for fuels and value-added bioproducts – A critical review. *Bioresource Technology, 344*, 126195. https://doi.org/10.1016/j.biortech.2021.126195.

Bayrakci, A. G., & Koçar, G. (2014). Second-generation bioethanol production from water hyacinth and duckweed in Izmir: A case study. *Renewable and Sustainable Energy Reviews, 30*, 306–316. https://doi.org/10.1016/j.rser.2013.10.011.

Bhatt, S. M., & Shilpa. (2014). Lignocellulosic feedstock conversion, inhibitor detoxification and cellulosic hydrolysis – A review. *Biofuels, 5*(6), 633–649. https://doi.org/10.1080/17597269.2014.1003702.

Bhattacharya, A., & Kumar, P. (2010). Water hyacinth as a potential biofuel crop. *EJEAFChe, 9*(1), 112–122.

Binod, P., Janu, K. U., Sindhu, R., & Pandey, A. (2011). Hydrolysis of lignocellulosic biomass for bioethanol production. In *Biofuels* (pp. 229–250). Academic Press. https://doi.org/10.1016/B978-0-12-385099-7.00010-3.

Biswas, B., & Biswas, A. S. (2022). Effects of some chemical nutrients on bio-ethanol production from water hyacinth (*Eichhornia crassipes*) hydrolyzed by heat and ethanol-resistant strain of Saccharomyces cerevisiae AB810. *Journal of the Indian Chemical Society, 99*, 100725. https://doi.org/10.1016/j.jics.2022.100725.

Cardona, C. A., & Sánchez, Ó. J. (2007). Fuel ethanol production: process design trends and integration opportunities. *Bioresource Technology, 98*(12), 2415–2457. https://doi.org/10.1016/j.biortech.2007.01.002.

Das, A., Ghosh, P., Paul, T., Ghosh, U., Pati, B. R., & Mondal, K. C. (2016). Production of bioethanol as useful biofuel through the bioconversion of water hyacinth (*Eichhornia crassipes*). *3 Biotech, 6*, 1–9. https://doi.org/10.1007/s13205-016-0385-y.

Devi, A., Bajar, S., Kour, H., Kothari, R., Pant, D., & Singh, A. (2022). Lignocellulosic biomass valorization for bioethanol production: A circular bioeconomy approach. *BioEnergy Research, 15*, 1820–1841. https://doi.org/10.1007/s12155-022-10401-9.

Dias, M. O., Junqueira, T. L., Jesus, C. D., Rossell, C. E., Maciel Filho, R., & Bonomi, A. (2012). Improving second-generation ethanol production through optimization of the first-generation production process from sugarcane. *Energy, 43*(1), 246–252. https://doi.org/10.1016/j.energy.2012.04.034.

Ganorkar, P. V., Jadeja, G. C., & Desai, M. A. (2022). Extraction of shikimic acid from water hyacinth (*Eichhornia crassipes*) using sonication: An approach towards waste valorization. *Journal of Environmental Management, 305*, 114419. https://doi.org/10.1016/j.jenvman.2021.114419.

Gebehiwot, H., Dekebo, A., & Endale, M. (2022). Chemical composition, pharmacological activities, and biofuel production of water hyacinth (*Eichhornia crassipes*): A review. *Journal of the Turkish Chemical Society, 9*(3), 849–866. https://doi.org/10.18596/jotcsa.1033493.

Gunnarsson, C. C., & Petersen, C. M. (2007). Water hyacinths as a resource in agriculture and energy production: A literature review. *Waste Management, 27*(1), 117–129. https://doi.org/10.1016/j.wasman.2005.12.011.

Hassan, S. S., Williams, G. A., & Jaiswal, A. K. (2018). Emerging technologies for the pretreatment of lignocellulosic biomass. *Bioresource Technology, 262*, 310–318. https://doi.org/10.1016/j.biortech.2018.04.099.

Hirphaye, B. Y., Mezgebe, F. H., Tura, A. M., & Fanta, G. M. (2022). Bio-ethanol production potential of water hyacinth (*Eichhornia crassipes*) as alternative energy feedstocks. *Research Square*. https://doi.org/10.21203/rs.3.rs-2356408/v1.

Holding, A. J. (2016). Ionic liquids and electrolytes for cellulose dissolution [Master's thesis]. https://doi.org/10.13140/RG.2.2.21504.23045.

Houfani, A. A., Anders, N., Spiess, A. C., Baldrian, P., & Benallaoua, S. (2020). Insights from enzymatic degradation of cellulose and hemicellulose to fermentable sugars – A review. *Biomass and Energy, 134*, 105481. https://doi.org/10.1016/j.biombioe.2020.105481.

Hu, G., Heitmann, J. A., & Rojas, O. J. (2008). Feedstock pretreatment strategies for producing ethanol from wood, bark, and forest residues. *BioResources, 3*(1), 270–294.

Ilo, O. P., Simatele, M. D., Nkomo, S. L., Mkhize, N. M., & Prabhu, N. G. (2020). The benefits of water hyacinth (*Eichhornia crassipes*) for Southern Africa: A review. *Sustainability, 12*, 9222. https://doi.org/10.3390/su12219222.

Juárez-Luna, G. N., Favela-Torres, E., Quevedo, I. R., & Batina, N. (2019). Enzymatically assisted isolation of high-quality cellulose nanoparticles from water hyacinth stems. *Carbohydrate Polymers, 220*, 110–117. https://doi.org/10.1016/j.carbpol.2019.05.058.

Kabir, M. M., Alam, F., Akter, M. M., Gilroyed, B. H., Didar-ul-Alam, Md., Tijing, L., & Shon, H. K. (2022). Highly effective water hyacinth (*Eichhornia crassipes*) waste-based functionalized sustainable green adsorbents for antibiotic remediation from wastewater. *Chemosphere, 304*, 135293. https://doi.org/10.1016/j.chemosphere.2022.13529.

Kheybari, S., Kazemi, M., & Rezaei, J. (2019). Bioethanol facility location selection using the best-worst method. *Applied Energy, 242*, 612–623. https://doi.org/10.1016/B978-0-444-63990-5.00006-2.

Kumar, A., Singh, L. K., & Ghosh, S. (2009). Bioconversion of lignocellulosic fraction of water-hyacinth (*Eichhornia crassipes*) hemicellulose acid hydrolysate to ethanol by Pichia stipitis. *Bioresource Technology, 100*(13), 3293–3297. https://doi.org/10.1016/j.biortech.2009.02.023.

Kumar, R., Kim, T. H., Basak, B., Patil, S. M., Kim, H. H., Ahn, Y., & Jeon, B. H. (2022). Emerging approaches in lignocellulosic biomass pretreatment and anaerobic bioprocesses for sustainable biofuel production. *Journal of Cleaner Production, 333*, 130180. https://doi.org/10.1016/j.jclepro.2021.130180.

Li, M., Jiang, B., Wu, W., Wu, S., Yang, Y., Song, J., Ahmad, M., & Jin, Y. (2022). Current understanding and optimization strategies for efficient lignin-enzyme interaction: A review. *International Journal of Biological Macromolecules, 195*, 274–286. https://doi.org/10.1016/j.ijbiomac.2021.11.188.

Li, S., Bashline, L., Lei, L., & Gu, Y. (2014). Cellulose synthesis and regulation. *The Arabidopsis Book*. American Society of Plant Biologists. https://doi.org/10.1199/tab.0169.

Limayem, A., & Ricke, S. C. (2012). Lignocellulosic biomass for bioethanol production: Current perspectives, potential issues, and prospects. *Progress in Energy and Combustion Science, 38*(4), 449–467. https://doi.org/10.1016/j.pecs.2012.03.002.

López-Ramírez, N., Volke-Sepulveda, T., Gaime-Perraud, I., Saucedo-Castañeda, G., & Favela-Torres, E. (2018). Effect of stirring on growth and cellulolytic enzymes production by *Trichoderma harzianum* in a novel bench-scale solid-state fermentation bioreactor. *Bioresource Technology, 265*, 291–298. https://doi.org/10.1016/j.biortech.2018.06.015.

Luo, X., Liu, J., Zheng, P., Li, M., Zhou, Y., Huang, L., & Shuai, L. (2019). Promoting enzymatic hydrolysis of lignocellulosic biomass by inexpensive soy protein. *Biotechnology for Biofuels, 12*(1), 1–13. https://doi.org/10.1186/s13068-019-1387-x.

Madadi, M., Tu, Y., & Abbas, A. (2017). Recent status on enzymatic saccharification of lignocellulosic biomass for bioethanol production. *Electronic Journal of Biology, 13*(2), 135–143.

Martín, M., & Grossmann, I. E. (2012). Energy optimization of bioethanol production via hydrolysis of switchgrass. *AIChE Journal, 58*(5), 1538–1549. https://doi.org/10.1002/aic.12735.

Mishima, D., Kuniki, M., Sei, K., Soda, S., Ike, M., & Fujita, M. (2008). Ethanol production from candidate energy crops: water hyacinth (*Eichhornia crassipes*) and water lettuce (*Pistia stratiotes* L.). *Bioresource Technology, 99*(7), 2495–2500. https://doi.org/10.1016/j.biortech.2007.04.056.

Nigam, J. N. (2002). Bioconversion of water-hyacinth (*Eichhornia crassipes*) hemicellulose acid hydrolysate to motor fuel ethanol by xylose–fermenting yeast. *Journal of Biotechnology, 97*(2), 107–116. https://doi.org/10.1016/S0168-1656(02)00013-5.

Osorio-González, C. S., Saini, R., Hegde, K., Brar, S. K., Lefebvre, A., & Ramírez, A. A. (2023). Crabtree effect on *Rhodosporidium toruloides* using wood hydrolysate as a culture media. *Fermentation, 9*, 11. https://doi.org/10.3390/fermentation9010011.

Pal, D. B., Tiwari, A. K., Prasad, N., Syed, A., Bahkali, A. H., Srivastava, N., Singh, R. P., & Gupta, V. K. (2023). Sustainable valorization of water hyacinth waste pollutant via pyrolysis for advanced microbial fuel investigation. *Chemosphere, 314*, 137602. https://doi.org/10.1016/j.chemosphere.2022.137602.

Panchapakesan, A., & Shankar, N. (2016). Fungal cellulases: An overview. In *New and Future Developments in Microbial Biotechnology and Bioengineering* (pp. 9–18). https://doi.org/10.1016/B978-0-444-63507-5.00002-2.

Rezania, S., Din, M. F. M., Mohamad, S. E., Sohaili, J., Taib, S. M., Yusof, M. B. M., Kamyab, H., Darajeh, N., & Ahsan, A. (2017). Review on pretreatment methods and ethanol production from cellulosic water hyacinth. *BioResources, 12*(1), 2108–2124.

Rocha-Meneses, L., Raud, M., Orupõld, K., & Kikas, T. (2017). Second-generation bioethanol production: A review of strategies for waste valorization. *Agronomy Research, 15*(3), 830–847.

Rocha-Meneses, L., Raud, M., Orupõld, K., & Kikas, T. (2019). Potential of bioethanol production waste for methane recovery. *Energy, 173*, 133–139. https://doi.org/10.1016/j.energy.2019.02.073.

Rojas-Sandoval, J., & Acevedo-Rodríguez, P. (2013). Water hyacinth (*Eichhornia crassipes*). *Centre for Agriculture and Bioscience International (CABI) Invasive Species Compendium*. https://doi.org/10.1079/cabicompendium.20544.

Sameni, J., Krigstin, S., & Sain, M. (2016). Characterization of lignins isolated from industrial residues and their beneficial uses. *BioResources, 11*(4), 8435–8456. https://doi.org/10.15376/biores.11.4.8435-8456.

Shahzadi, T., Mehmood, S., Irshad, M., Anwar, Z., Afroz, A., Zeeshan, N., & Sughra, K. (2014). Advances in lignocellulosic biotechnology: A brief review on lignocellulosic biomass and cellulases. *Advances in Bioscience and Biotechnology*. https://doi.org/10.4236/abb.2014.53031.

Shankar, K., Kulkarni, N. S., Sajjanshetty, R., Jayalakshmi, S. K., & Sreeramulu, K. (2020). Co-production of xylitol and ethanol by the fermentation of the lignocellulosic hydrolysates of banana and water hyacinth leaves by individual yeast strains. *Industrial Crops and Products, 155*, 112809. https://doi.org/10.1016/j.indcrop.2020.112809.

Sharma, S., Tsai, M. L., Sharma, V., Sun, P. P., Nargotra, P., Bajaj, B. K., Chen, C. W., & Dong, C. D. (2023). Environments. *Environments, 10*(6). https://doi.org/10.3390/environments10010006.

Sheng, Y., Lam, S. S., Wu, Y., Ge, S., Wu, J., Cai, L., Huang, Z., Le, V. Q., Sonne, C., & Xia, C. (2021). Enzymatic conversion of pretreated lignocellulosic biomass: A review on the influence of structural changes of lignin. *Bioresource Technology, 324*, 124631. https://doi.org/10.1016/j.biortech.2020.124631.

Soccol, C. R., Faraco, V., Karp, S. G., Vandenberghe, L. P., Thomaz-Soccol, V., Woiciechowski, A. L., & Pandey, A. (2019). Lignocellulosic bioethanol: Current status and future perspectives. In *Biofuels: Alternative Feedstocks and Conversion Processes for the Production of Liquid and Gaseous Biofuels* (pp. 331–354). https://doi.org/10.1016/B978-0-12-816856-1.00014-2.

Stickel, J. J., Elander, R. T., McMillan, J. D., & Brunecky, R. (2014). Enzymatic hydrolysis of lignocellulosic biomass. In *Bioprocessing of Renewable Resources to Commodity Bioproducts* (1st edn., pp. 77–103). John Wiley and Sons. https://doi.org/10.1002/9781118845394.ch4.

Suthar, S., Sharma, B., Kumar, K., Banu, J. R., & Tyagi, V. K. (2022). Enhanced biogas production in diluted acid-thermal pretreatment and cattle dung biochar mediated bio methanation of water hyacinth. *Fuel, 307*, 121897. https://doi.org/10.1016/j.fuel.2021.121897.

Talebnia, F., Karakashev, D., & Angelidaki, I. (2010). Production of bioethanol from wheat straw: an overview on pretreatment, hydrolysis, and fermentation. *Bioresource Technology, 101*(13), 4744–4753. https://doi.org/10.1016/j.biortech.2009.11.080.

Tanpichai, S., Biswas, S. K., Witayakran, S., & Yano, H. (2023). Water hyacinth: a sustainable lignin-poor cellulose source for the production of cellulose nanofibers. *ACS Sustainable Chemistry and Engineering*. https://doi.org/10.1021/acssuschemeng.9b04095.

Taouda, H., Chabir, R., Aarab, L., Miyah, Y., & Errachidi, F. (2017). Biomass and bio-ethanol production from date extract. *Journal of Materials and Environmental Science, 8*(9), 3391–3396.

Thu Ha Tran, T., & Nguyen, P. K. T. (2022). Enhanced hydrogen production from water hyacinth by a combination of ultrasonic-assisted alkaline pretreatment, dark fermentation, and microbial electrolysis cells. *Bioresource Technology, 357*, 127340. https://doi.org/10.1016/j.biortech.2022.127340.

Umai, R. D., Jacob, S., & Kumar, V. (2022). Deep eutectic solvent pretreatment of water hyacinth for improved holocellulosic saccharification and fermentative co-production of xylitol and lipids using *Rhodosporium toruloides* NCIM 3547. *Fermentation, 8*, 591. https://doi.org/10.3390/fermentation8110591.

Vanholme, R., Meester, B., Ralph, J., & Boerjan, W. (2019). Lignin biosynthesis and its integration into metabolism. *Current Opinion in Biotechnology, 56*, 230–239. https://doi.org/10.1016/j.copbio.2019.02.018.

Wang, Z., Zheng, F., & Xue, S. (2019). The economic feasibility of the valorization of water hyacinth for bioethanol production. *Sustainability, 11*(3), 905. https://doi.org/10.3390/su11030905.

Xin, F., & Geng, A. (2010). Horticultural waste as the substrate for cellulase and hemicellulase production by *Trichoderma reesei* under solid-state fermentation. *Applied Biochemistry and Biotechnology, 162*, 295–306. https://doi.org/10.1007/s12010-009-8745-2.

Yang, B., Dai, Z., Ding, S. Y., & Wyman, C. E. (2011). Enzymatic hydrolysis of cellulosic biomass. *Biofuels, 2*(4), 421–449. https://doi.org/10.4155/bfs.11.116.

Zhang, J., & Viikari, L. (2014). Impact of xylan on synergistic effects of xylanases and cellulases in enzymatic hydrolysis of lignocelluloses. *Applied Biochemistry and Biotechnology, 174*, 1393–1402. https://doi.org/10.1007/s12010-014-1140-7.

Zhang, Q., Wei, Y., Han, H., & Weng, C. (2018). Enhancing bioethanol production from water hyacinth by new combined pretreatment methods. *Bioresource Technology, 251*, 358–363. https://doi.org/10.1016/j.biortech.2017.12.085.

Zhao, D., Zhu, Y., Cheng, W., Chen, W., Wu, Y., & Yu, H. (2021). Cellulose-based flexible functional materials for emerging intelligent electronics. *Advanced Materials, 33*, 2000619. https://doi.org/10.1002/adma.202000619.

CHAPTER 5

Equitable Management and Sustainability in the Table Grape System in Northwestern Mexico: A Case Study

J. M. ROBLES-PARRA,[1] K. TERÁN-SAMANIEGO,[1] and J. L. PAZ-LUNA[2]

[1]*Coordinación de Desarrollo Regional, Centro de Investigación en Alimentación y Desarrollo, Carretera Enrique Astiazarán Rosas, C.P., Hermosillo, Sonora, Mexico*

[2]*Estudios Sociales, Revista de Alimentación Contemporánea y Desarrollo Regional, Centro de Investigación en Alimentación y Desarrollo, Carretera Enrique Astiazarán Rosas, C.P., Hermosillo, Sonora, Mexico*

ABSTRACT

The objective of this research is to show referential and empirical evidence of the relevance of incorporating equitable management in the processes of recruitment, promotion, and participation processes in an organization. With equitable management, it is easier to advance procedures focused on inhibiting labor discrimination and forming fairer and more equitable organizations. That is part of the intangible conditions that the market of table grape system organizations maintains in Northwestern Mexico. Those are related to the three components of sustainable development. The methodology is based on a semi-structured interview script that was applied to 14 table grape exporters. The research design is non-experimental, with a mixed approach, and it is intentional. The results show that in the system there are organizations that have already implemented strategies for the inclusion of people regardless of their physical appearance and living conditions. Other organizations are in the process of their incorporation. It is concluded that

Sustainable Agricultural and Food Systems Engineering: New Research, Practices, and Developments.
Jose Juan Buenrostro-Figueroa, Julio César Tafolla-Arellano, Leonardo Sepúlveda Torre, Porteen Kannan, A. K. Haghi (Eds.)

92% of table grape exporting organizations in the system in Northwestern Mexico tend to establish policies and means to achieve equitable management. This allows the inclusion of highly qualified employees and ensures the well-being of present and future generations. This position, which the market acquires, drives organizations to evolve in their organizational structures.

5.1 INTRODUCTION

The research has addressed the issue of equity in management with the objective of improving the performance of the organization and analyzing the need to sensitize society and those who lead organizations. The objective is to promote, in this way, fairer, more equitable, and inclusive work environments. Equitable management has become necessary to comply with the rights of those who have historically been poorly represented, such as ethnic groups, unions of different races, women, and immigrants, among others. The above-mentioned is crucial for an organization to focus its efforts on treating people well and using their talents and abilities, and not on aspects such as physical appearance and living conditions.

The organizations of the table grape system in Northwestern Mexico serve a specialized and knowledgeable market that raised demands related to non-visible and non-tangible issues in the product, to create consciousness among these organizations. For example, climate change, the loss of biodiversity, the depletion of natural resources, the application of pesticides, the use of fertilizers, child labor, unfair wages, and the lack of medical care for workers, without detriment to improving the economic capacity of the organization.

It is about maintaining the intangible market conditions, based on the concept of sustainable development and its three components: environmental, social, and economic. One more requirement is the eradication of labor discrimination at management levels and other positions in the organization. This is because the inclusion of people with knowledge, experience, and values, regardless of their appearance and condition, is beneficial for the organization, not only from an ethical and moral perspective but also from the point of view of business efficiency.

From what has been described, it is possible to realize the emergence of a group of consumers concerned about the implementation of practices related to sustainable development and equity as part of the social component of this principle. This is because they integrate a behavior that, in their perception, ensures the well-being of present and future generations. In addition, this is the stance that the market forces organizations to evolve their organizational structures.

This is how the implementation of the concept of equitable management contributes to sustainable development and equity in the organizations of the grape system of Northwestern Mexico and likewise contributes to the promotion of procedures that allow choosing a group of employees whose importance is based on being highly specialized and qualified.

5.2 LITERATURE REVIEW

5.2.1 EQUITABLE MANAGEMENT

The concept of equitable management has not been widely addressed; however, the roots of the words included in the concept are taken and analyzed in subsections.

5.2.1.1 EQUITY

Discrimination is a social event that violates the integrity, human rights, and basic autonomy of people. This arises unconsciously in social practices between people and authorities, in business, institutional, political, and public spaces. The prohibition of discrimination has become a fundamental aspect of human rights present in all international treaties, both universally nationally, and regionally (Giles & Méndez, 2019).

Inequity in opportunities occurs in the workplace when decisions about the selection, development, evaluation, or gratification processes are based on individual characteristics such as age, appearance, sex, religion, sexuality, public opinion, and skin color and not on academic training, skills, work experience, seniority, or professional development (Terán, Tiznado, Robles, & Paz, 2022).

Until today, the achievement of equity continues to be one of the main challenges in the world, especially in the countries of the Americas. The cause is due to labor discrimination that has prevailed in a context of social exclusion, which has created mistrust and impossibility for people to exercise their rights (Muñoz & Barrantes, 2016).

The concept of equity has been defined many times. Olvera & Arellano (2015) cite Karen S. Cook & Karen Hegtvedt (1983, p. 218) in their research and they mention that equity has been defined by them as the equality between input and result, for the people involved in the situation of a specific exchange.

The authors also cite the research of George Frederickson & Edmund C. Stazyk (2014, p. 11), where they explained that equity represents the conceptual unification of three values: justice, fairness, and equality. The concept of equity is based on three social principles: equality, fulfillment of rights, and justice. It is an ethical and just principle that defends the same conditions and opportunities for people without distinction, which is essential to develop their potential achieve a productive life, and realize oneself as a person (Terán, 2022).

5.2.1.2 MANAGEMENT

The word management (*gerentia*, in Latin) means the virtue of those who manage (Etymologies, 2023). The concept of management means administration, management, resource management, and business management. This concept refers to a system of planning activities, resources, and results in organizations, through systematized knowledge, techniques, methods, or tools to be applied in the processes of production, marketing, and expansion of companies (Viaña, 2018).

Guevara & Pérez (2021) comment that management can be assumed from a comprehensive vision, which aims to unify forces to define the aspects a company, organization, or institution must maintain to achieve greater competitiveness. There are three keywords in management: (i) strategy, to know where the company is headed; (ii) organization, to specify the strategies; and (iii) culture, with the aim of reinforcing the organizational structure.

Viaña (2018) mentions that the essential functions of management are: (a) establish interrelationship networks; (b) implement holistic strategies; (c) handle up-to-date information and communication technologies; (d) carry out informal leadership; (e) maintain a profile of impartiality; (f) promote innovation and creativity; and (g) promote the acquisition of knowledge to address the problems that surround management and the organization. Management must adapt to the new market requirements to achieve efficiency, competitiveness, and sustainability (Ferrín, 2018).

Management focuses on the use of resources with the purpose of generating profit and well-being in specific sectors of society. It also focuses on achieving the objectives with the best use of resources and aims to achieve a positive effect on society and organizations (Guerrero, Ballesteros, & Herrera, 2018). Therefore, the management process consists of a series of functions to be executed simultaneously and continuously, according to four essential aspects: planning, organization, direction, and control (Suárez, 2018).

5.2.1.3 THE FUSION OF EQUITY AND MANAGEMENT CONCEPTS

Castro & Álvarez (2016) mention that equity in management is like the inhibition of discrimination that arises in the context of corporate social responsibility and sustainable development, as well as a primary opportunity for organizations. Furthermore, it represents the need to guarantee the constant and equal participation of people, regardless of their condition, in all areas of a company, institution, and organization. The process of including equity in the practices of these entities responds to the need to create new models of organizational culture and leadership.

Teran et al. (2019) cite Yates (1993) in their research, who approaches the concept of equitable management. It is defined as a solution to the problems of the organization and an opportunity to push it towards a new direction. The new direction must be focused on joining efforts to achieve inclusion and thus prevent the rejection of people in the workplace due to race, color, religion, gender, or national origin. It is mentioned that equitable managers actively search for individuals capable of deserving opportunities to be prepared in service and work so that the barriers to their advancement are minimal.

Therefore, equitable management occurs in two stages. The first stage occurs when the company, organization, or institution promotes non-discrimination in the workplace by opening opportunities at managerial levels to those people who meet the required knowledge and values, regardless of gender, color, nationality, or sexual preference, among others. The second stage occurs when equitable managers work to inhibit discrimination by awarding opportunities in their reporting positions to applicants who also have the necessary skills, experience, values, and attitudes, regardless of gender, religion, origin, orientation, sexuality, or public opinion (Terán et al., 2020).

This is how equitable management is defined: it is the union of efforts to inhibit labor discrimination, through the recognition of competencies and merits of people for their inclusion in an organization. Equity in management should meet the conditions demanded by the market to create companies, institutions, and organizations that focus their efforts on the good treatment of people and on getting the most out of their talents and abilities (Terán, 2022).

5.2.2 INTANGIBLE MARKET CONDITIONS

For Solís, Robles, & Preciado (2017), a market is a group of potential customers who might be willing to exchange goods that are exchangeable among themselves. All of this is defined, mainly, by the conditions imposed by demand. The market is determined by the geographical area where the production and sale activities are carried out and, likewise, by the type of product, which is defined by its characteristics and ease of exchange.

For Huerta & Sandoval (2018), the market is the set of all real and potential buyers of a product or service, who share a specific need that can be satisfied through an exchange relationship. They argue that it is necessary to define the market, that is, to draw its limits so that substitute goods are included in this scenario and those that are not discarded.

The market defines the particularities of a product, for example, the price, quality, design, texture, hardness, and firmness, among others. For food products, these characteristics are related to the organoleptic, nutritional, and use characteristics, such as taste, smell, color, temperature, safety, and nutritional content. In addition, the market makes very specific demands on the organization. This corresponds to aspects that are not visible in the product, implicit in its production process, among other organizational practices. Consumer requirements are linked to their value system, ethical principles, habits, and ideologies (Barrera et al., 2022).

This is how the market not only demands attributes intrinsic to the product, but it also demands that they come from organizations that practice and promote two essential aspects: social responsibility in favor of workers, and environmental care. That is, they require attention to aspects that are not tangible in the product. And consumers maintain intangible market conditions (Wei et al., 2018).

Terán et al. (2019) mention these intangible demands and summarize them as a sustainability requirement that some market agents, directly or indirectly, impose on organizations, and that has led to important changes in them. These authors define intangible demands as market demands related to aspects that are not visible in the final product, and to social, environmental, and economic sustainability.

For this reason, the organizations of the table grape system in northwestern Mexico have been engaged in incorporating strategies to achieve equity and justice, as components of the social dimension of sustainable development (Hernández, 2020).

5.2.3 THE ORGANIZATIONS OF TABLE GRAPE SYSTEM IN NORTHWESTERN MEXICO

Table grape agricultural systems in Northwestern Mexico are a source of livelihood that provides employment and food opportunities (Orduño,

Kallas, & Ornelas, 2020). It is one of the most outstanding supports for the economy of both developed and developing countries (Asimeh, Nooripoor, Azadi, Eetvelde, & Sklenička, 2020).

Sonora contributes 89% of total production and 95% of Mexican table grape exports (Hurtado, Robles, Preciado, & Bañuelos, 2018; Contreras, 2019). The system is composed of 41 producers, located in Caborca, Pitiquito, Altar, Hermosillo, Carbó, San Miguel de Horcasitas, Guaymas, and Empalme, in the State of Sonora (Contreras, 2019).

Despite their contribution, senior managers of the organizations in the table grape system are concerned about the attention that must be given to market demands. These concerns are related to climate change, the loss of biodiversity, and the depletion of natural resources (Brun, Jeuffroy, Pénicaud, & Cerf, 2021), pesticide application, fertilizer use, and waste production (Schreefel, Schulte, de Boer, Schrijver, & van Zanten, 2020), child labor, unfair wages, and the lack of medical care for workers and their families (Terán, 2022). Organizations take these aspects into consideration without losing sight of their objective of improving their economic capacity (Shah et al., 2021).

Another market requirement for managerial levels is the eradication of labor discrimination by consumers, who demand that equitable management be implemented (Marrugo-Salas, 2016; Terán et al., 2019). Likewise, investors, employees, scholars, advisers, etc., have considered it necessary to include the participation of all people equally in management and other leadership positions. This is because their inclusion is beneficial for the organization, not only from the ethical and moral perspective but also from the perspective of business efficiency (Bañuelos et al., 2019).

Therefore, it is possible to demonstrate the emergence of a group of consumers concerned about the implementation of practices related to sustainable development and equity, as part of the social component. This conduct ensures the well-being of present and future generations (The Packer, 2014).

The position of the market forces organizations to advance in the protection of the environment, the care of workers, and the inclusion of people regardless of their sex, gender, age, religion, physical condition, nationality, or political preference (Terán, 2022). For this reason, evolution has been promoted in the organizational structures that are manifested in new organizational configurations and new managers (Solís, Robles, & Preciado, 2017).

Implementing the concept of equitable management contributes to sustainable development and equity in the organizations of the table grape system of Northwestern Mexico. At the same time, it contributes to the promotion of the necessary procedures that allow the selection of highly specialized and qualified employees, regardless of their physical appearance or life condition. They should be capable of solving organizational affairs from managerial levels, to replicate this comprehensive method in positions that depend on senior managers (Terán, 2022).

5.2.4 MARKET CHARACTERISTICS OF THE TABLE GRAPE SYSTEM OF NORTHWESTERN MEXICO

In the table grape system markets that currently exist worldwide, there are responsible and demanding consumers. Notably, those located in some European countries and the United States (Westbrook & Angus, 2021).

This consumer segment seeks brands that promote a positive change for society and/or the environment. Buyers identify more with organizations that follow these guidelines and prefer their products. They are even willing to pay high prices to acquire goods supported by practices related to sustainable development, in this case, focusing on equity and social justice (CSR Commitment, 2020).

To meet the expectations of this type of market, every year in the United States, The Packer Magazine generates a report on trends in fresh produce, including table grapes. They inform the sociodemographic characteristics of consumers who are inclined towards these products. Data are included such as the probability of buying table grapes in women (61%), in men (42%), the percentage of buyers who earn more than 1,00,000 dollars a year (54%), and the probability that people over the age of 59 buy grapes (73%) (The Packer, 2022).

That magazine also reports that the probability of purchasing in the United States, by region, shows a very similar percentage between them. However, the western zone stands out, where the states of Nevada, Utah, Idaho, Oregon, Washington, Arizona, and California are located. In that area, there is a 65% purchase probability, the highest percentage. Due to the geographic proximity of Sonora to California and Arizona, the Sonoran table grape plays an important role in the Anglo-Saxon market (The Packer, 2019).

A significant piece of data presented in The Packer Magazine (2014) is that North American consumers have confidence when buying products that come from areas such as Central America (Nicaragua and Honduras), Western Europe (Spain, Germany, and the Netherlands), South America (Chile, Brazil, and Ecuador) and Oceania (Australia and New Zealand). Around 77% of them feel even more comfortable buying products from

Mexico, which demonstrates the confidence that a large part of the US market has towards products from Mexico.

This report states that 44% of consumers surveyed would never buy products from China because they do not provide their customers with security in terms of food safety, policies, and practices that they implement, as well as the transportation of the product (The Packer, 2014). These factors affect the purchase decision since it is certain that 51% of those surveyed who purchased organic products did so to avoid consuming foods with chemicals. Around 47% responded that it was because of the taste of organic products, and 45% cited health reasons for being able to make or not the purchase (The Packer, 2014).

The market data discussed in the document by The Packer (2022) indicates a marked trend among US consumers for aspects related to food safety, nutritional quality, care for the environment, and attention to problems of workers and the community. This is because these are scenarios that are directly related to production and corporate image. The conditions mentioned have led to a notable change in the organizational forms towards increasingly sustainable, equitable, and fair structures, which consider the vision of the market, as well as the strategies of the organizations, in this case, table grape exporters.

In this section, it was possible to observe the description and characteristics of table grape consumers. This is a specialized segment, properly located and knowledgeable, that demands that organizations take responsibility for their operations and their production practices. This is achieved by implementing strategies that are related to sustainable development and equity as part of the social component.

5.3 METHODOLOGY

The present methodological scheme proposes to evaluate the question: Does equitable management exist, or not, in the organizations of the table grape system in Northwestern Mexico? It starts with a literature review to locate the state of the art and a semi-structured interview script, based on the components and subcomponents of the concept of equitable management (A la Torre, 2019; Contreras, 2019; Ibaceta & Villanueva, 2021; Terán, 2022).

The observation and evaluation units of this research are 41 table grape exporters, located in Hermosillo, Carbó, Caborca, and San Miguel de Horcasitas, municipalities of Sonora (Contreras, 2019). The informants are the managers, and they will be part of the investigation and will provide the data required to develop it because their hierarchical position is linked to decision-making regarding the process of non-discrimination at work (Caso et al., 2019).

The research design is non-experimental with a mixed and correlational approach. To determine the sample, a population of 41 organizations was used. The sample (n) is estimated over this size of N, with the following formula for finite populations:

$$n = \frac{Z^2(N)(p)(q)}{[E^2(N-1)]+[Z^2(p)(q)]} \quad (1)$$

where; "n" is the sample size; "N" is the population size; "Z" is the desired level of reliability; "E" is the estimation error; "p" is the probability of being chosen; and "q," the probability of not being.

Substituting we have:

$$n = \frac{1.96^2(41)(0.97)(0.03)}{[0.075^2(41-1)]+[1.96^2(0.97)(0.03)]}$$

The semi-structured interview was applied to obtain information related to equitable management and labor discrimination in the organization, for reasons of socioeconomic level, public opinion, religion, age, foreign origin, indigenous descent, national origin, sex, physical disability, growth in the organization, well-paid positions, general training, training in new needs, participation in meetings, and different sexual preferences in the organization's staff (Terán, 2022).

The data obtained from the 14 applied interviews were processed in the MaxQDA program (Kuckartz & Rädiker, 2019) and exported to Excel. The questions of equitable management related to the implementation of procedures for the non-discrimination of people at the time of entering the organization and/or during their employment were analyzed.

5.4 RESULTS

In the processes of hiring personnel and promotion in the table grape system, there is a tendency towards non-discrimination of people at different hierarchical levels (Figure 5.1). Out of the 16 attributes evaluated to determine the level of inclusion of each organization (origin, race, Indigenous descent, gender, religion, socioeconomic level, sexual preference, foreign origin, Mexican nationality, public or political opinion, pregnancy, age, physical disability, promotion, well-paid salaries, participation in meetings, and participation in training), only five organizations out of the 14 complied.

It is observed that organizations 4, 14, and 12 have only addressed 13 attributes, while 3, 2, and 1 have focused on 14 attributes. Organizations 13, 5, and 11 are considering 15 attributes in their processes. For organizations 8, 9, 10, and 7, it has been important to consider all 16 attributes in recruitment and promotion. Therefore, it is important to analyze and delve into the specific evidence corresponding to these organizations regarding compliance with the attributes.

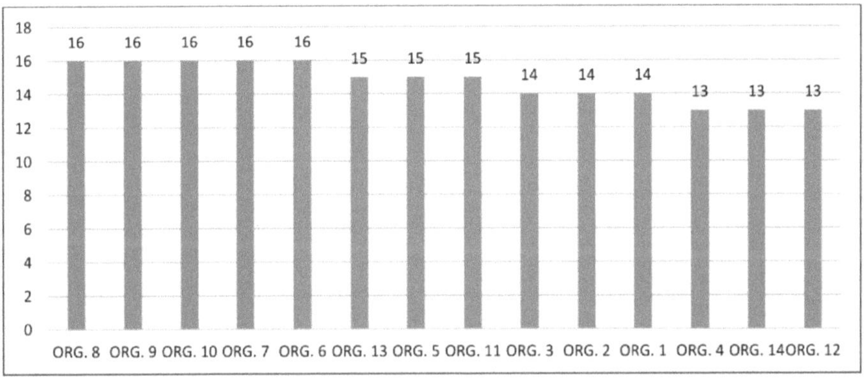

FIGURE 5.1 The score achieved for equitable management in each organization.

The 14 organizations interviewed reported that when hiring personnel, origin, indigenous descent, gender, socioeconomic level, and sexual preference are irrelevant attributes. In addition, they confirm that all personnel are considered for promotions, well-paid salaries, and participation in meetings, without their condition being an impediment. However, for the rest of the attributes (religion, participation in training, foreign origin, Mexican nationality, public or political opinion, pregnancy, age, and physical disability), the level of inclusion of people in the organization was not the same (Figure 5.2).

The interviewee from Organization 6 was asked: Does the origin of the people influence the hiring of management positions? And he replied: "No, no… we have, in fact, if I'm not mistaken… all of our managers come from the supervisor or day laborer level, field managers I mean." When asked the same question by the manager of Organization 12, he replied: "No, what is requested is the preparation, only… what the person offers, or what is knowledge, experience, study. It is requested for the position, the bachelor's degree profile, and the experience."

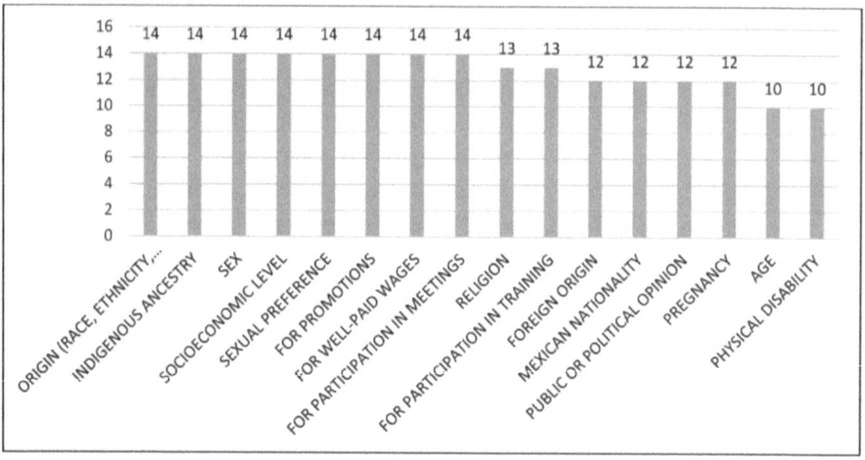

FIGURE 5.2 The importance of each equitable management attribute for organizations.

The manager of Organization 13 answered, in the interview, that Indigenous descent should not be an obstacle when hiring or promoting staff. In addition, he commented: "Well, in fact here, that's how many people improve; almost in all areas it happens that way; We have a boy who has just joined the administrative area, but he started working here as an entrance guard." Similarly, the manager of Organization 11 said that indigenous descent should not be considered when hiring or promoting personnel. He added: "In fact, there is an engineer of Yaqui origin here."

It was also asked in the interview if gender made any difference when hiring or promoting staff in the organization. The interviewee from Organization 3 commented: "No, definitely; and, above all, now with gender equality…we have to talk about it, here they are with all this, the truth." The manager of Organization 7 replied: "No, fortunately, in a man's world." On the other hand, the manager of Organization 11 said: "Gender matters less. Hey, I'm a woman, I swear, I'm the unique woman who goes to agricultural meetings."

In response to: Is the socioeconomic level considered when hiring or promoting someone? The manager of Organization 6 replied: "No… in fact the majority, I think one or two… but all the managers we have in the fields, most of them have come from a low level." The interviewee from Organization 12 said: "I say no because a former colleague who is in a managerial position; in that case, there was no influence. As I told you, it is the knowledge, the experience they have is considered."

They were asked if sexual preference was relevant when hiring staff. The manager of Organization 1 commented: "Not that kind of thing, in fact, we already have a couple of contracts... not me, I mean... I don't care." The interviewee from Organization 2 responded: "No, there are even several people who work in the company, and who have preferences towards the same sex. For example, the warehouse manager has been working with us for many years and he never has caused any problems for that."

In organization 8, the manager introduced her work team, and one of the coordinators answered the previous question: "I am openly gay, and I can tell you that the company is openly Catholic, very Catholic. However, in the five years that I have worked here, I have never felt discriminated against. On the contrary, I believe that if something keeps me here, I can tell you that I have always felt that I can be who I am, as I am here... what matters here is my work."

Also, everyone was asked: what is the procedure followed when already hired staff are offered an opportunity to move up to the management level? The interviewed manager of Organization 13 responded: "Here, I have seen people who are working in one activity and are promoted to another with greater responsibility. Basically, it is to meet the objectives of the position he held, they are not going to expect more than the one who was before, but that at least he does what he did, and that he meets the objectives."

The manager of Organization 6 said: "Whenever there is a vacancy, regardless of whether it is a managerial position or not, we look for internal talent. Movements always occur from people with low positions to middle positions in management. If we don't have internal talent, we look outside the organization."

In the interview, it was asked how they determine the salary, benefits, incentives, and rewards for positions, to find out if the work of managers and other collaborators is well paid, and also to find out if said remuneration corresponds to what is stipulated in the market. The manager of Organization 2 commented: "We do studies in the market to find out salaries, incentives, bonuses, and that kind of thing, and an average is made to create the tabulator that will guide us as a company. We offer a little more, we always have to offer a little more, precisely to avoid the problem of staff turnover, to retain the people we have, so that they don't leave to work in other fields, in other companies."

It was also questioned: what part of the subordinate personnel is invited to the meetings of managers and advisors? The manager of Organization 2 said: "It depends on the meeting. Sometimes we must discuss issues with the field manager or the field administrator. Sometimes there are issues that can be resolved with the food safety manager. Sometimes problems must be reported to the general manager; it depends on the matter that is going to be discussed, and considering that, those who can participate are invited." The interviewee from Organization 6 answered: "Depending on the theme of the meeting, staff from a certain area are invited… and they can range from an assistant to the general director."

The other question was: Does hiring for management levels influence people's religion? The interviewed manager of Organization 4 responded: "No, not 100%, but I tell you, it all adds up, right? So, well, you are free of whatever religion you want, but well…." In this regard, the rest of the interviewed managers assured that it was not an important attribute at the time of hiring.

It was also asked: How do you choose the people who will receive training, training, or development in new skills? The manager of Organization 1 answered: The first filter is that they must be permanently hired. The second filter, which perhaps should be the first, is that they want to train here with us. We have offered them training, and they have told us: "I'm not interested." To this same question, the manager of Organization 4 replied: "They ask us to work for at least one year… if you are going to take a training course, first aid, or search and rescue to be part of the security brigade, we have to be working for a year… And if we already received training, for example, in Sinaloa in an X or Y course, they make us sign to stay at least one year in the company."

In addition, it was asked: Is foreign origin a necessary condition for you to hire people at management levels? The manager of Organization 14 answered: "Yes because we have a distributor in San Diego, California, and Nogales, Arizona, USA, so yes." The response from the manager of Organization 1 was: "For my position, yes… in fact, it is not in the regulations, but it has always been a foreigner, from Chile."

Another question was: Does Mexican nationality influence your decision to hire a person who will occupy a management position? The manager of Organization 14 replied: "We are all Mexicans, there are no foreigners here in the offices; here, there is none." The manager of Organization 3 said: "Yes, and also for the rest of the workers because right now we are complying with certifications that require that."

The next question was: Is public or political opinion important for the hiring of the person who will occupy the position of manager? The manager of Organization 2 reported: "It's not important either… however, you don't want people in the company making political noise to staff, making noise, belonging to groups that make major demonstrations, and that is reflected here, no, I do not agree with that; they can have any public opinion, but that they do not influence the work." The manager of Organization 4 commented: Not openly, but it is secretly that we are PAN sympathizers… I think that perhaps that is why they hired me, he asked me openly: "Are you a PAN member?" "Yes, I've always said it, because I'm crazy, but I didn't know that they are PAN sympathizers."

Answers to the question: Do you consider the person's pregnancy status when hiring or promoting staff? They were extreme. The manager of organization 8 replied: "Negative… negative, because here is Alejandra, she is pregnant, and she was recently given the position of Coordinator." On the contrary, the manager of organization 12 said: At the time of hiring because I could hire them with one or two months of pregnancy. But if she is already going to give birth? … this situation is somewhat complicated because it is assumed that if someone is hired it is because someone is needed in the position, and that they are healthy… I mean, that they do not have an impediment, and that later does not leave the position vacated.

The next question was: is age a criterion that you consider when hiring or promoting staff? The manager of organization 13 responded: No… in my case, I started working here at 47 or 48 years of age… I thought that my age was going to be a limitation, and it was not like that. On the contrary, the manager of organization 14 said: "Now, age is considered"; and the manager of organization 2 commented: "It depends, for example, I am 46 years old… and if I hire a person over 60 years of age, I know it will be difficult for them to move." Likewise, the manager of organization 6 responded: "Perhaps it is a matter of how you see the person in the future… that is, there are jobs that are very demanding, and training someone takes years, that is, in one year you don't learn everything about the job… it takes three, four years to carry out all the work, because this job is very extensive, really." The manager of Organization 3 agreed and said: "For now, the age for hiring and promoting staff is considered."

The next question was: Is physical disability a criterion that you consider when hiring? The manager of Organization 13 replied: "Here, we have not seen physically disabled people… no." The manager of Organization 1 said: I have not hired a disabled person, but if I did, I would consider the type of disability. The manager of Organization 2 commented: "It depends, it is important to see specifically the needs of the position; You have to know if you can place a person with a physical disability in an office management position, or a field management position." On the contrary, the manager of Organization 14 answered: no because we do have disabled employees. Manager 7 said: "In the fields, people with disabilities have worked."

On the other hand, the manager of Organization 12 commented: "It depends on the disability." The manager of Organization 11 responded: "Perhaps I would consider him for the manager position... because of his availability, for those types of situations that perhaps require more care, mobilization. But neither has anyone with a disability applied for a job."

It should be mentioned that, of the 14 managers interviewed, 6 were women (representing 42.85% inclusion). Some of them started as assistants, which means professional growth in the organization, and permanence in it, regardless of pregnancy and/or maternity situations. The demographic data of each manager are specified, and the attributes for which each one is inclined (Table 5.1).

This is how organizations in the table grape system have worked on equitable management, by hiring personnel with an inclusive sense in the organization. They are currently addressing the non-discrimination of people in the hiring and promotion processes, where religion, foreign origin, Mexican nationality, public or political opinion, and pregnancy should not matter.

They also aim to ensure that once employees enter the organization, they have the security of participating in training. The managers of these organizations are aware that their efforts should be focused on providing facilities in their processes and facilities to achieve the inclusion of people with advanced age and physical disabilities.

Considering the above, it is crucial to analyze the level of non-discrimination that the system of table grape organizations has (Figure 5.3).

The percentage of compliance that the table grape system has, in terms of non-discrimination of people, when hiring and promoting personnel in the organization is 92%. What this means is that it is an inclusive system that maintains the tendency to implement the necessary adaptations to allow the participation of all people regardless of their condition, and primacy is given to abilities, knowledge, attitudes, values, and experience.

5.5 DISCUSSION

The fundamental attributes considered by the managers to achieve inclusion in the work areas and achieve equitable management in the organization were: origin-race, Indigenous descent, gender, socioeconomic level, sexual

TABLE 5.1 Demographic Data of the Managers, and the Most Important Attributes of Equitable Management for Each One of Them

Manager	Age	Sex	Profession	Position in the Organization	Years Working in the Organization	Most Important Equitable Management Attributes for Each Manager
1	45	M	Finance graduate	General manager	8	Leadership, responsibility, and empathy.
2	46	M	Horticulture engineer	General manager	23	Proactivity and teamwork.
3	68	M	Agricultural engineer	Manager in San Arturo field	45	Experience, skills, kindness, and empathy.
4	41	F	Degree in social work	Social responsibility manager	5	Experience, skills, proactivity, and empathy.
5	34	F	Degree in Accounting	General manager	1	Capabilities
6	38	M	Postgraduate in business administration.	Operations director	3	Leadership, experience, and kindness.
7	44	F	Degree in administration.	In charge of certifications and social responsibility department.	6	Capabilities, knowledge, and responsibility.
8	49	F	Participation in capacitation, training, and courses.	In charge of the departments of social responsibility, safety, environment, and safety and health.	5	Kindness, empathy, and teamwork.
9	38	F	Degree in Administration	In charge of the safety department	14	Capabilities
10	42	M	Bachelor of Administration	Administration manager	20	Capabilities
11	38	F	Marketing (not finished)	Administration manager	12	Good behavior, responsibility, and proactivity.
12	47	M	Degree in Accounting	In charge of the accounting department.	20	Skills, experience, and knowledge.
13	52	M	Degree in computer science	In charge of the safety and hygiene department.	14	Capabilities
14	55	M	Technical career	Purchasing manager	33	Responsibility and proactivity.

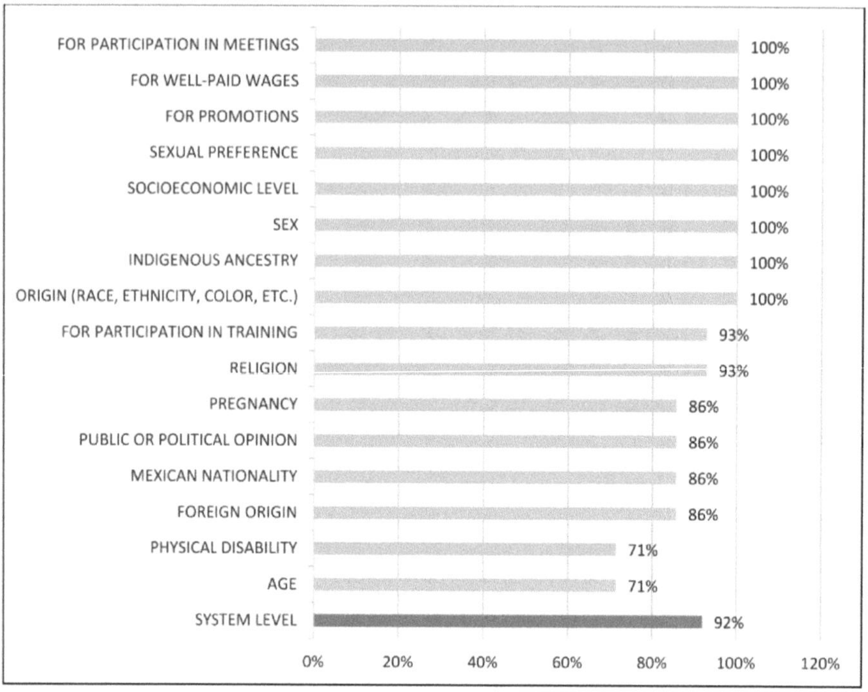

FIGURE 5.3 Level of non-discrimination reached by the system of table grape organizations in Northwestern Mexico.

preference, promotions, well-paid salary, and participation in meetings. Torres & Peri (2020) add elements to achieve equitable management and include gender as a substantial attribute that must be considered when including and/or promoting people in an organization.

Likewise, Bautista et al. (2019) include the obesity attribute, and Vela (2017) adds marital status. For the interviewees, the competence (capacity, ability, and skill) of the company's collaborators is significant and must be met and prevail over their physical appearance and their living conditions. This coincides with what Espinosa & Camaraco (2017) propose in their research, that competition at managerial levels is of great importance because it allows organizations to carry out processes of planning, direction, control, coordination, recruitment, selection, and training more successfully.

The aspects considered by the personnel of the managerial levels, and for the worker to be distinguished, are those related to responsibility, kindness, and the correct understanding of people. The manager must be a person with a good attitude, qualities, knowledge, and skills in the organizational field. In this way, it is expected to inspire and generate trust and credibility among its collaborators, since they are responsible for forming work teams and sharing the vision, mission, and objectives of the firm, through synergies, responsibilities, and labor commitments (García, Boom, & Molina, 2017).

This tells us that these organizations meet the requirements imposed by law. However, it is evident that more changes are required so that people with different physical abilities can access the company's facilities, be part of the organization, and attend to their responsibilities (CNDH, 2018).

5.6 CONCLUSION

Empirical evidence allows us to affirm that effectively, equitable management as an intangible market condition promotes changes in organizations towards more equitable and fair hiring, promotion, remuneration, and participation in meetings and training processes.

This verification also represents a conceptual, empirical, and methodological contribution to Management Theory and Market Theory, since the referential evidence was relatively limited with respect to studies in the agricultural industry (primary production sector), and more limited, even considering the theoretical-conceptual basis that guided this research.

The research work allows us to verify that market conditions force companies of the table grape system in northwestern Mexico to implement strategies related to equitable management, as a practice of the social component of sustainable development.

For future research, it is suggested that an additional instrument be prepared to collect more specific information on the inclusion practices that are being carried out by organizations of the table grape system in northwestern Mexico, both in their administrative and operational areas.

It is also suggested that equity be discussed at higher hierarchical levels of the organization, as well as other concepts that have been internalized in the organizations of the video mesa system, such as the competencies, ethics, and values of managers, and their effects on the structure of the organization and its functions.

This is an approximation work that represents the starting point for studies that can be carried out in greater depth and that allow a detailed explanation of the behavior of organizations that are interested in inhibiting discrimination in the workplace.

KEYWORDS

- **equitable management**
- **intangible market conditions**
- **sustainability**
- **table grape organizations**
- **table grape system**
- **video mesa system**

REFERENCES

A la Torre, M. A. (2019). Relación entre estilos de liderazgo y responsabilidad social empresarial en las organizaciones productoras de uva de mesa sonorense [Doctoral thesis, Centro de Investigación y Desarrollo (CIAD, A.C.)]. Hermosillo, Sonora, México.

Asimeh, M., Nooripoor, M., Azadi, H., Van Eetvelde, V., Sklenička, P., & Witlox, F. (2020). Agricultural land use sustainability in Southwest Iran: Improving land leveling using consolidation plans. *Land Use Policy, 94,* 104555.

Bañuelos-Flores, N., Robles-Parra, J. M., & Aranda-Figueroa, A. N. (2019). Los zarcillos que sostienen la producción de uva de mesa en Sonora: Testimonios orales de los líderes. *Ra Ximhai: Revista Científica de Sociedad, Cultura y Desarrollo Sostenible, 15*(2), 29–42.

Barrera-Rodríguez, A. I., Espejel, A., Pérez, M. G., & Ramírez-García, A. G. (2022). Atributos tangibles e intangibles y diferenciación sensorial de la vainilla mexicana. *Polibotánica, (54),* 241–255.

Brun, J., Jeuffroy, M. H., Pénicaud, C., Cerf, M., & Meynard, J. M. (2021). Designing a research agenda for coupled innovation towards sustainable agrifood systems. *Agricultural Systems, 191,* 103143.

Caso, A. M., Aguilar, A., Zizambo, J., & Sainz, J. (2019). Diversidad y discriminación en las empresas del sector servicios de la Ciudad de México. Consejo Nacional para Prevenir la Discriminación.

Castro, C., & Álvarez, B. (2016). La Igualdad en la Responsabilidad Social de las Empresas. Retrieved from: http://www.castello.es/web20/archivos/contenidos/61/D021_Igualdad_RSE.pdf (accessed on 30 May 2024).

CNDH. (2018). Guía con recomendaciones para considerar a las personas con discapacidad en protocolos de protección civil. Elaboración de un plan personal de emergencia. Retrieved from: http://appweb.cndh.org.mx/biblioteca/archivos/pdfs/01_Guia_DISCA_v_sept2018.pdf (accessed on 30 May 2024).

Compromiso RSE. (2020). Los consumidores, dispuestos a pagar más por productos de marcas sostenibles y transparentes. Retrieved from: https://www.compromisorse.com/rse/2020/01/22/los-consumidores-dispuestos-a-pagar-mas-por-productos-de-marcas-sostenibles-y-transparentes-/ (accessed on 30 May 2024).

Contreras, A. C. (2019). Certificaciones Agrícolas como Conducta Estratégica y el Alcance de Mercado de las firmas que forman el Sistema de Vid de Mesa Sonorense [Doctoral thesis, Centro de Investigación y Desarrollo CIAD]. Hermosillo, Sonora, México.

Espinosa, M. V., & Camargo, C. L. (2017). Habilidades gerenciales del contador público en las organizaciones modernas. *Gerencia Libre, 3,* 79–85.

Etimologías. (2023). *Gerencia.* Retrieved from: http://etimologias.dechile.net/?gerencia (accessed on 30 May 2024).

Ferrín-Schettini, H. M. (2018). La gerencia y sus nuevas tendencias: Un aporte al desarrollo empresarial. *Polo del Conocimiento, 3*(5), 237–249.

García-Payares, F., Boom-Cárcamo, E., & Molina-Romero, S. (2017). Habilidades del gerente en organizaciones del sector palmicultor en el departamento del Cesar-Colombia. *Visión de Futuro, 21*(2), 0–0.

Giles-Navarro, C. A., & Méndez-Mandujano, M. Á. (2019). La violación de los derechos humanos en México 2000-2018: Algunas características y tendencias a la luz de las estadísticas de la CNDH. Retrieved from: http://www.bibliodigitalibd.senado.gob.mx/bitstream/handle/123456789/4347/Nota_ViolacionesDHFinal46.pdf?sequence=1&isAllowed=y (accessed on 30 May 2024).

Guerrero, C. A., Ballesteros, L. G., & Herrera, J. B. (2018). Gerentes y equipo de trabajo efectivo: Empresa exitosa. *Dominio de las Ciencias, 4*(3), 416–430.

Guevara, L. R., & Pérez, D. V. (2021). Gerencia integral, tipos de liderazgos y su relación: Caso de estudio en el contexto empresarial local. *593 Digital Publisher CEIT, 6*(3), 107–121.

Hernández, F. G. (2020). Los objetivos de desarrollo sostenible a través de la responsabilidad social empresarial, la transformación organizacional y la innovación en una empresa manufacturera de Culiacán, Sinaloa México [Doctoral thesis, Universidad Juárez del Estado de Durango, Universidad Autónoma de Nayarit, Universidad Autónoma de Sinaloa].

Huerta-Dueñas, M., & Sandoval-Godoy, S. A. (2018). Sistemas de calidad como estrategia de ventaja competitiva en la agroindustria alimentaria. *Agricultura, Sociedad y Desarrollo, 15*(1), 19–28.

Hurtado-Bringas, B. A., Robles-Parra, J. M., Preciado-Rodríguez, J. M., & Bañuelos-Flores, N. (2018). Logística de transporte y desarrollo local en organizaciones exportadoras de uva de mesa sonorense. *Estudios Sociales, 28*(51), 0–0.

Ibaceta-Vergara, C. P., & Villanueva-Morales, C. F. (2021). Entornos virtuales de aprendizaje: Variables que inciden en las prácticas pedagógicas de docentes de enseñanza básica en el contexto chileno. *Perspectiva Educacional, 60*(3), 132–158.

Kuckartz, U., & Rädiker, S. (2019). *Analyzing Qualitative Data with MAXQDA: Text, Audio, and Video.* Springer International Publishing.

Marrugo-Salas, L. (2016). El acceso de la mujer a cargo de toma de decisiones en las empresas colombianas que cotizan en bolsa. *Entramado, 12*(1), 108–120.

Muñoz, B., & Barrantes, A. (2016). Equidad e inclusión social: Superando desigualdades. Retrieved from: http://www.oas.org/docs/inclusion_social/equidad-e-inclusion-socialentrega-web.pdf (accessed on 30 May 2024).

Olvera-Esquivel, J. G., & Arellano-Gault, D. (2015). El concepto de equidades y sus contradicciones: La política social mexicana. *Revista Mexicana de Sociología, 77*(4), 581–610.

Orduño, M. A., Kallas, Z., & Ornelas, S. I. (2020). Farmers' environmental perceptions and preferences regarding climate change adaptation and mitigation actions: Towards a sustainable agricultural system in México. *Land Use Policy, 99,* 105031.

Rojas, E. M. (2017). *Merchandising.* Bogotá, D.C., Colombia: Fondo editorial Areandino.

Shah, S. M., Liu, G., Yang, Q., Casazza, M., Agostinho, F., & Giannetti, B. F. (2021). Sustainability assessment of agriculture production systems in Pakistan: A provincial-scale energy-based evaluation. *Ecological Modelling, 455,* 109654.

Solís, D. D., Robles, J. M., & Preciado, J. M. (2020). Condiciones de mercado y vivienda sustentable. *Vivienda y Comunidades Sustentables, 7,* 62–76.

Suárez, M. J. (2018). Gestión Empresarial: Un paradigma del siglo XXI. *Revista Científica FIPCAEC (Fomento de la investigación y publicación en Ciencias Administrativas, Económicas y Contables), 3*(8), 44–64.

Terán, K. (2022). Equidad gerencial, como una demanda intangible de mercado: Hacia organizaciones sustentables (PhD thesis). Centro de Investigación en Alimentación y Desarrollo (CIAD, A.C.). Hermosillo, Sonora, México.

Terán, K., Robles, J. M., Preciado, J. M., & López, D. C. (2019). Equidad gerencial como una demanda intangible de mercado: hacia organizaciones sustentables. *Entre Ciencia e Ingeniería, 13*(26), 85–93.

Terán-Samaniego, K., Tiznado-Hernández, M. E., Robles-Parra, J. M., & Paz-Luna, J. L. (2022). The Influence of Managerial Equity on Corporate Social Responsibility: The Case of Agricultural Systems. *Estudios Sociales: Revista de Alimentación Contemporánea y Desarrollo Regional, 32*(60).

The Packer. (2014). Fresh Trends. Estados Unidos: Vance Publishing Corporation. Grapes. Retrieved from: http://www.thepacker.com/topics/grapes (accessed on 30 May 2024).

The Packer. (2019). Fresh Trends. Estados Unidos: Vance Publishing Corporation. Grapes. Retrieved from https://www.thepacker.com/magazines/fresh-trends-2019 (accessed on 30 May 2024).

The Packer. (2022). Fresh Trends. Estados Unidos: Vance Publishing Corporation. Grapes. Retrieved from: https://www.thepacker.com/magazines/2022-fresh-trends-magazine (accessed on 30 May 2024).

Torres, J. M., & Peris, Y. (2020). La discriminación por edad en el ámbito laboral. Fundaciópimec. Numitec. Retrieved from: https://www.transformacionsocialcompetitiva.com/wp-content/uploads/2021/10/Informe-NDPE_11-nov-20.pdf (accessed on 30 May 2024).

Vela-Barba, E. (2017). La discriminación en el empleo en México. Ciudad de México: Ediciones La Bilbioteca.

Viaña, F. (2018). Epistemología hacia la formación gerencial del siglo XXI. *Teorías, Enfoques y Aplicaciones en las Ciencias Sociales, 11*(23), 53–60.

Wei, J., Chen, Y., Tiemur, A., Wang, J., & Wu, B. (2018). Degradation of pesticide residues by gaseous chlorine dioxide on table grapes. *Postharvest Biology and Technology, 137,* 142–148.

Westbrook, G., & Angus, A. (2021). Las 10 principales tendencias globales de consumo para 2021. Euromotor Internacional. Retrieved from: https://mexwco.com/2021/11/10/las-10-principales-tendencias-de-consumo-para-el-2021-parte-ii/ (accessed on 30 May 2024).

CHAPTER 6

Development and Characterization of Sweet Bread with Jackfruit Bulb Flour

X. RUELAS-CHACÓN,[1] D. BELLEZA-PACHECO,[1] R. ROJAS-MOLINA,[2] O. N. REBOLLOSO-PADILLA,[3] J. D. CORONA-FLORES,[4] J. I. GARCÍA-LÓPEZ,5 N. CAMPOSECO-MONTEJO,[5] A. FLORES-NAVEDA,[5] and P. ÁLVAREZ-VAZQUEZ[6]

[1]*Food Science and Technology Department, Universidad Autónoma Agraria Antonio Narro, Calzada Antonio Narro, Buenavista, Saltillo, Coahuila, C.P., Mexico*

[2]*Agronomy Faculty, Universidad Autónoma de Nuevo Léon, Francisco I. Madero, ExHacienda el Cañada, General Escobedo, Nuevo Léon, C.P., Mexico*

[3]*Animal Production Department, UAAAN, Saltillo, Mexico*

[4]*Planning Direction, UAAAN, Saltillo, Mexico*

[5]*Plant Breeding Department, UAAAN, Saltillo, Mexico*

[6]*Natural and Renewable Resources Department, UAAAN, Saltillo, Mexico*

ABSTRACT

The jackfruit (*Artocarpus altilis*) is a very perishable fruit with a short life, becoming unacceptable to consumers as a fresh fruit. The food bakery development is an alternative to reduce post-harvest losses by using flours obtained from various parts of the fruits and provides an alternative for regular consumers. The objective of the present research was to evaluate the quality characteristics of sweet bread based on jackfruit bulb flour. Methodology: Jackfruit bulb flour was obtained by drying at 60?C for 24

Sustainable Agricultural and Food Systems Engineering: New Research, Practices, and Developments. Jose Juan Buenrostro-Figueroa, Julio César Tafolla-Arellano, Leonardo Sepúlveda Torre, Porteen Kannan, A. K. Haghi (Eds.)

hours and was used to make bread with different percentages of whole-wheat flour (WWF), jackfruit flour (YF), and wheat flour (WF). Three formulations were analyzed: Treatment 1 (50% WWF–50% WF; control treatment), Treatment 2 (50% WF–25% WWF–25% YF), and Treatment 3 (50% WF–37.5% WWF–12.5% YF). The AOAC methods were used to evaluate moisture, crude protein, crude fiber, fat, ash, carbohydrates, caloric value, and textural properties. The antioxidant properties were analyzed with the DPPH and ABTS, and iron-reducing power methods. Total polyphenols were quantified by the Folin-Ciocalteu Reagent. Microbial analysis was carried out for aerobic mesophiles, fungi, total coliforms, and *Salmonella spp*. Sensory evaluation was carried out with 40 semi-trained panelists using a hedonic test. Results: An ANOVA was calculated for the proximal analyses, and functional parameters, and in case a difference was detected, a Tukey's test was applied. Significant differences were observed in quality parameters, showing that the use of *jackfruit flour* increased protein and fiber content. There were significant differences between treatments according to the texture profile: greater hardness, elasticity, and chewiness for the control; greater cohesiveness for treatment-1; greater gumminess for treatment-2. There were no significant differences in the content of functional components. Microbiological analyses for aerobic mesophiles, fungi, total coliforms, and *Salmonella* spp. showed acceptable results according to NOM-247-SSA1-2008. There were no significant differences between treatments in the sensory evaluation. Conclusions: It was possible to develop sweet breads using 12.5% and 25% jackfruit flour, with high fiber and protein content, and high scores in the hedonic test.

6.1 INTRODUCTION

The importance of functional foods has a great impact on the human diet today; therefore, it is sought to have a product with functional characteristics and high nutritional value. To obtain these products, innovation plays an essential role at all times, from formulation to obtaining the final product, taking into account the tests to be analyzed.

Functional foods can be divided into two broad categories. The first category consists of functional foods that naturally contain a component that offers additional benefits to the consumer. The other category of functional foods consists of processed foods in which the component is added to the food to give it additional benefits (De Carrasco & Encina, 2008; Fu, Shiau, & Chang, 2014; Akmeemana et al., 2022).

The cereal industry is one of the most important globally since cereal is the basis of human nutrition. The baking area is the one that stands out the most, where we can find different types of bread according to the method of preparation and the type of flour used. The combination of ingredients for the preparation of bread is very extensive in both sweet and savory breads, which has led to the introduction of ingredients that have properties for the benefit of health.

Wheat has various uses in the food industry and is the main ingredient in making bread, cookies, pastries, and pasta. However, a variety of wheat considered to be of good quality for making bakery products could be considered of poor quality for making cookies or some other non-bakery product. This is due to the fact that the main groups of products made with wheat (baking, biscuits, confectionery, and pastification) differ from each other both in their formulation and processing characteristics and in those of the finished product (moisture, texture, density, flavor, etc.), requiring specific grain quality characteristics for each product group (De Carrasco & Encina, 2008; Güemes, Totosaus, Hernández, Soto, & Aquino, 2009; Bassett et al., 2013). One of the ingredients in sweet bread is fruits, which give desirable organoleptic attributes to the product, acting as antimicrobial and antioxidant functional foods. By using a mixture of integral flours such as wheat and fruit flour, a sweet bread can be developed that meets the nutritional requirements and is functional.

Artocarpus altilis is a leafy tree native to Indonesia and Polynesia, from where it has spread to all tropical regions of the world. Using this fruit as flour in the preparation of sweet bread due to its exotic flavor results in an innovative product that provides cardiovascular properties to the body. By adding the large amount of fiber provided by whole wheat flour, we can have a purely functional product (Barros-Castillo et al., 2021; Cheng et al., 2023).

It is known that the majority of the global population consumes sweet bread, but most of them do not consume whole wheat type. In the present work, the development of sweet bread with jackfruit (*Artocarpus altilis*) flour, wheat flour, and whole wheat flour is described. This bread is a unique food that provides nutrients to complement the human diet, with better sensory attributes compared to conventional bread. It is important to develop bakery products that contain functional ingredients, such as whole wheat flour, which provides fiber to improve intestinal transit and digestion (Cheng et al., 2023; Olivas, Nevárez, & Gastélum, 2009; Islam, Biswas, Esham, Yoy, Khan, & Hasan, 2023).

Artocarpus altilis is used because it is a fruit rich in starches, with a high carbohydrate content considered a good source of vitamin B, a moderate source of vitamins A and C, and rich in calcium, iron, potassium, and phosphorus. The caloric value and protein content of the fruit are 75 to 80 calories per 100 g and 1.5%, respectively. It is also considered a primary food source, making it possible to incorporate flour into bread that provides all these nutritional qualities that promote health (Cheng et al., 2023; Islam, Biswas, Esham, Yoy, Khan, & Hasan, 2023).

It is important to consume foods that contain antioxidant compounds in moderation. Research in the medical field has shown that flavonoids isolated from *A. altilis* exhibit strong cytotoxic activity against leukemia in histological cultures, thus offering great benefits. The sap is reported to be used in medicinal preparations in Polynesia. The manufacturing process is easy, resulting in a product different from conventional sweet bread (Cheng et al., 2023; Islam, Biswas, Esham, Yoy, Khan, & Hasan, 2023; Konsue, Bunyameen, & Donlao, 2023).

Due to the above, the objective of the research was to design and evaluate the quality characteristics of sweet bread based on jackfruit bulb flour (*Artocarpus altilis*).

6.2 MATERIALS AND METHODS

The jackfruit was obtained from a commercial establishment in Saltillo, Mexico. Whole-wheat flour, wheat flour, oil, butter flavor, whole milk, eggs, sugar, and baking soda were purchased from a local grocery store.

6.2.1 PREPARATION OF BREAD

The bread was based on flour percentages. The measured ingredients (Table 6.1) were placed in the order of liquids and dry ingredients into a Breadman Dream Machine TR3000 (Salton/Maxim Housewares, Inc., Mt. Prospect, Ill., U.S.A.). Mold trays (cupcake) with 6.5 cm upper diameter, 4 cm lower diameter, and 3.5 cm height layers were prepared for each sample. The bread was baked in a rotary oven at a temperature of 200?C for approximately 15 min. Once baked, they were demolded, cooled, and bagged, stored for later characterization.

TABLE 6.1 Formulations for Jackfruit Sweet Bread (WWF: Whole Wheat Flour; WF: Wheat Flour; and YF: Jackfruit Flour)

Ingredients (%)	Treatment 1 (Control) 50%WWF–50%WF	Treatment 2 50%WF–25%WWF–25%YF)	Treatment 3 50%WF–37.5%WWF–12.5%YF
Whole wheat flour	12.63	6.31	9.47
Wheat flour	12.63	12.63	12.63
Jackfruit flour	0	6.31	3.15
Oil	16.84	16.84	16.84
Whole milk	21.05	21.05	21.05
Egg	13.68	13.68	13.68
Vanilla	2.10	2.10	2.10
Butter flavor	2.10	2.10	2.10
Sugar	17.89	17.89	17.89
Baking soda	1.05	1.05	1.05

6.2.2 *PROXIMATE AND CHEMICAL ANALYSES*

Protein content (%) was determined by the Kjeldahl method using 1 g of ground sample, a correction factor of 5.7, typical of flour mixtures, was used for the calculation. Fat content (%) and ash content (%) were determined utilizing a Soxhlet extractor (Velp Scientifica, Monza-Brianza, Italy) on 5 g of ground sample and diethyl ether as a solvent, and the muffler furnace method, respectively (AOAC International, 2015). Moisture was determined with the IR-200 Moisture Analyzer (Denver Instrument Co., Arvada, Colo., U.S.A.). Carbohydrate concentration of the bread samples was calculated by difference, and fiber content was determined by the AOAC (2015). All analyses were carried out in triplicate for each flour type. Three loaves of each bread type were analyzed.

6.2.3 *TEXTURE OF BREAD*

Texture analysis was performed in a texture analyzer (model TA-XT-2iHR, Stable Microsystems, Ltd., Surrey, U.K.) containing a 5 kg load cell. Calibrations were carried out with a 2 kg load cell. Data were analyzed using Exponent Software supplied with the instrument. Upper and lower crusts of bread balls (n = 6) were subjected to a penetration depth of 30 mm through a two-cycle sequence using a spherical probe (25 mm in diameter) (Cyl. Perspex P/25, Stable Microsystems, Ltd.) at a crosshead speed of 1 mm/s. The following texture parameters were extracted from the measured force deformation curves: hardness (N) = maximum force required to compress the sample (peak force during the first compression cycle); gumminess = height that the sample recovers during the time that elapses between the end of the first compression and the start of the second; cohesiveness = extent to which the sample could be deformed before rupture (A1/A2, A1 being the total energy required for the first compression and A2 the total energy required for the second compression); chewiness (N) = the work needed to chew a solid sample to a steady state of swallowing (hardness × cohesiveness × gumminess).

6.2.4 MICROBIAL ANALYSES

Microbiological determinations of aerobic mesophiles, total coliforms, fungi, and Salmonella spp. All this is based on the Official Mexican Norm NOM-247-SSA1-2008 (2018).

6.2.5 DETERMINATION OF ANTIOXIDANT PROPERTIES

6.2.5.1 2′2-AZINO-BIS-3-ETHYLBENZTHIAZOLINE-6-SULFONICACID-2NH4 (ABTS)

A stock solution of Trolox was prepared at a concentration of 200 ppm, and the following calibration curve was made. An aqueous solution of ABTS+ (2′2-Azino-bis-3-ethylbenzthiazoline-6-sulfonic acid-2NH_4) was prepared at a concentration of 7 mM while preparing a solution of potassium persulfate ($K_2S_2O_8$) at a concentration of 2.45 mM. The ABTS+ solution and the potassium persulfate solution were mixed in a ratio of 2:1, respectively, which was left to oxidize for 12 hours before being used. Dilute solutions of the oxidized ABTS+ mixtures were made in ethanol until an absorbance value of 0.700 ± 0.002 nm was obtained. About 50 μL of each concentration of the calibration curve was pipetted directly onto each well, as well as 50 μL of the analyzed sample directly onto the well (considering the dilution factor of the sample), after which 950 μL of the reagent ABTS+ was pipetted, previously adjusted, at each concentration of the calibration curve and sample analyzed. Finally, the reading was taken at a wavelength of 734 nm.

6.2.5.2 2,2-DIPHENYL-1-PICRYLHYDRAZYL (DPPH) FREE RADICAL-SCAVENGING ACTIVITY TEST

Five grams of bread were extracted as previously reported, with a methanol:water (70:30, v/v) mixture. Then, 200 μL of the extract was mixed with 2.6 mL of methanol and 2 mL of DPPH. The absorbance was recorded at 517 nm by a UV-visible Genesys spectrophotometer after an incubation time of 30 minutes at room temperature. A blank was prepared and analyzed following the same procedure. Radical-scavenging activity was calculated as follows: I% = [(Abs0 – Abs1)/Abs0] × 100, where Abs0 was the absorbance of the blank and Abs1 was the absorbance of the sample. The antioxidant capacity value (μM Trolox g^{-1}) of samples was obtained from the calibration curve by measuring the absorbance at 517 nm of Trolox methanolic solutions at different concentrations. The cupcakes of each treatment were analyzed in triplicate.

6.2.5.3 FERRIC REDUCING ANTIOXIDANT POWER ASSAY (FRAP)

A stock solution of gallic acid was prepared at a concentration of 200 ppm, and the following calibration curve was made. A phosphate buffer solution at pH 7 was prepared by mixing a solution of 61.5 mL of 1M dibasic potassium phosphate (K_2HPO_4) + 38.5 mL of 1M monobasic potassium phosphate (KH_2PO_4) and diluted to 1 L with distilled H_2O. Next, a solution of potassium ferrocyanide ($C_6N_6FeK_4$) was prepared at a concentration of 1%, as well as a solution of trichloroacetic acid (CCl_3COOH) at a concentration of 10% and a solution of ferric chloride ($FeCl_3$) at a concentration of 0.1%. Test tubes for the samples were labeled, and the concentrations of the calibration curve were pipetted into their corresponding test tubes, along with 50 μL of the sample to be analyzed (considering the dilution factor of the sample), 120 μL of the phosphate buffer solution at each concentration of the calibration curve and sample analyzed, and 220 μL of the potassium ferrocyanide solution at each concentration of the calibration curve and sample to be analyzed. It was homogenized for later incubation in a water bath for 20 minutes at a temperature of 50?C. After that, 120 μL of the trichloroacetic acid solution was pipetted at each concentration of the calibration curve, and the sample was analyzed. About 450 μL of distilled H_2O was added at each step, along with 100 μL of the ferric chloride solution at each concentration of the calibration curve and sample analyzed. Finally, it was read at a wavelength of 700 nm.

6.2.5.4 TOTAL POLYPHENOL ANALYSIS

For this determination, a gallic acid solution was prepared at a concentration of 200 ppm, and a calibration curve was made. A sodium carbonate (Na_2CO_3) solution was prepared at a concentration of 75 g/L, the test tubes were previously labeled according to the samples analyzed and concentrations of the calibration curve. Subsequently, 250 μL of each concentration of the calibration curve was pipetted into their corresponding test tubes, 250 μL of the sample analyzed in their corresponding test tubes (considering the dilution factor of the sample), 250 μL of Folin-Ciocalteu reagent at each concentration of the calibration curve and sample analyzed, 250 μL of the sodium carbonate solution at each concentration of the calibration curve and sample analyzed. It was homogenized and incubated in a water bath for 30 minutes at 40?C and then pipetted 2 mL of distilled H_2O at each concentration of the calibration curve and the sample was analyzed. It was read at a wavelength of 750 nm.

6.2.6 BREAD SENSORIAL EVALUATION

One day after preparation, breads were submitted to a sensory analysis in order to evaluate appearance, texture, crust color, flavor, and degree of satisfaction. Samples were placed in codified recipients using three digits randomly. One cupcake of each treatment was given to each taster. The breads were submitted to a hedonic test, applying a 5-point scale. Sensory data were statistically analyzed considering the Kruskal-Wallis test at a 5% probability.

6.3 RESULTS AND DISCUSSION

6.3.1 PROXIMAL ANALYSES

Table 6.2 shows the evaluated parameters of the proximal analysis in the control and the treatments. Of the parameters evaluated from the proximal analysis, it was found that treatment 3 presents less moisture; therefore, the amount of dry matter is more for this formulation than the rest. It is very interesting that the control, composed of wheat flour, presented a humidity and amount of dry matter similar or almost equal to treatment two, which is composed of a mixture of jackfruit and wheat flour.

TABLE 6.2 Proximal Analyzes of the Three Samples of Sweet Bread Made with Jackfruit Flour

Proximal Parameters (%)	Sweet Bread		
	Treatment 1 (Control)	Treatment 2	Treatment 3
Moisture	72.3a	72.2a	69.1a
Dry matter	27.7a	27.9a	30.9a
Protein	8.8b	10.7a	5.7c
Fat	30.3a	26.9a	27.1a
Ashes	2.7a	3.4a	2.9a
Fiber	22.4a	16.3b	17.5b
ELN*	35.8c	42.7b	46.8a
Total	100.0	100.0	100.0
Energy (Kcal)	9062a	5798c	6311b

*ELN: Nitrogen-free extract consists mainly of carbohydrates. Means followed by different letters differ among them as analyzed by the Tukey test, at 5% probability.

The protein, as a result, is favorable for the formulation that presents a greater amount of jackfruit flour. This means that by adding *Artocarpus altilis* flour to wheat flour, the effect is reflected in the increase in protein. Wheat flour by itself presents a fairly good percentage of protein. According to Codex Standard 152–1985, the minimum percentage of protein in wheat flour is 7%. The values obtained for the control, which is made only from wheat flour, comply with said parameter, as well as treatment 3, which presents 50% wheat flour in the formulation. Treatment 3 presents a very low amount of protein compared to the other formulations; it is detected that by presenting a smaller amount of jackfruit flour, the effect decreases in the amount of protein. The most favorable formulation for the percentage of protein is undoubtedly treatment 2. Fat is a fundamental parameter in the product. Here, animal fat has been used at the same concentration. It can be observed that the jackfruit flour used in treatments 2 and 3 has a favorable effect on the product, since compared to the control, the amount of raw fat is less. There is a higher ash content for treatment 1 and a lower value for the control. Product or treatment 2 is rich in the ash parameter. According to Codex Standard 152–1985, there is no maximum or minimum limit for the ash content in wheat flour, so the final bread product will be even richer in ashes with the addition of jackfruit flour.

It is well known that wheat has a lot of fiber, so it is reflected in the study, that the control that is made from wheat and whole wheat flour has a greater amount of this parameter. In treatment 2, where the concentration of whole wheat flour is lower compared to treatment 3, it can be seen that this is how the values obtained are. The greater the amount of whole wheat flour, the higher the fiber percentage. In the bromatological analysis of the nutritional and functional evaluation study of *Artocarpus altilis* starch by Ngoc et al. (2023), the jackfruit analyzed by itself presents a protein percentage of 12.29% compared to wheat, which does not even present half of its value.

The nitrogen-free extract present in the control is minimal compared to treatment 2, where it is known that the nitrogen-free extract is mainly composed of carbohydrates. The use of jackfruit flour, which is a fruit where the amount of carbohydrates is good, helps the ELN parameter to be high. Sensorially, it is appreciated that the sweetness is characteristic of bread that contains jackfruit flour (Kuskoski, Asuero, Troncoso, Mancini, & Fett, 2004; Kumar et al., 2021; Ortega-González et al., 2022).

The control formulated with wheat flour provides us with a greater amount of energy, where it is known that this cereal provides a large amount of energy to the body. It can be deduced that treatment 1, which has the least amount of wheat flour, is poorer in the amount of gross energy it provides as a product. Therefore, we can say that by mixing jackfruit flour and wheat flour to formulate a sweet bread, jackfruit flour will not have an effect of increasing gross energy. The greater the amount of jackfruit flour and the less wheat flour, the percentage of gross energy will be lower.

6.3.2 TEXTURE ANALYSES

Table 6.3 shows the average values obtained from the evaluated parameters of the texture profile, where it is observed that there was a significant difference between the formulations in hardness, cohesiveness, and gumminess. All the components used for the formulation were the same; the only variation was the concentration and the type of flour.

TABLE 6.3 Texture Analysis of the Jackfruit Bread Treatments

Treatments	Texture Parameters					
	Hardness	**Elasticity**	**Cohesiveness**	**Gumminess**	**Chewability**	**Strength**
Treatment 1	23.5537a	0.8537a	0.4757b	11.1753a	9.5403a	0.2233a
Treatment 2	17.3500c	0.8327a	0.5980a	10.3867b	8.6500a	0.2570a
Treatment 3	20.3277b	0.8367a	0.5697a	11.5787a	9.6840a	0.2467a

Note: Means followed by different letters differ among them as analyzed by the Tukey test, at 5% probability.

Statistically, there is a significant difference in the hardness parameter (Table 6.3), with a higher value for the control and a minimum value for treatment 2. This is due to the fact that the control formulation is composed only of wheat and whole wheat flour. Whole wheat flours tend to be harder because they contain gluten and because the type of whole wheat grain used is hard and not soft. Therefore, treatment 2, which has a greater amount of jackfruit flour and a lower amount of whole wheat flour, tends to be softer. The amount of fat present in each treatment is a very important factor in the hardness of the bread. A greater amount of fat has been obtained in the control, which makes its hardness more evident compared to the rest of the treatments.

Table 6.3 shows the elasticity results, where there is no significant difference for the formulations. Wheat flour proteins, specifically gluten proteins, give the dough a unique functionality that sets it apart from other cereal flour. From a rheological point of view, wheat flour dough behaves like a fluid. Viscoelastic, this property makes the dough elastic and extensible (Ortega-González et al., 2022; Osuna, Judis, Romero, & Bertola, 2013).

The cohesiveness values in Table 6.3 show a significant difference between the formulations. Being an important parameter related to strength, chewiness, and gumminess, statistically the results are not the same, but visually it can be seen that the control presents a lower value. It can be said that the jackfruit flour modifies this property in treatments 2 and 3, which are visually taller.

Gumminess, defined as the product of hardness by cohesiveness, like hardness, showed significant differences for this parameter. The hardness, gumminess, and chewiness are increased by adding whole wheat (Ortega-González et al., 2022; Torres, Torres, Acevedo, & Gallo, 2015), which is rich in fiber. The formulations have fiber due to the flours used, which helps to ensure that this evaluated parameter is the same for treatments 1, 2, and 3 (Table 6.3).

The chewiness parameter did not present a significant difference; statistically, the three formulations show similarity in this parameter (Table 6.3).

Sensorially, it is defined as the force required to disintegrate a solid food until it is ready to be swallowed. It is determined as the product of firmness x cohesiveness (defined as the ratio between A2/A1) x elasticity (Osuna, Judis, Romero, & Bertola, 2013). Extrapolating the results of this parameter, the three formulations statistically require the same strength to be swallowed by the consumer, depending on the chewing time as well as the hardness of the bread.

The resistance parameter does not present a significant difference, so it can be said that the evaluated parameter is statistically the same for the three formulations. Table 6.3 shows that the resistance of bread in its three formulations has low values compared to other foods based on wheat flour, such as cookies and whole grain bars.

6.3.3 *MICROBIAL ANALYSES*

Table 6.4 shows the results obtained from the different microorganisms evaluated in the final product, following the Official Mexican Standard NOM-247-SSA1-2008 (2018), Products and Services. Cereals and their products. Cereals, cereal flour, meal, or semolina. Foods based on cereals, edible seeds, flour, semolina, or their mixtures. Bakery products. Provisions and sanitary and nutritional specifications.

TABLE 6.4 Microbial Counts of Aerobic Mesophiles, Total Coliforms, Fungi, and *Salmonella* spp. in Sweet Bread Samples

Treatments	Microbial Analyzes (CFU/g)			
	Aerobic Mesophiles	**Total Coliforms**	**Fungi**	**Salmonella spp. in 25 g**
Treatment 1	5,760	0	10	Negative
Treatment 2	5,760	20	20	Negative
Treatment 3	5,760	0	40	Negative

Note: Means followed by different letters differ among them as analyzed by the Tukey test, at 5% probability.

Treatments 2 and 3, as well as the control based on the Official Mexican Standard NOM-247-SSA1-2008 (2018), do not exceed the number of colony-forming units, so the bread presents a tolerable amount of aerobic mesophiles. The jackfruit flour used does not affect aerobic mesophiles, since in all cases for the formulations, the values obtained are the same.

Treatment 2 presents a high value of 20 CFU/g, which does not exceed the value established by the standards. The rest of the trials are 100% free of total coliforms, resulting in a stable product with a longer shelf life under appropriate conditions. The preparation and handling during and after treatment 2 are reflected in the value given by it. It is attributed that contamination is easier during handling and not in the process, since the temperature at which the bread was baked is high, destroying all microorganisms.

The amount of fungi present in the tests is minimal compared to the value established by the standards of 300 CFU/g, coupled with the temperatures used during the process that helped reduce fungal colonies that could be present before baking.

Salmonella spp. does not present in any of the tests; therefore, the ingredients used were of good quality, and the process of elaboration of the product, and the handling were adequate.

6.3.4 ANTIOXIDANT PROPERTIES

6.3.4.1 ABTS

Table 6.5 shows the antioxidant capacity of the three formulations studied; treatment 3 (24.28%) has the highest antioxidant capacity compared to the others. It is followed by treatment 2 (22.77%) and finally the control with a value of 16.15% as a percentage of inhibition of the ABTS radical. The antioxidant capacity is directly related to the pigment content of the fruit (De Carrasco & Encina, 2008; Spada et al., 2020), as well as to the components present in the same food.

TABLE 6.5 Antioxidant Capacity of the Functional Components in the Sweet Jackfruit Breads Formulated

Treatments	ABTS (mg EAG/L)	DPPH (mg EAG/L)	FRAP (mg EAG/L)	Total Polyphenol (mg EAG/L)
Treatment 1	18.28b	2.68a	9.63b	16.57b
Treatment 2	25.95a	1.13b	17.91a	19.64a
Treatment 3	27.69a	2.09a	16.64a	17.80a

Note: Means followed by different letters differ among them as analyzed by the Tukey test, at 5% probability.

The antioxidant capacity of a food depends on the nature and concentration of the natural antioxidants present in it. *Artocarpus altilis* presents a low concentration of antioxidant capacity as a natural fruit, but the antioxidant potential evidenced by the ethanolic and aqueous extracts of *A. altilis* (Rivas, Muñoz, Pino, & Balcázar, 2015) from the leaves of the plant is highly significant compared to other medicinal plants.

Treatments 2 and 3 have 25% and 12% jackfruit flour, respectively. It can be observed that the greater the amount of jackfruit flour in this bread formulation, the lower the result in percent inhibition as well as the milligrams of Trolox equivalents when compared to when a lower amount of flour is added.

The antioxidant capacity of a food is due to the antioxidant activity of its different compounds, among which are phenolic compounds, carotenes, anthocyanins, ascorbic acid, etc. (Jurado et al., 2016). Therefore, it can be said, from the results obtained, that there are phenolic compounds, carotenes, and ascorbic acid present in *A. altilis* in a minimal way, which would have some antioxidant potential. This potential has a synergistic effect between the bioactive compounds that make up the fruit (Kuskoski, Asuero, Troncoso, Mancini, & Fett, 2004; Rivas, Muñoz, Pino, & Balcázar, 2015; Pineda, Salucci, Lázano, Maiani, & Ferro, 1999).

6.3.4.2 DPPH

Table 6.5 shows the mg EAG/L of the different formulations studied, for control (2.68) presents the highest value of the rest, treatment 2 with a value of 1.13 mg EAG/L which is low, and treatment 3 presents 2.09 mg EAG/L. Compared to other foods, this does not present great antioxidant capacity in general, it is also mentioned that high temperatures directly influence the content of bioactive compounds (Kuskoski, Asuero, Troncoso, Mancini, & Fett, 2004; Rivas, Muñoz, Pino, & Balcázar, 2015; Pineda, Salucci, Lázano, Maiani, & Ferro, 1999).

6.3.4.3 FRAP

Table 6.5 shows each of the values obtained for metals ferric ion, expressed in mg of gallic acid equivalents per liter of extract. The FRAP technique shows that the bread with the highest concentration of *A. altilis* flour, which is treatment 2, has a ferric ion metal reduction level of 17.91 mg EAG, this being the highest compared to treatment 3 and the control. Generally, this food has a minimal antioxidant capacity, the addition of antioxidant compounds to the different formulations will depend on whether a study is required with foods with high antioxidant capacity (Kuskoski, Asuero, Troncoso, Mancini, & Fett, 2004; Rivas, Muñoz, Pino, & Balcázar, 2015; Jurado et al., 2016).

6.3.4.4 TOTAL POLYPHENOL CONTENT

There are several classes and subclasses of polyphenols that are defined based on the number of phenolic rings that they possess and the structural elements that these rings present. The main groups of polyphenols are phenolic acids (derived from hydroxybenzoic acid or hydroxycinnamic acid), stilbenes, lignans, phenolic alcohols, and flavonoids (Konsue, Bunyameen, & Donlao, 2023; Kumar et al., 2021). Table 6.5 shows the results of the quantification of total phenols by the Folin-Ciocalteu method, whose results are expressed in mg/Eq of gallic acid per liter of extract. It is observed that the ethanolic extract of sweet bread based on *A. altilis* and wheat flour presented for treatment 2 a higher amount of total phenols (19.64 mg/Eq. Gallic acid/L of extract). The control (16.57 mg/Eq of gallic acid per liter of extract) presents a lower amount of total phenols compared to treatments 2 and 3, it is reflected that jackfruit flour in its different concentrations influences the values obtained for phenols. We can say that the higher the percentage of jackfruit flour incorporated into the formulation, the greater the amount of phenols, not counting the loss due to high temperatures. It is important to mention that the phenolic compounds found in fruits and vegetables play a fundamental role in health; since they decrease the risk of chronic diseases, such as cardiovascular disease and cancer (Kuskoski, Asuero, Troncoso, Mancini, & Fett, 2004; Rivas, Muñoz, Pino, & Balcázar, 2015; Kalse & Swami, 2022).

6.3.5 SENSORY EVALUATION OF SWEET JACKFRUIT BREAD

Table 6.6 shows the average values obtained from the evaluated parameters of the sensory evaluation. It is observed that there was no statistically significant difference. This is due to the fact that when evaluating the bread, the panelist judges were consumers, which means that they could not identify the differences between attributes of the formulations. It is not a bad result, since the consumers are not trained, and the different treatments are the same for them.

TABLE 6.6 Sensory Analysis of the Odor, Color, Flavor, Texture, and Appearance Attributes Obtained from 40 Semi-Trained Panelists

Treatments	Attributes Evaluated				
	Odor	**Color**	**Flavor**	**Texture**	**Appearance**
Treatment 1	4.25a	4.10a	4.05a	3.88a	4.08a
Treatment 2	4.13a	4.22a	4.25a	4.17a	4.30a
Treatment 3	4.13a	4.08a	4.30a	4.00a	4.30a

Means followed by different letters differ among them as analyzed by Kruskal-Wallis, at 5% probability.

6.4 CONCLUSION

The proximal analysis presents a significant difference in the study variables of protein and fiber, with these being greater in treatments 2 and 3. The antioxidant components present in the treatments are minimal between treatments 2 and 3. The variables analyzed in the texture profile that showed significant differences were hardness, cohesiveness, and gumminess, due to the jackfruit flour. The bread is safe according to the Official Mexican Standard NOM-247-SSA1-2008, Products and Services. For the participating panelists, the bread is the same in color, flavor, smell, texture, and appearance, so no significant difference was found in the data analyzed in the sensory evaluation, but the scores given were high in all attributes.

KEYWORDS

- **antioxidants**
- **Artocarpus altilis flour**
- **food development**
- **jackfruit bread**
- **quality parameters**
- **sensory analysis**
- **sensory evaluation**

REFERENCES

Akmeemana, Ch., Wickramasinghe, I., Wanniarachchi, P. Ch., & Vithanage, Th. (2022). Effect of drying and frying pre-treatments on nutrient profile, antioxidant capacity, cooking time, and sensory acceptability of easy-to-cook jackfruit seeds. *Appl. Food Res., 2*, 100234. DOI: 10.1016/j.afres.2022.100234.

AOAC International. (2015). Métodos oficiales de análisis de AOAC 2015: Análisis Proximal. [Online] Retrieved from: https://www.aoac.org/AOAC_Prod_Imis/AOAC_Member/Default.aspx?WebsiteKey=2e25ab5a-1f6d-4d78-a498-19b9763d11b4&hkey=8fc2171a-6051-4e64-a928-5c47dfa25797 (accessed on 30 May 2024).

Barros-Castillo, J. C., Calderón-Santoyo, M., Cuevas-Glory, L. F., Pino, J. A., & Ragazzo-Sánchez, J. A. (2021). Volatile profiles of five jackfruit (*Artocarpus heterophyllus* Lam) cultivars grown in the Mexican Pacific area. *Food Res. Int., 139*, 109961. DOI: 10.1016/j.foodres.2020.109961.

Bassett, M. N., Pérez-Palacios, T., Cipriano, I., Cardoso, P., Ferreira, I. M. P. L. V. O., Samman, N., & Pinho, O. (2013). Development of bread with NaCl reduction and calcium fortification: Study of its quality characteristics. *J. Food Qual., 37*, 107–116. DOI: 10.1111/jfq.12079.

Cheng, M., He. J., Wang, H., Li, Ch., Wu, G., Zhu, K., Chen, X., Zhang, Y., & Tan, L. (2023). Comparison of microwave, ultrasound, and ultrasound-microwave assisted solvent extraction methods on phenolic profile and antioxidant activity of extract from jackfruit (*Artocarpus heterophyllus* Lam.) pulp. *LWT, 173*, 114395. DOI: 10.1016/j.lwt.2022.114395.

Codex Standard 152–1985, Roma, Italia.

De Carrasco, R. R., & Encina, C. R. (2008). Determinación de la capacidad antioxidante y compuestos bioactivos de frutas nativas peruanas. *Rev. Soc. Quím. Perú, 74*, 108–124.

Fu, J. T., Shiau, S. Y., & Chang, R. C. (2014). Effect of calamondin fiber on rheological, antioxidative, and sensory properties of dough and steamed bread. *J. Texture Stud., 45*, 367–376. DOI: 10.1111/jtxs.12087.

Güemes, N. V., Totosaus, A., Hernández, J. F., Soto, S., & Aquino, E. N. (2009). Propiedades de textura de masa y pan dulce tipo "concha" fortificados con proteínas de suero de leche. *Ciênc. Tecnol. Aliment. Campinas, 29*, 70–75.

Islam, M. R., Biswas, M. M. H., Esham, M. K. H., Yoy, P., Khan, S. M. K., & Hasan, S. M. K. (2023). Jackfruit (*Artocarpus heterophyllus*) by-products a novel source of pectin: studies on physicochemical characterization and its application in soup formulation as a thickener. *Food Chemistry Advances, 2*, 100273. DOI: 10.1016/j.focha.2023.100273.

Jurado, B. T., Aparcana, I. M., Villarreal, L. S., Ramos, E., Calixto, M. R., Hurtado, P. E., & Del Carmen, K. M. (2016). Evaluación del contenido de polifenoles totales y la capacidad antioxidante de los extractos etanólicos de los frutos de aguaymanto (*Physalis peruviana* L.) de diferentes lugares del Perú. *Rev. Soc. Quím. Perú, 82*, 272–279.

Kalse, S. B., & Swami, S. B. (2022). Recent application of jackfruit waste in food and material engineering: a review. *Food Biosci., 48*, 101740. DOI: 10.1016/j.fbio.2022.101740.

Konsue, N., Bunyameen, N., & Donlao, N. (2023). Utilization of young jackfruit (*Artocarpus heterophyllus* Lam) as a plant-based food ingredient: Influence of maturity on chemical attributes and changes during in vitro digestion. *LWT, 180*, 114721. DOI: 10.1016/j.lwt.2023.114721.

Kumar, M., Potkule, J., Tomar, M., Punia, S., Singh, S., Patil, S., Singh, S., Illkiya, T., Kaur, C., & Kennedy, J. F. (2021). Jackfruit seed slimy sheath, a novel source of pectin: Studies on antioxidant activity, functional group, and structural morphology. *Carbohydr. Polym. Tech. & Appl., 2*, 100054. DOI: 10.1016/j.carpta.2021.100054.

Kuskoski, E. M., Asuero, A. G., Troncoso, A. M., Mancini, J., & Fett, R. (2004). Aplicación de diversos métodos químicos para determinar actividad antioxidante en pulpa de frutos. *Ciênc. Tecnol. Aliment., 25*, 726–732.

Ngoc Le, T. A., Lin Lee, J. J., & Chen, W. N. (2023). Stimulation of lactic acid production and Lactobacillus plantarum growth in the coculture with Bacillus subtilis using jackfruit seed starch. *J. Funct. Foods, 104*, 105535. DOI: 10.1016/j.jff.2023.105535.

Norm Official Mexican NOM-247-SSA1-2008, Productos y servicios. (2018). Cereales y sus productos. Cereales, harinas de cereales, sémolas o semolinas. Alimentos a base de: cereales, semillas comestibles, de harinas, sémolas o semolinas o sus mezclas. Productos de panificación. Disposiciones y especificaciones sanitarias y nutrimentales. Métodos de prueba (accessed on 30 May 2024).

Olivas, R., Nevárez, G. V., & Gastélum, M. G. (2009). Las pruebas de diferencia en el análisis sensorial de los alimentos. *Tecnociencia, 3*, 1–7.

Ortega-González, L., Güemes-Vera, N., Piloni-Martini, J., Quintero-Lira, A., & Soto-Simental, S. (2022). Substitution of wheat flour by jackfruit (*Artocarpus heterophyllus* Lam) seed flour: Effects on dough rheology and deep-frying doughnuts texture and sensory analysis. *Food Sci., 30*, 100612. DOI: 10.1016/j.ijgfs.2022.100612.

Osuna, M., Judis, M. A., Romero, A., & Bertola, N. (2013). Características texturales y de color en panes con mezclas de harinas. *Heladería-Panadería Latinoamericana, 224*, 62–68.

Pineda, D., Salucci, M., Lázano, R., Maiani, G., & Ferro, A. (1999). Capacidad antioxidante y potencial de sinergismo entre los principales constituyentes antioxidantes de algunos alimentos. *Rev. Cubana Aliment. Nutr., 13*, 104–111.

Rivas, K. E., Muñoz, D. L., Pino, C. N., & Balcázar, M. N. (2015). Actividad antioxidante, contenido fenólico total y citotoxicidad de extractos polares obtenidos de plantas antidiabéticas colombianas. *Rev. Cubana Plantas Medic., 20*, 277–289.

Spada, F. P., Mandro, G. F., da Matta Junior, M. D., & Canniatti-Brazaca, S. G. (2020). Functional properties and sensory aroma of roasted jackfruit seed flours compared to cocoa and commercial chocolate powder. *Food Biosci., 37*, 100683. DOI: 10.1016/j.fbio.2020.100683.

Torres, J. D., Torres, R., Acevedo, D., & Gallo, L. A. (2015). Evaluación instrumental de los parámetros de textura de galletas de limón. *Rev. Vector, 10*, 14–25.

CHAPTER 7

Characterization of the Spherification of Alginate, Xanthan Gum, and Soursop Pulp

X. RUELAS-CHACÓN,[1] I. MORALES-IBARRA,[1] L. O. FUENTES-LARA,[2] O. N. REBOLLOSO-PADILLA,[3] J. D. CORONA-FLORES,[4] J. I. GARCÍA-LÓPEZ,[5] N. CAMPOSECO-MONTEJO,[5] A. FLORES-NAVEDA,[5] and P. ÁLVAREZ-VAZQUEZ[6]

[1]*Food Science and Technology Department, Universidad Autónoma Agraria Antonio Narro, Calzada Antonio Narro, Buenavista, Saltillo, Coahuila, C.P., Mexico*

[2]*Animal Nutrition Department, UAAAN, Saltillo, Mexico*

[3]*Animal Production Department, UAAAN, Saltillo, Mexico*

[4]*Planning Direction, UAAAN, Saltillo, Mexico*

[5]*Plant Breeding Department, UAAAN, Saltillo, Mexico*

[6]*Natural and Renewable Resources Department, UAAAN, Saltillo, Mexico*

ABSTRACT

Spherification is a technique derived from molecular gastronomy consisting of a controlled gelation of a liquid, submerged in a vessel containing a solution of calcium chloride. Liquid forms small spheres with food inside. In molecular gastronomy, various ingredients are replaced in prepared foods trying to provide the same chemical response as the original ingredients but with new flavors, textures, and functional properties that satisfy consumers. An advantage of this technique is to provide healthier and dietary foods. The

Sustainable Agricultural and Food Systems Engineering: New Research, Practices, and Developments.
Jose Juan Buenrostro-Figueroa, Julio César Tafolla-Arellano, Leonardo Sepúlveda Torre, Porteen Kannan, A. K. Haghi (Eds.)

aim of this research was to obtain and characterize a functional food with soursop by spherification with alginate and xanthan gum. The experimental design for physicochemical characterization was a completely randomized design with three replications and for sensory analysis evaluation, was a randomized complete block. For the characterization of the new product, AOAC methods were used and for sensory analysis, a hedonic test was applied. The physicochemical variables analyzed were total dry matter (77.75%), moisture (22.25%), total nitrogen (12.08%), fat (0.26%), ash (13.01%), fiber (9.4%), total sugars (71.31 mg/L), and calorific value (3.28 kcal/g). Product stability was estimated by visual analysis of moisture loss at 30, 60, and 90 minutes with an analytical balance. The soursop spherifications were microbiologically analyzed (aerobic mesophilic bacteria, fungi, and yeasts) for 5 days. On the first day, in the counting of mesophilic aerobic bacteria, a microbial load of 4.5×10^4 CFU/g was obtained. On the fifth day, there was 7.5×10^4 CFU/g. Due to the presence of fungi and yeasts, there was a development of 2.0×10^3 CFU/g on the first day and 7.0×10^3 CFU/g on the fifth day. For sensory evaluation, 16 trained judges participated, and the attributes evaluated were: overall appearance, odor, flavor, texture, and overall acceptance. A hedonic test with a five-point scale was used. The results were analyzed using analysis of variance, and if any significant difference was found, the Fisher test was performed to compare means at a significance level of 5%. The characteristics of significant differences ($P>0.05$) were presented in overall appearance and texture. The soursop spherification product can be used as a new product due to its nutritional content and can be consumed by all kinds of consumers.

7.1 INTRODUCTION

In gastronomy, a new trend has emerged to prepare food in a more attractive way for the consumer, something innovative that attracts the attention of the consumer with its appearance and can be perceived with the 5 senses. This trend has a name and is known as molecular gastronomy, from which different techniques emerge, where classical methodologies have been modified through the application of new technologies. They began to combine science with the kitchen, where scientists started to delve into culinary issues (Dorantes, 2016).

We are currently in a time when levels of competitiveness are increasing. Science and technology advance rapidly to achieve improvement in the quality of life of human beings in all areas. Fortunately, the world of gastronomy is not far behind. Using all the scientific-technical advances and their applications in the kitchen, we see how new cooking techniques or preparation of dishes begin to be developed (Escandell, 2015).

Spherification is a technique in which different foods can be encapsulated. It is one of the most viable ways that be responsible for offering fresh food with a rich nutritional contribution, safeguarding the sensory characteristics of food, and competing in the same way with the most industrialized foods.

This technique is based on the controlled gelation of a liquid that, when immersed in a calcium chloride bath, forms spheres, constituting a semi-rigid mesh formed by chemical bonds responsible for a liquid taking consistency and forming perfect spheres (Lupo, González, & Maestro, 2012; Ramírez, Jerónimo, Manzano, & Hernández, 2016).

Imeson (2010) highlights the prebiotic effect of low molecular weight alginates, the benefits of their intake as daily fiber for the reduction of blood sugar and cholesterol levels, as well as the ability to prolong the shelf life of products. Viscosity is the main characteristic of alginate solutions and together with its reactivity against polyvalent cations, it is what generates characteristics such as thickening, stabilizing, and gelling.

This spherification technique is intended to make food that, combined with the soursop pulp, is nutritious, provides energy, and, on top of that, is attractive to the consumer (Avendaño, López, & Palou, 2013). The soursop fruit is used in this study since there are investigations where it is alluded to for having different healing properties and for its nutritional contribution.

Currently, there is very little information related to molecular gastronomy, mainly about the spherification technique. The most important thing is to know what we consume and what benefits it brings to health. SAGARPA (2016) points out that soursop grows in tropical, subtropical, and humid areas. It is highlighted that the natural distribution area of this fruit covers the tropical region of southern Mexico, Central America, northern South America, and the Caribbean islands. It is affirmed that in Mexico, more than 16,000 tons of soursop are produced annually, with Nayarit, Colima, and Michoacán being the main producers of this fruit. More than 2,000 hectares are dedicated to cultivating it, and its production value exceeds 102 million pesos. Currently, Mexico exports soursop pulp to the United States and producers in Nayarit sell their fresh fruit to self-service stores in the country. However, soursop fruits are very perishable. Four to five days after harvest, at room temperature, they show softening and alterations in flavor and external and internal color. Fruit production per tree is low compared to other fruit trees, so the commercial success of soursop depends on an adequate selection of genotypes with higher production and quality fruits. Despite the fact that one of the centers of origin of this fruit is southern Mexico and part of Guatemala, it's genetic improvement and intensive plantations in the state of Morelos, Mexico, are almost nil (Evangelista, Cruz, Pérez, Mercado, & Dávila, 2002).

As a consequence, 60% of postharvest losses have been reported due to the perishable nature and fragility of the physical damage of this fruit. Therefore, the export of soursop must be carried out on the day of harvest and transport must be by air at 13?C, which is expensive (Tovar, Mata, García, & Montalvo 2011).

The soursop, for its nutritional contribution, has been studied lately, since it has different healing properties which makes it very interesting to use in various foods. One of the consuming countries is the United States where the natural fruit is consumed in the form of pulp, nectar, and jelly (Márquez, 2009; Coria, Montalvo, Yahia, & Obledo, 2016).

Currently, the fruit is used to eat fresh as a whole fruit or in fruit salad, it is used as a raw material to prepare juices, ice creams, desserts, and homemade cakes. As processed fruit, natural or frozen soursop pulp, concentrate, jam, nectar, jellies, and puree are marketed. The fruit is highly appreciated for beverages and the producing countries export juice in industrial presentations or cans for final consumption. Mexico exports canned soursop with seeds. It is also used for mixtures with liquors. In Indonesia, the unripe fruit is cooked as a vegetable and used in soups, and in Brazil, portions of soursop with skin are roasted or fried (Coria, Montalvo, Yahia, & Obledo, 2016; FAO, s.f.). Since spherification is a modern culinary technique, it allows a new way of presenting this food to attract the consumer. The purpose of this work is to obtain a soursop spherification with alginate and xanthan gum, which is nutritious and attractive for the consumer.

7.2 MATERIALS AND METHODS

The soursop was obtained from a commercial establishment in Saltillo, Mexico. Xanthan gum and alginate were the chemical reagents purchased from a local chemical supplier.

7.2.1 PREPARATION OF THE SPHERIFICATION OF SOURSOP PULP

It began with the washing and disinfection of the soursop; subsequently, the peel of the fruit was removed, also removing the seeds from it, thus obtaining only the pulp of the soursop. This was crushed with a blender (Nutribullet 2.0, 600 W, Nutribullet, C.V. Mexico). Afterwards, two formulations were continued to obtain the spherifications. To the first formulation, 65% water, 33% pulp, 0.7% Splenda, and 0.8% alginate were added, and 0.5% xanthan gum was added to generate thickness and thus prevent liquid leakage through the membrane. The same was added to the second formulation 2, with the difference that 0.05 ml of lemon essential oil was added to increase the flavor of the soursop.

An 8% w/v calcium chloride solution was then prepared in 1,000 mL of drinking water. Immediately, with the help of a spoon, a small amount of the formulation was introduced into a calcium chloride solution, since it was dropped by drop, there was no specific time when removing them; approximately it was between 3 and 5 minutes. Later, when removing them, they were placed in a container containing 1 liter of drinking water to eliminate traces of calcium particles that did not dissolve well. When performing the sensory evaluation with these two samples, there was no significant difference in odor, and it was decided to only carry out the analysis proximal to the first formulation.

7.2.2 CHARACTERIZATION OF THE SPHERIFICATION OF ALGINATE, XANTHAN GUM, AND SOURSOP PULP

7.2.2.1 DETERMINATION OF SPHERE SIZE

The average thickness of the spheres was determined using a vernier, measuring the length and width of a total of 20 spheres.

7.2.2.2 EVALUATION OF PARTIAL DRY MATTER

The most widely used method to determine partial dry matter is the elimination of free water by means of circulating heat followed by the determination of the weight of the residue. This technique is based on subjecting the inputs to temperatures between 55 and 60?C. The temperature is regulated to achieve maximum drying and to avoid a minimum loss of volatile substances and other decomposing substances. For this, 473.8 grams of fresh samples were placed in an aluminum tray and left in an oven (Flamineta Premiere brand) at 65?C for 14 hours (AOAC, 1985).

7.2.2.3 TOTAL DRY MATTER ANALYSIS

Total dry matter analysis consists of a portion that can be burned since it is made up of substances that contain carbon or organic matter and provide energy to the food. The other incombustible portion is made up of substances that cannot be burned, and the waste they form is ashes when subjected to calcination. The already completely dry sample was crushed in a mortar, and later 2 g of this sample was placed in a crucible at a constant weight. This was then placed in an oven (Thelco Model 27 Brand) at 100?C for 12 hours (AOAC, 1985).

7.2.2.4 DETERMINATION OF MOISTURE PERCENTAGE

Moisture was determined with the IR-200 Moisture Analyzer (Denver Instrument Co. Arvada, Colo., U.S.A.).

7.2.2.5 PHYSICAL STABILITY ASSAY

Stability was estimated by analyzing the moisture loss of the spheres using an analytical balance (Mettler Toledo model). About 10 grams of fresh sample were used and it was checked at 30, 60, and 90 minutes. As time passed, their shape was observed, and the spheres were slightly pressed to check if they kept liquid inside.

7.2.2.6 DENSITY DETERMINATION

An indirect method was used. Initially, the spheres were weighed on an analytical balance to find their mass. To obtain their volume, the spheres were completely submerged, and carefully, in a test tube that has an exact volume of water (V0); then the final volume (Vf) was read, the volume of the solid corresponding to the difference (Vf–V0). The density of the spheres was calculated by dividing the mass by the volume.

7.2.2.7 PROXIMATE AND CHEMICAL ANALYSES

Protein content (%) was determined by the Kjeldahl method using 1 g of ground sample. Fat content (%) and ash content (%) were determined utilizing a Soxhlet extractor (Velp Scientifica, Monza-Brianza, Italy) on 5 g of ground sample and diethyl ether as a solvent, and muffler furnace method, respectively (AOAC, 1985). Moisture was determined with the IR-200 Moisture Analyzer (Denver Instrument Co., Arvada, Colo., U.S.A.).

Carbohydrate concentration of the samples was calculated by difference, and fiber content was determined by the AOAC (1985). All analyses were carried out in triplicate for each treatment.

7.2.3 MICROBIAL ANALYSES

One milliliter of each sample for initial characterization, after processing and during shelf-life, respectively, was serially diluted in 0.1% sterile peptone water. The PDA culture media (fungi and yeasts) and standard count (aerobic mesophiles) were prepared, and the materials to be used such as tubes, bottles, and plastic spikes, and each of the media (PDA and standard count) were sterilized. Once the media was sterilized, they were poured into Petri dishes and allowed to cool. The boxes were marked according to the medium and identifying the dilutions. All Plate Count Agar (Difco, Becton, Dickinson & Company, Sparks, MD, USA) of these dishes were incubated at 35?C for 48 hours, after which bacterial loads were counted. Samples were also pour-plated into Potato Dextrose Agar (Difco, Becton, Dickinson & Company, Sparks, MD) to enumerate yeast and molds. Dishes were incubated at 21?C for 7 days, after which time the colonies were counted.

7.2.4 SENSORY EVALUATION

Two formulations were made, the first with alginate, xanthan gum, and soursop pulp: the second contained alginate, xanthan gum, soursop pulp, and lemon essence. Around 16 trained judges participated, aged between 20 and 23 years, who were provided with the two formulations, and a hedonic test was applied with a five-point scale using two scales: verbal and texture; and the attributes evaluated were general appearance, smell, taste, texture, and overall acceptance. The results were analyzed using analysis of variance, and a Fisher test was performed to compare the significant differences, and the means were compared with a significance level of 5%.

7.2.5 STATISTICAL ANALYSIS

Analysis of variance (ANOVA) was carried out using Infostat V. 2018 (Infostat Software Estadístico, Argentina). Differences between treatments were analyzed using the Fisher test ($p < 0.05$).

7.3 RESULTS AND DISCUSSION

7.3.1 DETERMINATION OF THE SIZE OF THE SPHERE

Within the parameters for the determination of size, the length, width, and weight of the spheres were considered. Using a total of 20 spheres. Given that the spheres were irregularly shaped due to the method used to make them, there is some variation in their length. The comparisons of the length of the samples where the average was 1.24 cm with a standard deviation of 0.19. The width parameter of the samples, the mean was 0.88 cm with a standard deviation of 0.09. The average in terms of weight was 0.45 g and the standard deviation was 0.05.

In the extrusion drip method used by Boon-Beng et al. (2013), the alginate solution is extruded through a capillary and allowed to separate from the nozzle as a droplet under the influence of gravitational force or external force. They mention that the production of pearls, as they call them, to have the desired size and spherical shape often required some trial-and-error relative to the liquid formulation and experimental setup, for example, solution viscosity or surface tension, tip size, pickup distance, etc. In soursop spherifications, what affected the same size in all was the distance of fall, and this is due to its irregular shape since the method that was used is not recommended for mass production. The use of Pasteur pipettes is recommended to obtain homogeneous spheres and use the extrusion method.

7.3.2 DETERMINATION OF PHYSICAL STABILITY

This determination was used to visually assess the moisture loss. The stability of the spheres was between 60 and 90 minutes since at this time the spheres still showed a certain amount of liquid inside, and visually at 80 minutes they could still be seen in good shape. Rivera (2013) records an average stability in the alginate spheres of 7,236 seconds, which is equivalent to 2 hours.

7.3.3 DENSITY DETERMINATION

Since density is the scalar magnitude referring to the amount of mass in a certain volume of a substance, the result according to the median in the spherification was 0.48 g/cm^3, and the standard deviation was 0.29. There was a great difference between the minimum density data, which is 0.44 g/cm^3, and the maximum data, which is 0.96 g/cm^3. According to Rivera (2013), the density in the alginate spherifications was 0.88 g/ml and he mentions that the density of the spheres varied by the gelation time. Another important factor may be due to the mass of the spherifications since it was different; due to the very varied size in which the spheres came out because of the distance of the fall, it was not the same, and the total amount that was taken to make the spheres varied.

7.3.4 PROXIMATE AND CHEMICAL ANALYSES

7.3.4.1 PROTEIN ANALYSIS

According to the literature, the amount of protein in soursop is 1 g per 100 g of soursop pulp. Table 7.1 shows the results of the percentage of total protein, the spherification obtains a median of 0.89%, a relatively high percentage compared to the data of the strawberry spherification where it obtains 0.37% protein; and compared to the fresh soursop fruit, a bit low.

TABLE 7.1 Means of Parameters of Soursop, Alginate, and Xanthan Gum Spherification

Parameters Analyzed	Soursop, Alginate, and Xanthan Gum Spherification
Protein (%)	0.89
Fat (%)	0.03
Total sugars (g/kg)	0.07
Fiber (%)	1.15
Caloric value (Kcal/100 g)	3.28

7.3.4.2 FAT DETERMINATION

In this determination, a total average of the 3 repetitions that were made of the sample with 0.03% fat, shown in Table 7.1 with a standard deviation of 0.01, was obtained; in the natural fruit, it has 1 g of fat for every 100 g of soursop pulp according to García (2010).

In the strawberry spherification of Aguirre (2016), he obtained a total of 0.05% lipids in his sample. A higher percentage than ours since compared to the natural strawberry fruit, it reports 0.3% fat, a lower value than that of the natural soursop fruit.

7.3.4.3 TOTAL SUGAR CONTENT

To determine this analysis, a sucrose curve was carried out, which can be seen in the annexes section, in order to obtain the slope equation and with this, obtain the results with the absorbances obtained in the sample of soursop spherifications. Table 7.1 shows the summary measurements of the three repetitions, that were performed, the results are presented in g/kg. Ramírez & Pacheco (2011) determined the total sugars in the soursop pulp with a total of 52.04 g/100 g (dry basis), obtaining a greater amount than the spherification carried out; this could probably be due to the soursop since the origin of the one used in this study is not known exactly and it can vary according to the region.

Barahona (2013) obtained a total of 67.9 g/kg of total sugars in the soursop pulp. This value is much higher because the natural pulp is very rich in sugars. One of the important aspects for the soursop spherifications to obtain a lower sugar value was the amount of water used in its preparation, and it was sweetened with sucralose, a sweetener that does not provide calories.

7.3.4.4 FIBER ANALYSIS

The percentage of crude fiber had a standard deviation of 0.18 in the three repetitions with a median of 1.15% compared to normal fruit and 8.55% higher than spherifications. Table 7.1 also shows that the standard deviation is high since it shows that the minimum data is 0.91% and the maximum is 1.27%. There is a big difference, this may be due to some error in the process of this analysis, such as the weight of the samples.

In Table 7.1, the results of the caloric value obtained in the spherifications are presented. Compared to Aguirre (2016), we obtain a lower amount since in the strawberry spherifications a total of 73.49 Kcal/100 g is obtained. This is due to the fact that in this they use a greater quantity of ingredients and in a greater proportion. It uses 10% fructose, 7% sugar, and lesser amounts of red dye 40, citric acid, ascorbic acid, sodium benzoate, potassium sorbate, sodium citrate, aminoquelated iron, and xanthan gum. Barahona (2013) obtained a total of 59.74 Kcal/100 g in the natural soursop, which is low because it is a fruit that has a higher percentage of moisture. In the soursop spherification, only 3.28 Kcal/100 g is obtained, which can be seen in Table 7.1. This is because a lower percentage of ingredients is used than the strawberry spherifications, and it also has a high percentage of water.

7.3.5 MICROBIOLOGICAL ANALYSES

In Table 7.2, the growth of aerobic mesophilic bacteria on the first day was 4.5×10^4 CFU/g, and on the fifth day, there was a growth of 7.5×10^4

CFU/g. Since there are no standards establishing a limit for the microbial load in spherifications, it was compared with a sanitary norm (Ministerio de salud República del Perú, 2008) that sets microbiological criteria for the sanitary quality and safety of food and beverages for human consumption in Peru. This norm applies to foods prepared without heat treatment such as raw salads, mayonnaise, potato sauces, dressings, desserts, juices, and homemade yogurt, among others. The norm specifies a minimum limit of 10^5 mesophilic aerobics and a maximum limit of 10^6 CFU/g. By comparing these standards with the results obtained in the soursop spherification, it can be seen that the data fall within these limits.

TABLE 7.2 Means of the Microorganisms Count CFU/G of the Soursop, Alginate, and Xanthan Gum Spherification During the First and Fifth Day of Storage

Microbial Counts (CFU/g)	Soursop, Alginate, and Xanthan Gum Spherification	
	1st Day	**5th Day**
Aerobic mesophilic bacteria	4.5×10^4	7.5×10^4
Fungi and yeast	2.0×10^3	7.0×10^3

Table 7.2 shows the growth of fungi and yeasts also had a significant difference on the first day of 2.0×10^3 CFU/g with the fifth day of 7.0×10^3 CFU/g. Because there is no specific standard for this product, the permissible limits for fungi and yeast cannot be verified. In the case of fruit pulp, according to the Ministry of Health and Social Protection of Colombia (2013), the microbiological requirements for fruit pulp without frozen heat treatment or not, must have a maximum allowable index to identify a good quality level of 1×10^3 in the count of molds and yeasts CFU/g and a maximum permissible index to identify an acceptable level of quality of 3.0×10^3 CFU/g.

7.3.6 SENSORY EVALUATION

The results of this analysis were concentrated in Table 7.3 where it is shown that there is a significant difference in global appearance and texture. And in the other attributes they are not significantly different; despite the fact that in the observations of the judges the comments that the smell of lemon was noticeable in formulation B where lemon essence was added predominated. However, the highest means are found for formulation A, to which all the bromatological analysis was done.

TABLE 7.3 Means Values of the ANOVA of the Sensory Evaluation of the 2 Treatments of Spherification

Attributes	Spherification 1 (Soursop, Alginate, and Xanthan Gum)	Spherification 2 (Soursop, Alginate, Xanthan Gum, and Lemmon Essential Oil)
Overall appearance	4.13a	3.5b
Odor	4.13a	3.81a
Flavor	3.19a	3.13a
Texture	3.25a	2.13b
Global Acceptance	3.44a	3.38a

Note: Means with a common letter are not significantly different ($P > 0.05$).

7.4 CONCLUSION

The spherification was achieved using soursop as the main ingredient according to the classification of spherifications by Aguirre (2016), a basic spherification consisting of liquid food plus alginate with xanthan gum introduced into a calcium chloride solution. Obtaining a new innovative product for molecular gastronomy and other applications. It is deduced that the best formulation was Spherification 1, in which all the ingredients were stable for the formation of the sphere. The formulation that obtained the best acceptance was A, which did not include lemon oil. Due to the bromatological characteristics of Spherification 1, it can be consumed by all types of consumers due to its nutritional content, since it provides 0.90% protein, 0.07 g/kg of total sugars, 0.03% fat, a low percentage of carbohydrates of 3.28 Kcal/100 g and 1.11% soluble fiber, in addition to the medicinal properties attributed to it. It allows a new way of presenting this food to attract the consumer. The purpose of this work is to obtain a soursop spherification that is nutritious for the consumer and attractive.

KEYWORDS

- **alginate**
- **carbohydrates**
- **medicinal properties**
- **molecular gastronomy**
- **sensory evaluation**
- **soursop pulp**
- **spherification**

REFERENCES

Aguirre, A. (2016). Esferificación inversa de bebida nutricional sabor fresa, desarrollada para el beneficio de la niñez intermedia y determinación de su tiempo de vida útil en condiciones adecuadas. Guatemala de la Asunción.: Universidad Rafael Landívar.

AOAC. (1985). Official Methods of Analysis. 16th ed. Assoc. Off. Anal. Chem.; Arlington, VA.

Avendaño, G., López, A., & Palou, E. (2013). Propiedades del alginato y aplicaciones en alimentos. *Temas selectos de Ingeniería de Alimentos, 7*, 87–96.

Barahona, V. (2013). Evaluación de la actividad antioxidante y valor nutracéutico de las hojas y frutos de la guanábana (*Annona muricata*). Riobamba, Ecuador.: Escuela superior politécnica de Chimborazo.

Boon-Beng, L., Pogaku, R., & Eng-Seng, C. (2013). Size and shape of calcium alginate beads produced by extrusion dripping. *Chem. Engineering Technol., 10*, 1627–1642.

Coria, A., Montalvo, E., Yahia, E., & Obledo, E. (2016). Annona muricata: A comprehensive review of its traditional medicinal uses, phytochemicals, pharmacological activities, mechanisms of action, and toxicity. *Arabian J. Chem., 2–10*, 1–30.

Dorantes, L. (2016). Los componentes de la gastronomía molecular: sabor, ciencia y arte. Simposio gastronomía molecular. Presentado en el V Congreso Internacional de Ingeniería Bioquímica.

Escandell, J. (2015). Gastronomía molecular: de la empírica a la innovación científica. *Rev. Cien. Farm. Alim., 1*, 1–8.

Evangelista, S., Cruz, J., Pérez, S., Mercado, E., & Dávila, G. (2002). Producción y calidad frutícola de guanábanos (*Annona muricata* L.) provenientes de semilla de Jiutepec, Morelos, México. *Rev. Chapingo Serie Hort., 9*, 69–79.

FAO. (s.f.). Fichas técnicas. Productos frescos de frutas. Recuperado de http://www.fao.org/3/a-au173s.pdf (accessed on 30 May 2024).

García, M. (2010). La guanábana (*Annona muricata* L.). Propiedades y usos. *Rev. Citri. Frut., 27*, 69–70.

Imeson, A. (2010). Food stabilizers, thickeners, and gelling agents. United Kingdom: Wiley Blackwell.

Lupo, B., González, A., & Maestro, A. (2012). Microencapsulación con alginato en alimentos. Técnicas y aplicaciones. *Rev. Venezolana Ciencia y Tecnol. Alim., 3*, 130–151.

Márquez, C. (2009). Caracterización fisiológica, físico-química, reológica, nutraceútica, estructural y sensorial de la guanábana (*Annona muricata* L. cv. Elita). Universidad Nacional de Colombia SEDE Medellín Facultad de Ciencias Agropecuarias.

Ministerio de Salud República del Perú. (2008). Norma sanitaria que establece los criterios microbiológicos de calidad sanitaria e inocuidad para los alimentos y bebidas de consumo humano. Recuperado de http://www.saludarequipa.gob.pe/desa/archivos/Normas_Legales/alimentos/RM591MINSANORMA.pdf (accessed on 30 May 2024).

Ministerio de Salud y Protección Social. (2013). Reglamento técnico sobre los requisitos sanitarios que deben cumplir las frutas y las bebidas con adición de jugo (zumo) o pulpa de fruta o concentrados de fruta, clarificados o no, o la mezcla de éstos que se procesen, empaquen, transporten, importen y comercialicen en el territorio nacional. República de Colombia. Recuperado de https://www.minsalud.gov.co/Normatividad_Nuevo/Resoluci%C3%B3n%203929%20de%202013.pdf (accessed on 30 May 2024).

Ramírez, A., & Pacheco, E. (2011). Composición química y compuestos bioactivos presentes en pulpas de piña, guayaba y guanábana. *Interciencia, 36*, 71–75.

Ramírez, C., Jerónimo, L., Manzano, M., & Hernández, L. (2016). Encapsulamiento alimentario a base de alginato sódico y cloruro cálcico para evitar la oxidación del mango. Recuperado de http://www.academia.edu/7751821/_firus_frut_nutritivamente_divertido_encapsulamiento_alimentario_a_base_de_alginato_sodico_y_cloruro_calcico_para_evitar_la_oxidaci%c3%93n_del_mango (accessed on 30 May 2024).

Rivera, E. (2013). Evaluación de la capacidad de esferificación de quitosano como sustituto de alginato en la elaboración del falso caviar. Trabajo de investigación para optar por el grado de Químico de Alimentos. Carrera de Química de Alimentos. Quito: UCE. 94p.

SAGARPA. (2016). Guanábana: exótica, deliciosa y saludable. Recuperado de http://www.gob.mx/sagarpa/articulos/guanabana-exotica-deliciosa-y-saludable#acciones (accessed on 30 May 2024).

Tovar, B., Mata, M., García, H., & Montalvo E. (2011). Efecto de emulsiones de cera y 1-metilciclopropeno en la conservación poscosecha de guanábana. *Rev. Chapingo Serie Hort., 17*, 53–61.

CHAPTER 8

Polysaccharides as Wall Materials in Microencapsulation of Bioactive Compounds for Food Preservation

JUAN MANUEL TIRADO-GALLEGOS,[1] JUAN C. BUSTILLOS-RODRÍGUEZ,[2] MAGALI ORDÓÑEZ-GARCÍA,[2] EDMUNDO JUÁREZ-ENRIQUEZ,[3] DAVID NEDER-SUÁREZ,[3] MARÍA J. RODRÍGUEZ-ROQUE,[4] and NDAHITA DE DIOS-AVILA[4]

[1]*School of Animal Sciences and Ecology, Autonomous University of Chihuahua, Chihuahua, Mexico*

[2]*National Technological of Mexico, Technological Institute of Cd. Cuauhtemoc, Chihuahua, Mexico*

[3]*School of Chemistry, Autonomous University of Chihuahua, Chihuahua, Mexico*

[4]*School of Agrotechnological Sciences, Autonomous University of Chihuahua, Chihuahua, Mexico*

ABSTRACT

Many bioactive compounds present in foods or used as additives are susceptible to degradation during processing and preservation. A potential solution to overcome this drawback is microencapsulation, which involves trapping a substance of interest inside another substance (wall material), thus producing millimeter-sized particles. This procedure guarantees protection against oxygen, light, humidity, temperature, etc. In addition, the microcapsules obtained behave as controlled delivery systems in the food formulations; therefore, it offers a wide field of application in food technology. Several

Sustainable Agricultural and Food Systems Engineering: New Research, Practices, and Developments.
Jose Juan Buenrostro-Figueroa, Julio César Tafolla-Arellano, Leonardo Sepúlveda Torre, Porteen Kannan, A. K. Haghi (Eds.)

encapsulation techniques include spray drying, coacervation, extrusion, lyophilization, etc. Various types of wall materials have been used for microencapsulation, but the most popular ones are proteins and carbohydrates. Polysaccharides are widely used since they are of natural origin, abundant, and inexpensive. The particle size, efficiency, and stability of the encapsulation process depend on several factors, mainly the encapsulation method and the wall material. This review highlights the use of native and modified polysaccharides in encapsulating bioactive substances for their application in food preservation.

8.1 INTRODUCTION

Preservatives are substances added to foods to prevent or avoid spoilage caused by microbial growth, enzymatic activity, oxidation, light, pH changes, etc. (Calderón-Oliver & Ponce-Alquicira, 2022). The basic requirements for this type of additives, apart from their activity as such, are that they have no toxicity for the consumer's health and do not affect the sensory attributes of the food to which they are added. Most additives used for this purpose are synthetic; however, the number of artificial additives allowed has decreased due to their potential toxic effects (Baldin et al., 2016). In addition, at present, there is a growing demand from consumers for chemically synthesized substances (Wu & Zhou, 2021; Buljeta, Pichler, Šimunović, & Kopjar, 2022). With consumers' preference for natural additives and their concern about the harmful health effects of synthetic preservatives, the food industry has been encouraged to consider natural alternatives (Đorđević et al., 2015; Iñiguez-Moreno, Ragazzo-Sánchez, & Calderón-Santoyo, 2021). Different bioactive compounds can be used as natural additives to replace synthetic food preservatives. These compounds have antioxidant or antimicrobial properties, or both; the most important are polyphenols present in plant extracts, essential oils of plants and spices, and bacteriocins (Calderón-Oliver & Ponce-Alquicira, 2022).

Another characteristic of bioactive compounds is that their activity is not only limited to functioning as antimicrobials or antioxidants; they can also promote beneficial effects on consumer health, which has increased interest in consuming foods rich in these compounds (Castro-Alatorre et al., 2021; Onwulata, 2013). However, many of these substances, once extracted from biological materials, are sensitive to environmental factors and processing conditions, such as exposure to oxygen, light, pH variations, and high temperatures (Đorđević et al., 2015; Choudhury, Meghwal, & Das, 2021; Mehta et al., 2022). A strategy implemented by the food industry to preserve bioactive compounds in foods is microencapsulation, which consists of a process by which these compounds are encapsulated in different wall materials to protect them from degradation and loss of their functional activity. Additionally, this technique can enhance the physicochemical and sensorial properties and increase acceptance of the products (Mehta et al., 2022; Yan et al., 2022).

Microencapsulation processes use biodegradable and biocompatible polymeric materials and various techniques to obtain an adequate size of microcapsules to ensure the delivery of the bioactive compound to the specific site (Choudhury, Meghwal, & Das, 2021; Parente, Sousa, Marques, Forte, & Tavares, 2022). Among the most commonly used polysaccharides as wall materials are maltodextrin, gum Arabic, and native and modified starches (Comunian et al., 2020; Farias-Cervantes et al., 2018; Gómez-Contreras et al., 2021; Jafari, Ghanbari, Dehnad, & Ganje, 2018; Sapper, Palou, Pérez-Gago, & Chiralt, 2019). However, other researchers have combined them with other biopolymers such as hydroxypropyl methylcellulose, inulin, gelan gum, agave fructans, pectin, and sodium alginate (Sapper, Palou, Pérez-Gago, & Chiralt, 2019; Asensio, Paredes, Martin, Allemandi, Nepote, & Grosso, 2017; Yoplac, Vargas, Robert, & Hidalgo, 2021). These studies have shown that the use of biopolymers can substantially improve the efficiency of the encapsulates when used in combinations.

This chapter will address recent advances in encapsulating active substances through different microencapsulation techniques. The focus will be on those works that used polysaccharides as wall material and were applied in food preservation.

8.2 MICROENCAPSULATION TECHNIQUES

Many factors that influence the selection of microencapsulation techniques, type of bioactive compounds, wall materials, their interaction, and working conditions, should be noticed by researchers to produce microparticles with desirable characteristics (Huang, Yuan, & Baojun, 2021). Different techniques for microencapsulation have been reported, and their main difference is the means of the combination of the wall material and the core. Different techniques have been developed to be used in the microencapsulation of bioactive compounds, for example, extrusion, spray drying, freeze drying, emulsification, coacervation, and ionic gelation, to mention the most employed (Montoya-Yepes et al., 2023).

8.2.1 EXTRUSION

The extrusion technique is a mode of encapsulation of bioactive compounds. Extrusion used in the encapsulation of food ingredients is not the same as extrusion cooking to produce puffed and/or texturized food products. Extrusion microencapsulation has been successfully applied to entrap unstable (oxidation-prone) volatiles (flavor compounds) in several carbohydrate amorphous matrices (Lazou & Krokida, 2017). The extrusion uses mild parameters; however, the capsules or microbeads produced from extrusion encapsulation are often larger and more varied compared with other techniques. The inconsistent particle size could be caused by various factors such as the diameter of the extruder equipment (How, Lai, Pui, & In, 2022). Extrusion techniques might be identified according to the equipment used (electrostatic extrusion, coaxial airflow extrusion, jet extrusion or vibrating nozzle, and extrusion with rotational atomizer discs). This technique consists of mixing the bioactive compound with the encapsulant material and passing it through the extrusion nozzle of the equipment, resulting in droplets that are immediately transformed into capsules in a solidification bath. In this technique, there is an absence of extreme temperatures and solvents (Bamidele & Emmambux, 2021).

8.2.2 SPRAY DRYING

Spray drying is extensively employed in the food industry. It presents numerous benefits, such as fast drying, high throughput, and continuous operation. This technique consists of the feed mix (wall cover and interest substance) being sprayed into drops in a flow of heated air during the drying process (conversion of a solution into a powder in a chamber). Due to heat and mass transfer, the droplets are dried instantly. The final product can be obtained in the form of fine powder, conglomerates, or small compact particles with low moisture content and with high physical, chemical, and microbial stability. Spray drying is a common and cost-effective method for protecting a variety of substances, making it the most used encapsulation technique for bioactive compounds. Their use in these compounds could open up new possibilities for the controlled delivery of beneficial compounds to the host, as well as a potentially cost-effective method to prevent and/or treat a wide range of diseases (Akther, Jothi, Badsha, Rahman, Das, & Alim, 2023; Jafari et al., 2022).

8.2.3 FREEZE DRYING (LYOPHILIZATION)

Freeze drying is a technique employed for dehydrating thermolabile products rich in bioactive compounds, which refers to the elimination of moisture from food molecules that have been frozen, through sublimation (Kandasamy & Naveen, 2022). The process includes the complete freezing of samples, ice sublimation (primary drying), and desorption of remaining unfrozen (secondary drying), it is necessary that capsule temperature storage is 2–3?C below the temperature level of collapse, since, can lose its macroscopic structure, in this sense, due to most bioactive compounds are chemically unstable and heat-sensitive, freeze-drying is assumed to be the most suitable technique for their encapsulation (Buljeta, Pichler, Šimunović, & Kopjar, 2022).

8.2.4 EMULSIFICATION

Emulsification includes the dispersion of one liquid into another (two immiscible liquids) in the form of droplets and has been shown to be an excellent technique for the encapsulation, controlled release, and targeted delivery of bioactive substances. Improved emulsions can be designed by combining emulsifiers with distinct properties to meet requirements. For example, the formation of multilayer interfaces around oil droplets using layer-by-layer self-assembly in oil-in-water emulsions can improve the quality and stability of many compounds (Yan, Zhang, Zhu, Qi, & Li, 2022). Due to some compounds presenting low water solubility, it is necessary to improve bioavailability through the development of emulsions as encapsulation and delivery systems. It is necessary to explore new active ingredient carriers and investigate the steady state of the emulsion system (Tang, Wang, Yu, Song, & Lin, 2020).

8.2.5 COACERVATION

Coacervation, often called "phase separation," results from the electrostatic interactions of oppositely charged biopolymers in an aqueous medium.

This technique allows the incorporation of hydrophobic bio-functional compounds directly into food products, attracting interest in the food and pharmaceutical industry (Wang, Akanbi, Agyei, Holland, & Barrow, 2018). Both simple and complex techniques of coacervation can be used. In simple coacervation, a nonsolvent or a more water-soluble polymer is used. In complex coacervation, the capsule is formed by the ionic interaction of two oppositely charged polymers, commonly the positive charges on protein molecules and anionic macromolecules. A large number of coating materials have been evaluated for coacervation microencapsulation, but the most used coating system is the gelatin/gum acacia system (Poshadri & Aparna, 2010). Coacervation is very versatile and is suitable for the protection of compounds against degradation, to prevent the loss of volatiles, besides allowing modulated release and increased stability of the encapsulated material. Different factors such as polymer concentration, pH, ionic strength, polymer mixing ratio, and thermal treatments have a significant effect on the formation of complex coacervates, but the most critical point in the complex coacervation process is defining the optimum condition (Ferreira, Nicoletti, & Dragosavac, 2023).

8.2.6 IONIC GELATION

Ionic gelation is a simple microencapsulation method through the formation of complexes occurring when an aqueous polymeric solution containing active material is dripped on anionic salt, generally on a calcium ion solution at appropriate concentrations. The stability and structure of microstructures in ionic gelation are determined by the appropriate concentration of anionic salt and biopolymer, core load, the extent of cross-linking, and drying conditions. As wall materials, alginates, chitosan, carboxymethyl cellulose, and pectin are commonly used polymers for the formation of gel systems (Mehta et al., 2022). In addition, the ionic gelation technique is used for microencapsulating active compounds because it can be technically regarded as low-cost and does not require advanced equipment, high temperatures, or organic solvents. Particulate types of gel have many uses, including structuring, strengthening, and texturizing agents in food matrices and the potential to enhance visual product acceptability (Mohammadalinejhad & Kurek, 2021). Table 8.1 describes the advantages and disadvantages of the different microencapsulation techniques described above.

TABLE 8.1 Advantages and Disadvantages of Main Microencapsulation Techniques Applied to Bioactive Compounds

Technique	Advantages	Disadvantages
Extrusion	• Continuous process. • Absence of extreme temperatures. • Encapsulate hydrophilic and hydrophobic compounds.	• Obtaining large particles. • Capsules with porous structure. • Limited wall materials.
Spray drying	• Quick process. • Versatility. • High yield.	• Use of elevated temperatures. • Agglomerates formation (variated particle size). • Clogs in the atomization system.
Freeze drying (lyophilization)	• Excellent retention of bioactive compounds. • Encapsulation of thermosensitive compounds. • Long-term preservation of bioactive compounds.	• Long process. • Expensive process. • Expensive final product storage.
Emulsion	• Controlled release of the encapsulated compounds. • Release of multiple bioactive ingredients. • Flavor-masking by trapping bitter compounds.	• Unstable to environmental conditions. • Droplet aggregation. • Expensive production.
Coacervation	• High encapsulation efficiency. • Variety of biopolymers as wall materials. • Protection against degradation.	• Expensive process. • Long process. • Complex process.
Ionic gelation	• Controlled release. • Low cost. • Small particle size.	• Long process. • Low scale. • Low polydispersity.

8.3 POLYSACCHARIDES AS WALL MATERIAL FOR MICROENCAPSULATION

The generation and use of synthetic films in food packaging have grown quickly over the past decades, resulting in serious environmental concerns. Edible and biodegradable polysaccharide-based films could be an alternative to synthetic packaging materials in several food applications and could offer new opportunities to develop novel food packaging systems. Also, biodegradable films can reduce environmental problems associated with food packaging (Cazón, Velazquez, Ramírez, & Vázquez, 2017). The selection of adequate wall materials is critical for encapsulation efficiency and microcapsule stability. Generally, the materials should be food-grade, biodegradable, and able to form a barrier between the internal phase and its surroundings (Lazou & Krokida, 2017). The selection of wall material is based on physicochemical properties, like glass transition, solubility, molecular weight, diffusibility, crystallinity, and emulsifying and film-forming ability (Kandasamy & Naveen, 2022).

Polysaccharides are currently the most often employed biopolymers as wall materials in the microencapsulation of bioactive compounds due to their favorable properties such as biocompatibility, efficient yield, and costs (Halahlah, Piironen, Mikkonen, & Ho, 2022). In this sense, polysaccharides are widely available and present different structures with different advantages including being easily digested, non-toxic, biocompatible, and able to remain stable over a wide range of pH and temperatures. Given the above, polysaccharides are regarded as a promising biopolymer for the encapsulation and delivery of bioactive compounds. For encapsulation purposes, the most common polysaccharides are native and modified starch and derivatives, alginate, pectin, cellulose and derivatives, gum Arabic, chitosan, carrageenan, inulin, and seaweed polysaccharides (Zhang, Belwal, Li, Lin, Xu, & Luo, 2020).

8.3.1 STARCH AND DERIVATES

Starch is an abundant material. Common sources of starch are corn, cassava, potato, rice, and wheat, among others. This polymer has received attention in some industries like food, plastics, and pharmaceuticals due to its properties of gelling ability, film formation, and biodegradability. However, the industrial application of native starch is limited due to high viscosity, instability under thermal and shear stresses, and tendency to rapidly retrograde. Hence, physical and chemical starch modifications improve their viability and expand their range of applications in different industries such as encapsulation of bioactive compounds (Montoya-Yepes et al., 2023). Chemically modified starches are one of the most common wall or carrier materials that most closely reproduce the functional properties of gum Arabic. Hydrolyzed starches like maltodextrins are available in dextrose equivalent (DE) and offer good protection against oxidation and high total solid contents (Poshadri & Aparna, 2010).

8.3.2 ALGINATE

Alginate is an unbranched anionic biopolymer extracted from brown algae and can also be obtained from *Pseudomonas* bacteria. It is soluble in water, and the most important functional attribute of alginate is gelling, which occurs in the presence of cations through the cross-linking between ions and uronates. For microencapsulation applications, sodium alginate is one of the preferred materials because it forms granules immediately in the presence of polyvalent cations. An advantage of this polysaccharide is that gel formation occurs at ambient temperature, allowing cells to easily spread and be trapped. However, it also has some limitations, such as low barriers and instability toward heat treatment (Nezamdoost-Sani, Khaledabad, Amiri, & Khaneghah, 2023).

8.3.3 PECTIN

Pectin is a complex polysaccharide that is widely found in the middle layer of plant cell walls, mainly in fruits. Pectin forms gelled beads in the presence of calcium cations, being more stable than Ca-alginate beads. It presents tolerance to a wide range of conditions, including pH and temperature. However, the high-water content and porosity of the beads require strategies like the insertion of polysaccharide fillers to overcome low retention and improve the physical properties of pectin for microencapsulation purposes (Silveira, Almeida, Andreola, Alvim, & Prata, 2023).

8.3.4 CELLULOSE

Cellulose is the most widespread polymer in nature. It is the main component of the cell walls of higher plants. There are several cellulose derivatives, including ethyl cellulose, cellulose acetate, cellulose acetate butyrate, cellulose acetate phthalate, and trimethylsilyl cellulose. Industrial usage of cellulose derivatives is widespread, and encapsulation is no exception. Microencapsulation shells made from cellulose and cellulose derivative polymers are becoming popular because of their biodegradability and functionality. Interest in cellulose and its numerous derivatives is caused by the set of unique properties that can be used in different application fields, including food, paper, bio-material, and pharmaceutical industries (Aleksanyan, 2023; Peng et al., 2023). Cellulose and derivatives have been used as a material in bioactive compound encapsulation due to its high stiffness, strength, compaction, and large surface area (Amalraj, Gopi, Thomas, & Haponiuk, 2018).

8.3.5 GUM ARABIC

Gum Arabic is a complex polysaccharide, also known as gum acacia, and is a plant exudate originating from some plants in the *Acacia* family. Exudation of the gum occurs naturally as a response to tissue damage or environmental stress and is widely used as a wall material in different microencapsulation processes, especially in the food industry. The above is due to its high encapsulation efficiency, good ability to form interfacial layers, excellent stability against temperature, pH, ionic strength, emulsifying power, and high solubility. These properties traditionally make it the most popular wall material for spray-drying and, when paired with gelatin, can also be used for complex coacervation (Akinci & Holthaus, 2023; Alves et al., 2023).

8.3.6 CHITOSAN

Chitosan is a polysaccharide that can be obtained from chitin via the deacetylation process. Chitin is most commonly extracted from cell walls in fungi and the exoskeletons of arthropods, such as crustaceans and insects. These last ones emerge as a great importance source of this polysaccharide. Chitosan has various biological activities, especially antimicrobial, film-forming, and antioxidant properties, which have allowed the use and application of chitosan in various industries. This occurs due to a crosslinking process, where the properties of this polymer are modified, leading to the formation of hydrogels that are an excellent material for encapsulation, allowing the absorption of ingredients and their controlled release (Kulka & Sionkowska, 2023; Yang et al., 2023).

8.3.7 CARRAGEENAN

Carrageenans are sulfated polysaccharides obtained from seaweed by hot alkali separation. There are six types of carrageenans, all soluble in water. They have been explored as controlled release vehicles, gelling agents, and encapsulating agents in the delivery systems of bioactive components. This polysaccharide exhibits outstanding advantages compared with other encapsulation materials, such as high stability and electronegativity. Due to this, carrageenan can improve the retention rate of bioactive ingredients (Chakraborty, 2017). Carrageenan is often employed as a food additive, including as a gelling agent (most common application), thickening agent, and stabilizer. As a packaging material, carrageenan can provide a more stable environment for the delivery of bioactive components, which is conducive to the delivery and maximizes the value of bioactive ingredients. Due to its versatility, carrageenan has been employed in the formulation of nanoparticles, complexes, hydrogels, emulsions, and microcapsules (Dong, Wei, & Xue, 2021).

8.4 PROTECTION OF BIOACTIVE COMPOUNDS THROUGH MICROENCAPSULATION

Many encapsulation methods have been developed to protect vulnerable bioactive components (Cheng, Xie, Qi, Liu, Guo, Sun, & Liu, 2017; Fredes, Becerra, Parada, & Robert, 2018; Yamashita et al., 2017). This section will describe some research on the protection and preservation of bioactive compounds by microencapsulation.

Tomsone et al. (2020) studied other wall materials to obtain microcapsules of Horseradish (*Armoracia rusticana* L.) juice by spray drying, showing that maltodextrin and a combination of maltodextrin/gum Arabic were the most effective wall materials for the retention of bioactive compounds, confirming the efficiency of microencapsulation for preserving bioactive compounds. One of the major technological difficulties in the pharmaceutical, food, and chemical use of oils and bioactive components is their preservation. Lescano et al. (2014) showed that microencapsulation by complex coacervation of *Acrocomia aculeata* (Jacq.) Lodd. ex Mart. oil using gelatin and gum Arabic as wall materials was efficient and helped to preserve carotenoids and antioxidant activity. Rezvankhah et al. (2022) encapsulated hemp seed (*Cannabis sativa* L.) oil via freeze-drying using maltodextrin and gum Arabic as wall materials. They reported that hemp seed oil was physically entrapped during freeze-drying and these wall materials showed preservation effects on the encapsulation of sensitive bioactive components of the oil. Similarly, Ng et al. (2013) observed positive effects when kenaf (*Hibiscus cannabinus* L.) seed oil was microencapsulated using spray drying with sodium caseinate, maltodextrin, and soy lecithin as wall materials. The microcapsules of kenaf seed oil were resistant to lipid oxidation and degradation, as well as the loss of bioactive compounds.

On the other hand, most essential oils from plants and spices have limitations for incorporation into food due to factors such as low solubility in water. Citral essential oil, for example, has antimicrobial activity but low stability at room temperature. Therefore, Calderón-Santoyo et al. (2021) evaluated the thermal stability of citral essential oil microencapsulated in maltodextrin, gum Arabic, and sodium alginate by spray drying. Thermal analysis of both microencapsulated and unencapsulated citral essential oil showed that microencapsulation provided thermal protection to the oil from 25–175?C. Furthermore, the microencapsulation process resulted in a material with high solubility (83%) and low wettability time (27 s). These results suggest that the powders obtained are stable and easy to handle on a large scale.

All this demonstrates the importance of preserving bioactive compounds in food matrices through various methods, especially microencapsulation, since, in addition to preserving the integrity, they contribute to extremely valuable aspects such as their nutritional properties and contribution to health.

8.5 RELEASE OF BIOACTIVE COMPOUNDS FROM MICROCAPSULES

Controlled release involves the distribution of bioactive compounds at a specific time interval when encountering a particular stimulus (Boostani & Jafari, 2020). The advantages of these controlled release systems are that the active ingredient is released at controlled rates over prolonged times and can avoid degradation of the compounds during processing and storage (Jeyakumari, Zynudheen, & Parvathy, 2016).

Delivery systems can be designed to react to one or a combination of environmental factors (relative humidity, temperature, UV radiation, etc.) to achieve the desired release rate (Rezvankhah, Emam-Djomeh, & Askari, 2020). Many studies have focused on obtaining information on the factors that influence the release of bioactive compounds from complex matrices, this being the main purpose in food industries (Zhang et al., 2022). According to the literature, different mechanisms involved in the release of microencapsulated compounds can be recognized (Figure 8.1):

1. **Diffusion:** This mechanism is governed by the solubility of the bioactive substances, due to which it creates a concentration gradient in the particle as a driving force for diffusion, and by the permeability of the microcapsule wall materials (Zhang et al., 2022). The release rate of compounds can be limited by the characteristics of the bioactive compounds (polarity, water vapor pressure, and molecular weight), properties of the carrier matrix (such as physical state,

interactions, and even its rheological properties), the particle properties (shape, size, and structure), and the concentration gradient of the bioactive in the microcapsule and the surrounding environment (McClements, 2014).

2. **Dissolution:** This process is governed by the solubility of the wall material in the surrounding medium (Choudhury, Meghwal, & Das, 2021). According to Ko & Gunasekaran (2014), when active substances are encapsulated in slowly dissolving wall materials (proteins and carbohydrates), the release rate depends on the solubility of the encapsulated substances and the thickness of the wall material. On the other hand, when the wall material includes non-polar components such as waxes, the release rate depends on the penetration rate of the dissolution fluid into the matrix.

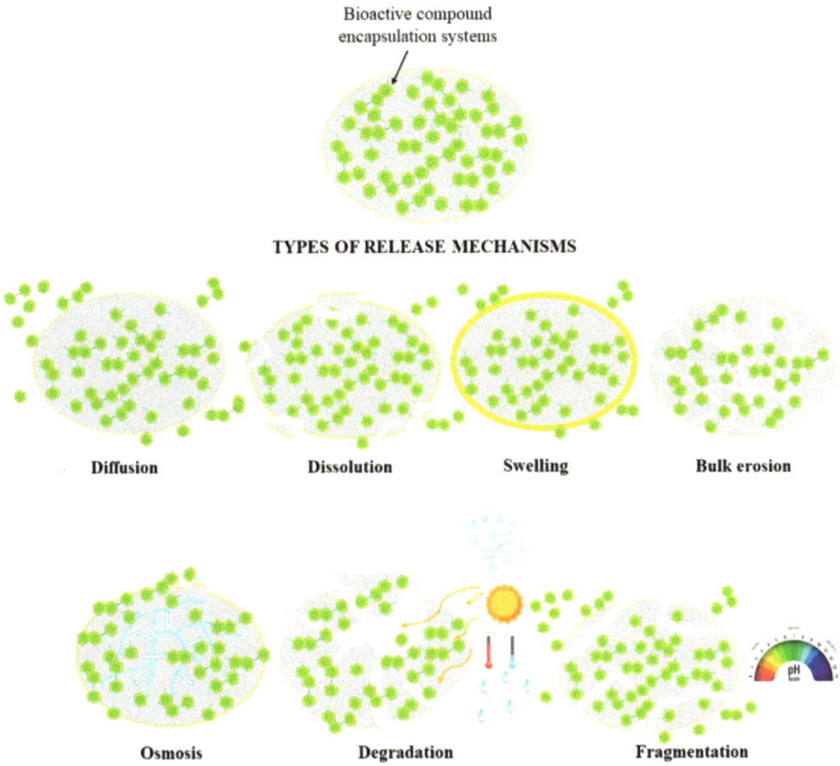

FIGURE 8.1 Controlled release of bioactive substances from encapsulation systems.

3. **Swelling:** The bioactive compounds can be released from their encapsulating matrix when they swell due to solvent absorption. The release rate will depend largely on the wall material's swelling rate and the time required for the bioactive to diffuse throughout the swollen system (McClements, 2014). This release mechanism can be controlled by proper selection of the polymeric matrix and control of environmental conditions such as temperature and pH (Boostani & Jafari, 2020).
4. **Bulk Erosion:** This mechanism can occur due to physicochemical factors such as high temperatures, exposure to strong acids/bases, and enzymatic reactions, which can cause wall degradation and promote the release of the bioactive compound (Jafari, Esfanjani, Katouzian, & Assadpour, 2017).
5. **Osmosis:** When there is an osmotic pressure difference between the inside and outside of the microcapsule, the wall behaves as a semi-permeable membrane. It allows the passage of the substance trapped inside to the outside through cracks or fractures in the wall (Choudhury, Meghwal, & Das, 2021; Jafari, Esfanjani, Katouzian, & Assadpour, 2017).
6. **Degradation:** The release of the active ingredient occurs due to the biodegradation of the encapsulating polymer by the combined action of microorganisms and abiotic factors that fragment the encapsulating matrix into small fractions allowing the release of the encapsulated substances (Zhang et al., 2022).
7. **Fragmentation:** In this mechanism, the bioactive ingredients are released from their containing capsule when they crack or break due to various environmental parameters such as shear force or pressure, pH changes, and enzymatic actions (Boostani & Jafari, 2020). The release rate of the bioactive components will depend on the size and shape of the particles obtained (McClements, 2014). The bioactive agent may eventually be released from the fragments through diffusion, dissolution, or erosion mechanisms with a faster release profile due to the formation of smaller particle sizes and increased surface area (Jafari, Esfanjani, Katouzian, & Assadpour, 2017).

Microencapsulated bioactive compounds are also used to formulate active food packaging materials (Zhang, Yu, Regenstein, Xia, & Dong, 2021; Bahrami, Delshadi, Assadpour, Jafari, & Williams, 2020). According to Zhang et al. (2021), the release rate of active agents in the bioactive films to the product is influenced by three factors: (i) microstructure (porosity, number of microchannels); (ii) intramolecular interactions between materials (hydrogen bridges, dipole-dipole interactions, etc.); and (iii) other factors (mainly temperature and moisture).

8.6 APPLICATION OF ENCAPSULATED BIOACTIVE COMPOUNDS FOR FOOD PRESERVATION

8.6.1 APPLICATION IN MEAT PRODUCTS

The quality of meat and its derivatives depends to a great extent on the processes of oxidation and the growth of microorganisms. In the case of oxidation, the most sensitive components are lipids and myoglobin. The following describes some of the most recent research on using bioactive compounds encapsulated in polysaccharides that have been applied to preserve meat and meat products.

As mentioned in other sections, the bioactive compounds most evaluated as preservatives are obtained from plants and therefore have also been the most evaluated in meat products. Rocchetti et al. (2022) obtained elderberry (*Sambucus nigra* L.) extracts microencapsulated in maltodextrins (13.5 g/100 g of extract) by spray drying. The extracts were evaluated at two concentrations (ME2.5 = 2.5 g/kg and ME5 = 5 g/kg) and compared with sodium erythorbate (0.5 g/kg), a synthetic antioxidant (ERY); on the shelf life of hamburger meat packaged under modified atmosphere (80% O_2 and 20% CO_2) for 13 days storage at 2?C. In all treatments, the concentration of MDA/kg increased with storage time. However, at the end of the experiment, MDA concentrations did not show significant differences for the control (4.68 MDA/kg), the ERY treatment (5.39 MDA/kg), and the ME2.5 treatment (2.57 MDA/kg), while ME5 patties recorded the lowest value (2.09 MDA/kg). Similar behavior was observed while monitoring protein oxidation; the total carbonyls concentration (nmol carbonyls/mg protein) for the control, ERY, ME2.5, and ME5 treatments were 4.69, 4.35, 2.87, and 2.39, respectively. These results suggest that elderberry extracts can be used as natural antioxidants instead of ERY for this product type. Recently, other sources of natural substances with antioxidant properties have been explored, such as propolis. This resin is produced by bees (*Apis mellifera* L.) for hive protection against insects and microorganisms. Reis et al. (2017) obtained propolis extracts (20% soluble solids) microencapsulated by spray drying in CAPSUL® (commercially modified waxy corn starch) with an encapsulation efficiency of 76.86%. The efficiency of encapsulated propolis (0.3 g/kg) on the stability of hamburger meat stored at –15?C was evaluated and compared against commercial antioxidant ERY (0.1 g/kg). According to the results, the hamburger with propolis presented color, appearance, and texture very close to the control and ERY hamburgers; however, its overall acceptability was 63% since the odor and flavor impacted negatively on consumer acceptability. On the other hand, compared to the burgers control and ERY, the malonaldehyde concentrations per kg of meat (MDA/kg meat) in the burgers with microencapsulated propolis were lower during all sampling (0, 7, 14, 21, and 28 days). At the end of the experiment, the samples with encapsulated propolis presented 0.95 MDA/kg meat, while the commercial antioxidant registered 1.21 MDA/kg meat.

With respect to pork meat, Chen et al. (2021) used a mixture of quinoa protein and gum Arabic to microencapsulate eugenol by coacervation. The optimal protein-gum mixture for obtaining the coacervates was a 4:1 ratio at pH 4. The microencapsulates were applied for minced pork preservation. According to the microbiological analysis, in comparison with the meat with free eugenol (0.362% w/w) and the meat with microencapsulated eugenol (0.6% w/w), the control treatment (meat without eugenol) presented a higher total bacterial count (TBC) from day 3 of storage and exceeded the limit (6 log10 CFU/g) acceptable for pork meat on day 6 (6.26 log10 CFU/g). For meat with free eugenol and microencapsulated eugenol, these samples remained within the TBC limit until day 9 with TBCs of 5.09 and 4.55 log10 CFU/g, respectively. Treatments with microencapsulated eugenol had the lowest values ($p < 0.05$) for TBC, pH, and total volatile base nitrogen (TVB-N) throughout the experiment. In other research, Cunha et al. (2018) evaluated the effect of microencapsulated pitaya (*Hylocereus costaricensis* [F.A.C. Weber] Britton & Rose) skin extracts (100 and 1,000 ppm) on the color, texture, and oxidative stability of pork patties pressurized (500 MPa for 10 min) refrigerated at 4?C for 9 days. The extracts were encapsulated in maltodextrins by spray drying with a phenolic compound retention of 81%. The pH of the samples without antioxidants (control) presented the highest pH values. In contrast, the treatments with commercial antioxidants (100 ppm of butylated hydroxytoluene (BHT)) and microencapsulated extracts (100 and 1,000 ppm) presented the lowest values, with no significant differences between them. In general, pitaya extracts and BHT increased the oxidative stability of the product and retarded protein oxidation, and the latter contributed to maintaining the texture with minimal changes during storage. Later, in a similar study by the same authors (Cunha et al., 2021), pitaya skin extracts were microencapsulated in maltodextrin in a 1:2 ratio (maltodextrin: total solids of the extract). The authors evaluated the effect of microcapsules at different concentrations (100, 1,000, and 2,000 ppm) on the oxidation parameters of ground pork patties subjected to UV-C radiation. Compared with the commercial antioxidant BHT (100 ppm), microencapsulated extracts of pitaya showed the same effect on preventing protein oxidation. On the other hand, the samples with BHT presented the lowest values for TBARS. However, it was observed that the higher concentration of microencapsulation (2,000 ppm) prevented lipid oxidation under UV-C radiation up to 6 days of refrigerated storage and color change of pork patties in a similar way as the BHT treatment.

The effect of adding microencapsulated bioactive compounds has also been evaluated in different types of cooked meats. For example, Wang et al. (2018) microencapsulated clove essential oil (*Syzygium aromaticum* L.) (CEO) in a mixture of starch and β-maltodextrin (1:5 starch: β-maltodextrin ratio) by spray drying with an efficiency of 84.5%. To evaluate the effect of microencapsulated CEO on the preservation of cooked chicken, cooked pork, and cooked beef, cubes (20 x 20 x 20 mm) of cooked meat were cut and immersed in aqueous solutions with different concentrations of microencapsulated CEO (0, 0.100, 0.090, 0.080, 0.070, 0.060, 0.050, 0.040, 0.030, 0.020, 0.010 wt.%) for 0.5 h and then monitored in an incubator (37?C and RH = 75%) for different times (24, 48, 72, and 96 h). To evaluate the effect of heat on the microencapsulates, the procedure described above was repeated, but the immersion of the cooked meat was in the microencapsulate solutions boiled for 30 min and cooled to 27?C. The results showed that the microencapsulated CEO presented antifungal activity when applied at concentrations higher than 0.070%. The antifungal activity was presented at concentrations higher than 0.080% in microencapsulates subjected to heat treatment, evidencing the high thermal stability of the microencapsulates.

Apart from functioning as antioxidants and antimicrobial agents, some extracts can also be used simultaneously as natural colorants. Baldin et al. (2016) microencapsulated Jabuticaba (*Myrciaria cauliflora* Berg) extracts with maltodextrin (260 g of maltodextrin/kg of extract) by spray drying and evaluated on fresh sausage as a natural dye with antioxidant and antimicrobial activity at two concentrations (2% and 4%). During refrigerated storage for 15 days, oxidative stability evaluated by TBARS was higher in sausages with 2% and 4% microcapsules, with MDA values lower than 0.1 mg/kg sample, while the control (sausages without colorant) and sausages with carmine (0.006%) presented the highest values (from 0.3 to 0.6 MDA/kg sample). Regarding microbiological analysis, the sausages with the microencapsulated extracts had the lowest microbial counts. However, a higher concentration of microencapsulation (4%) negatively affected consumer acceptability, while the samples with 2% of microcapsules showed the same acceptability as the control and the treatments with carmine.

On the other hand, there are also reports of processed meat products. In this regard, Huq et al. (2014) encapsulated nisin, a polypeptide with antimicrobial activity, in alginate microspheres and cellulose nanocrystals. Three concentrations of nisin (16, 31, μg/mL, and 63 μg/mL) encapsulated and unencapsulated (16, 31, μg/mL, and 63 μg/mL) were evaluated on the growth of *Listeria monocytogenes* in refrigerated hams for 28 days. Fourier transform infrared spectroscopy (FTIR) studies showed interactions between the nisin molecules and the wall material, which allowed a higher availability of viable nisin during storage. The *in vitro* study (BHI-agar deep-well model) during storage at 4?C showed that the available nisin over time increased with the concentration of nisin in the microencapsulation, while the viable nisin in the unencapsulated treatments decreased dramatically from the first day of storage to 2.1, 6, and 11 μg/mL when using 16, 31, and 63 μg/mL of unencapsulated nisin, respectively. On the other hand, the highest concentration of free nisin (63 μg/mL) presented only 2.61 μg/mL of viable nisin after 14 days; when the same concentration of nisin was encapsulated, the availability by day 28 of storage was 31.26 μg/mL Regarding the application of nisin in hams inoculated with L. monocytogenes (~3 log colony-forming units (CFU)/g) vacuum-packed and stored under refrigeration, the color and pH were not affected by the addition of nisin. Compared to free nisin, the *L. monocytogenes* counts were reduced by 2.65, 1.50, and 3.04 log CFU/g after 28 days of storage. These results suggest the great potential of microencapsulated nisin for this product type.

In another study, Xu et al. (2019) evaluated the effect of microencapsulated mulberry polyphenols on the oxidation properties of dried minced pork slices during thermal processing (baked for 3 min at 150?C) and storage (vacuum-packed in plastic bags) at room temperature. Mulberry polyphenols were encapsulated in gum Arabic by freeze-drying with the best polyphenol:wall material ratio at 1:90. The polyphenols were included in the product formulation as individual unencapsulated (1.01 g/kg mixture), encapsulated (93.20 g/kg mixture), or the mix of both (0.51 and 46.60 of unencapsulated and encapsulated polyphenols, respectively). Although polyphenol content decreased with heat treatment and days of storage (20 days), the concentration of polyphenols was higher when encapsulated polyphenols were applied than with unencapsulated polyphenols. In all treatments, lipid oxidation was higher in the control (mixture without antioxidants) than in the other treatments. Similar behavior was observed in protein oxidation. A synergistic effect between free and encapsulated polyphenols was observed in both oxidation parameters.

8.6.2 APPLICATION IN DAIRY PRODUCTS

Nowadays, the developing interest of consumers towards natural ingredients has resulted in the use of bioactive compounds in dairy applications. Milk and dairy products are among the most important foods used in our daily diet and can be a good option to provide bioactive compounds to the body, since they not only provide basic nutrition to the human body but also contain a significant source of bioactive ingredients. However, many of these compounds can become degraded, both due to environmental conditions and during the intestinal digestion phase (Adinepour, Pouramin, Rashidinejad, & Jafari, 2022; Mohite & Waghmare, 2020). Due to this, it has been necessary to look for alternatives that allow these compounds to remain active. Microencapsulation is one of the most used and attractive options for the protection, stability, and controlled release of these bioactive compounds (Comunian & Favaro-Trindade, 2016).

Dairy fortification has been receiving a lot of attention, especially in functional dairy foods like probiotic yogurts, butter, cheese, ice cream, and other fermented milk products (Shamil, Nautiyal, & Omar, 2023). Bioactive compounds provide better quality and stability to food products, increasing the demand for functional foods among consumers day by day. Currently, there are multiple cases of the use of microencapsulation and its application in the preservation and improvement of dairy products.

8.6.2.1 MILK

Milk is a priceless and indispensable liquid that has enormous long-term advantages for individuals of all ages and society. This food is obtained from different sources, like domesticated cows, goats, sheep, and camels, since thousands of years ago (Shamil, Nautiyal, & Omar, 2023). The milk market follows the same trend as the food market with regard to fortification and functional appeal. The use of bioactive compounds in dairy foods has high nutritional and health relevance (De Oliveira et al., 2021). For example, the microencapsulation of fatty acids helps to increase the nutritional value of dairy products, or by incorporating Omega-3 microcapsules in whole milk powder, it remains stable for a longer time (Calderón-Oliver & Ponce-Alquicira, 2022). Milk proteins are an essential medium for probiotic microflora and due to their physicochemical and structural properties, these can be used as a probiotic distribution system (Abbas et al., 2022). In this sense, De Oliveira et al. (2021) showed that the incorporation of *Spirulina sp.* LEB-18 microencapsulated in maltodextrin and soy lecithin and added to chocolate milk presented an increase in protein content and reduced total lipids, besides showing good suspension and an increase of DPPH and phenolic compounds. On the other hand, Patrignani et al. (2017) evaluated the potential of high-pressure homogenization for the microencapsulation of two probiotic lactic acid bacteria, *Lactobacillus paracasei* A13 and *Lactobacillus salivarius* subsp. *salivarius* CET 4063 to produce functional fermented milks, in the presence of sodium alginate and vegetable oil. They found that microencapsulation conditions increased the resistance to the simulated digestion processes and improved functionality and sensory properties.

8.6.2.2 CHEESE

Cheese is a dairy product widely appreciated all over the world and can be distinct from other dairy products in the sheer volume of varieties that are available (Bir, Widmar, Thompson, Townsend, & Wolf, 2020). Within the types of cheese, Cheddar cheese has the distinction of being a good carrier of probiotic microorganisms due to its properties as a relatively high pH (Abbas et al., 2022). However, bioactive compounds have been added to different types of cheese in order to improve the functional properties of the product. For example, Ribeiro et al. (2016) used a rosemary aqueous extract microencapsulated as a functional ingredient for cottage cheese and observed that the samples added with the microencapsulated extracts maintained their antioxidant properties more efficiently throughout storage, thus increasing the bioavailability upon ingestion. On the other hand, Pérez-Soto et al. (2021) showed that the addition of *Opuntia oligacantha* (C. F. Först) microcapsules to fresh cheese did not affect the physicochemical parameters of the cheese, while total coliforms decreased, as well as aerobic mesophiles and mold-yeast counts. Regarding the functional properties, an increase in total phenols was observed, and adding microcapsules influenced the antioxidant activity more positively. In this sense, Moghanjougi et al. (2020) elaborated a bacterial cellulose film with microencapsulated *Lactobacillus acidophilus* and *Bifidobacterium animalis* in sodium alginate and pectin by emulsification. According to the results, the bacterial cellulose film had the potential for inhibition of fungal growth in white brined cheese and keeping intact the chemical properties of cheese, contributing to its preservation. Kavas et al. (2021) applied the extrusion microencapsulation method using sodium alginate as a coating material to prevent the loss of viability of *Lactobacillus paracasei* and *Bifidobacterium longum* during the storage period of white cheese from goat milk stored for 180 days. They showed that microorganisms were available in cheese samples containing microcapsules after the storage time, and this cheese could be considered as probiotic.

8.6.2.3 FERMENTED DAIRY

Fermented dairy foods and beverages are obtained from the fermentation of milk, through the action of microorganisms. They were among the first "processed" food products made by the addition of bacteria to usually heat-treated animal milk, followed by incubation to reduce the pH, with or without coagulation pretreatment. The most common example of fermented milk is yogurt (Savaiano & Hutkins, 2021). The addition of bioactive compounds to fermented dairy could confer health benefits and consider them as functional foods. There are microbes with beneficial contributions to human health called "probiotics," such as *Lactobacillus* and *Bifidobacterium*. Among the health benefits of probiotics are included inhibition of the development of harmful or pathogenic bacteria, supporting a healthy gut microflora, among others. However, there are some factors affecting probiotics during their functionality period like the gastric conditions. Hence, the protection of living probiotic cells became a serious issue. One way to solve this is the application of microencapsulation technology, which is used to integrate functional substances into dairy products and beverages to boost their health-promoting characteristics (Abbas et al., 2022).

The incorporation of bioactive ingredients in fermented dairy foods is an objective of the research and food industry and is necessary to improve and preserve these products. In a study conducted by Thakur et al. (2022), lyophilized microencapsulated phenolic extract powder from the peel of wild pomegranate fruit was utilized for the enrichment of yogurt, and it was observed that the antioxidant properties of yogurt can be enhanced after the incorporation of the powder without sensorial alterations. Similarly, microencapsulated anthocyanin-rich extracts have been added to fermented or unfermented dairy beverages to increase their nutritional and sensorial values (Lima, Madalão, dos Santos, Bernardes, Saraiva, & Silva, 2019). In this sense, yogurt added with grape pomace and flaxseed oil in micro-encapsulated form was investigated. The incorporation of these bioactive compounds in microencapsulated form caused a significant decrease in pH, syneresis, and a significant increase in acidity, water holding capacity, and viscosity. Likewise, yogurts with the microencapsulated form showed the highest amount of phenolics and antioxidant activity (Saberi, Saremnezhad, Soltani, & Faraji, 2023). On the other hand, de Moura et al. (2022) evaluated the microencapsulation by double emulsion (ionic gelation) of anthocyanins from *Hibiscus sabdariffa* L. extract applied in natural yogurt, obtaining a great encapsulation efficiency of bioactive compounds, besides promoting color and functional appeal to yogurt.

8.6.3 APPLICATION IN FRUITS AND VEGETABLE PRODUCTS

Fruits and vegetables suffer postharvest losses during storage caused by improper handling, physiological disorders, cold damage, pests, and microbial growth. These factors cause severe damage to their appearance, which is the most critical quality trait for selecting fresh products (Bahrami, Delshadi, Assadpour, Jafari, & Williams, 2020; Ahmad & Ali, 2019). On the other hand, many fruits are harvested seasonally, perishable, and exhibit high respiratory rates, making them extremely susceptible to mechanical damage and easy spoilage, which leads to a loss of their nutritional and beneficial properties for health. For this reason, different industrial technologies have been employed to promote their conservation (Leyva-Porras et al., 2021). One of the strategies to counteract postharvest losses is using packaging to control interactions between food and the environment, protecting and helping to maintain product quality (Nogueira et al., 2020). Currently, given the environmental problems caused by petroleum-derived plastics and the demands of consumers for increasingly healthier and safer foods, films and coatings have been developed from biodegradable materials (proteins, carbohydrates, and lipids) that incorporate bioactive components (antioxi-dants and antimicrobials) for the production of active packaging that can be applied to fruits and vegetables (Bahrami, Delshadi, Assadpour, Jafari, & Williams, 2020; Nogueira et al., 2020; Ban et al., 2020).

Active packaging and coatings are intended to reduce the risk of food-borne pathogens, prolong shelf life, and improve the quality and safety of postharvest fruit and vegetable products (Iñiguez-Moreno, Ragazzo-Sánchez, & Calderón-Santoyo, 2021). In this regard, Ban et al. (2020) mentioned that microencapsulation of essential oils (EO) is a suitable delivery system for food preservation due to its antioxidant and antimicrobial properties. In this sense, Ban et al. (2020) obtained microcapsules from the essential oil of Ginger (*Zingiber officinale* L.), using chitosan (Q) and carboxymethylcellulose (CMC) (1:1 ratio) as wall material by lyophilization. The microcapsules were evaluated for 15 days to preserve jujube fruit (*Ziziphus jujuba* Mill.). The results showed that using microencapsulated ginger essential oil (GEO) on jujube fruits delayed senescence, changes in total soluble solids, titratable acidity, and ripening (prevented the oxidation of anthocyanins during the process of fruit development and ripening). In addition, the microcapsules promoted greater disease resistance and increased antioxidant activity compared to the control. Overall, GEO microencapsulation maintained postharvest quality and prolonged the shelf life of jujube fruit.

There are other studies in which microencapsulated bioactive compounds are added to the formulation of biodegradable films to evaluate their *in vitro* antimicrobial activity. For example, Calderón-Santoyo et al. (2021) assessed the performance of pectin films added with citral microcapsules on phytopathogenic fungi. The results showed that the films could control the growth of *Penicillium italicum, Colletotrichum gloeosporioides*, and *Aspergillus niger*, suggesting that these materials provide a basis for developing active coatings with potential applications in postharvest management of phytopathogenic fungi. In a recent study, Calderón-Santoyo et al. (2022) obtained microcapsules from citral by spray-drying oil-in-water emulsions. The wall materials used were gum Arabic and sodium alginate (7.5 and 0.4% w/v, respectively) and miglycol 812 (5% v/v) as an emulsifying agent. The microcapsules presented sizes between 1 and 17 μm, with a citral retention percentage greater than 42%. Finally, the microcapsules were added to the formulation of coatings for application on bananas (*Musa paradisiaca* L. cv. portalimon) to evaluate their antifungal activity against the phytopathogen *Fusarium pseudocircinatum* (O'Donnell & Nirenberg), one of the main causal agents of postharvest fungal diseases (rots). The coating reduced by 56% the disease caused by the phytopathogen on banana fruits; therefore, this could be an alternative to control postharvest rots in fruits and extend their shelf life. In another study, Cheng et al. (2019) encapsulated carvacrol (CAR) in β-cyclodextrin (βCD) for the formulation of sodium alginate-based active films to evaluate their *in vitro* activity against *Trichoderma* sp. In addition, the coating was applied to extend the shelf life of White mushrooms

(*Agaricus bisporus*). The results show that the microcapsules obtained exhibited *in vitro* antifungal activity against Trichoderma at a concentration of 15 g/L. At the same time, the antifungal properties of the films were more effective when the microcapsules were incorporated at 30 g/L. In addition to inhibiting the growth of *Trichoderma* sp., the films provided a good barrier against CO_2 and water. Moreover, packaging with the active films increased the activity of superoxide dismutase, catalase, and ascorbate peroxidase enzymes, thus decreasing oxidative damage during storage at 4?C for 3 days and delaying senescence. In a recent study, Shao et al. (2021) developed a bioactive paper for the preservation of edible fungi by combining cinnamon essential oil microcapsules with paper-based materials and starch, the cover paper provided an effective packaging material for the preservation of edible fungi and prolonged shelf life.

The addition of microencapsulated active substances has also been explored in processed products. For example, Lavelli et al. (2016) obtained microcapsules from methanol extracts of grape skins using maltodextrin (DE 12) as wall material by spray drying. The microencapsulated extracts were added to apple puree, which was stored for 30 days. The results indicate that adding the microcapsules promoted greater stability to the puree during storage. In addition, these microcapsules have the potential to be used as a functional ingredient that provides natural colorants.

8.6.4 APPLICATION IN OTHER FOODS

Some foods are very prone to oxidation, such as meat, a complex food with a high-fat content. When lipid oxidation occurs, the development of odors and deterioration of characteristics like color, texture, and nutritional value are inevitable. To preserve these aspects, Gómez et al. (2018); Ariestya et al. (2016) evaluated the addition of coconut shell liquid smoke microencapsulated with dextrin to maintain the quality of tilapia meat during storage. They concluded that coconut shell liquid smoke microencapsulate contains bioactive compounds and organic acids that can have an antibacterial impact on tilapia meat.

Bakery products are part of a healthy diet; however, they are characterized by their short shelf life. Like other foods, these products are also subject to microbiological spoilage (Axel, Zannini, & Arendt, 2017), so the bakery industry is also looking for natural alternatives to synthetic preservatives. In this regard, Gonçalves et al. (2017) encapsulated thyme (*Thymus vulgaris* L.) essential oil (TEO) through complex coacervation, using bovine skin gelatin and gum Arabic as wall material. The microcapsules showed an encapsulation efficiency of 85%. The antimicrobial properties of microencapsulated TEO were evaluated *in vitro* against various food pathogens, demonstrating a minimum inhibitory concentration (MIC) ranging from 0.010 to 0.030 mg/mL against fungi. For the *in-situ* study, a cake was prepared with microcapsules containing 0.60 g of thymol. After 15 days, the control treatment (cake without preservatives) showed fungal growth, while the cake with thymol remained free of fungal growth throughout the study (30 days).

8.7 CONCLUSION

In this chapter, we have explored the importance of microencapsulation as a promising strategy to maintain the viability of bioactive compounds from different natural sources; we specifically focus on those compounds that were microencapsulated using polysaccharides as wall material to be applied in food preservation. Within the wall materials, polysaccharides are natural polymers with a relatively low cost and great versatility to be used in all microencapsulation techniques. Their microcapsules have good retention percentages and maintain the viability of bioactive compounds over time. The majority of substances that have been encapsulated and applied in food are of plant origin. The results suggest that incorporating encapsulated bioactive compounds has great potential to replace antimicrobials and synthetic antioxidants. They can give a plus to consumer health by turning food into functional food. However, there is still more future research, especially in assessing the sensory attributes of food in which these substances can be applied.

KEYWORDS

- **antimicrobials**
- **antioxidants**
- **biopolymers**
- **food preservation**
- **fungal growth**
- **microcapsules**
- **shelf-life**

REFERENCES

Abbas, M. S., Saeed, F., Afzaal, M., Jianfeng, L., Hussain, M., Ikram, A., & Jabeen, A. (2022). Recent trends in encapsulation of probiotics in dairy and beverage: A review. *Journal of Food Processing and Preservation, 46*(7), e16689.

Adinepour, F., Pouramin, S., Rashidinejad, A., & Jafari, S. M. (2022). Fortification/enrichment of milk and dairy products by encapsulated bioactive ingredients. *Food Research International*, 111212.

Ahmad, A., & Ali, A. (2019). Improvement of postharvest quality, regulation of antioxidants capacity, and softening enzyme activity of cold-stored carambola in response to polyamines application. *Postharvest Biology and Technology, 148*, 208–217.

Akinci, E., & Holthaus, D. (2023). Hydrocolloids and gums as encapsulating agents. In R. Sobel (Ed.), *Microencapsulation in the Food Industry* (2nd ed., pp. 309–323). Elsevier.

Akther, S., Jothi, J. S., Badsha, M. R., Rahman, M. M., Das, G. B., & Alim, M. A. (2023). Drying methods affect on bioactive compounds, phenolic profile, and antioxidant capacity of mango powder. *Journal of King Saud University-Science, 35*(1), 102370.

Aleksanyan, K. V. (2023). Polysaccharides for biodegradable packaging materials: Past, present, and future (brief review). *Polymers, 15*(2), 451.

Alves, E. S., Ferreira, C. S. R., Souza, P. R., Bruni, A. R. S., Castro, M. C., Saqueti, B. H. F., Santos, O. O., Madrona, G. S., & Visentainer, J. V. (2023). Freeze-dried human milk microcapsules using gum Arabic and maltodextrin: An approach to improving solubility. *International Journal of Biological Macromolecules*, 124100.

Amalraj, A., Gopi, S., Thomas, S., & Haponiuk, J. T. (2018). Cellulose nanomaterials in biomedical, food, and nutraceutical applications: A review. *Macromolecular Symposia*. Wiley Online Library, 1800115.

Ariestya, D. I., Swastawati, F., & Susanto, E. (2016). Antimicrobial activity of microencapsulation liquid smoke on tilapia [*Oreochromis niloticus* (Linnaeus, 1758)] meat for preservatives in cold storage (±5?C). *Aquatic Procedia, 7*, 19–27. https://doi.org/10.1016/j.aqpro.2016.07.003.

Asensio, C. M., Paredes, A. J., Martin, M. P., Allemandi, D. A., Nepote, V., & Grosso, N. R. (2017). Antioxidant stability study of oregano essential oil microcapsules prepared by spray-drying. *Journal of Food Science, 82*(12), 2864–2872.

Axel, C., Zannini, E., & Arendt, E. K. (2017). Mold spoilage of bread and its biopreservation: A review of current strategies for bread shelf life extension. *Critical Reviews in Food Science and Nutrition, 57*(16), 3528–3542. https://doi.org/10.1080/10408398.2016.1147417.

Bahrami, A., Delshadi, R., Assadpour, E., Jafari, S. M., & Williams, L. (2020). Antimicrobial-loaded nanocarriers for food packaging applications. *Advances in Colloid and Interface Science, 278*, 102140.

Baldin, J. C., Michelin, E. C., Polizer, Y. J., Rodrigues, I., de Godoy, S. H. S., Fregonesi, R. P., & Trindade, M. A. (2016). Microencapsulated jabuticaba (*Myrciaria cauliflora*) extract is added to fresh sausage as a natural dye with antioxidant and antimicrobial activity. *Meat Science, 118*, 15–21.

Bamidele, O. P., & Emmambux, M. N. (2021). Encapsulation of bioactive compounds by "extrusion" technologies: A review. *Critical Reviews in Food Science and Nutrition, 61*(18), 3100–3118.

Ban, Z., Zhang, J., Li, L., Luo, Z., Wang, Y., Yuan, Q., Zhou, B., & Liu, H. (2020). Ginger essential oil-based microencapsulation as an efficient delivery system for the improvement of Jujube (*Ziziphus jujuba* Mill.) fruit quality. *Food Chemistry, 306*, 125628. https://doi.org/10.1016/j.foodchem.2019.125628.

Bir, C., Widmar, N. O., Thompson, N. M., Townsend, J., & Wolf, C. A. (2020). US respondents' willingness to pay for Cheddar cheese from dairy cattle with different pasture access, antibiotic use, and dehorning practices. *Journal of Dairy Science, 103*(4), 3234–3249.

Boostani, S., & Jafari, S. M. (2020). Controlled release of nanoencapsulated food ingredients. In *Release and Bioavailability of Nanoencapsulated Food Ingredients* (pp. 27–78). Elsevier.

Buljeta, I., Pichler, A., Šimunović, J., & Kopjar, M. (2022). Polysaccharides as carriers of polyphenols: comparison of freeze-drying and spray-drying as encapsulation techniques. *Molecules, 27*(16), 5069.

Calderón-Oliver, M., & Ponce-Alquicira, E. (2022). The role of microencapsulation in food application. *Molecules, 27*(5), 1499.

Calderón-Santoyo, M., Iñiguez-Moreno, M., & Ragazzo-Sánchez, J. A. (2021). Microencapsulation of citral and its antifungal activity into pectin films. *Biointerface Research in Applied Chemistry, 12*, 7488–7502.

Calderón-Santoyo, M., Iñiguez-Moreno, M., Barros-Castillo, J. C., Miss-Zacarías, D. M., Díaz, J. A., & Ragazzo-Sánchez, J. A. (2022). Microencapsulation of citral with Arabic gum and sodium alginate for the control of Fusarium pseudocircinatum in bananas. *Iranian Polymer Journal, 31*(5), 665–676. https://doi.org/10.1007/s13726-022-01043-4.

Castro-Alatorre, N. C., Gallardo-Velázquez, T., Boyano-Orozco, L. C., Téllez-Medina, D. I., Meza-Márquez, O. G., & Osorio-Revilla, G. (2021). Extraction and microencapsulation of bioactive compounds from muicle (*Justicia spicigera*) and their use in the formulation of functional foods. *Foods, 10*(8), 1747.

Cazón, P., Velazquez, G., Ramírez, J. A., & Vázquez, M. (2017). Polysaccharide-based films and coatings for food packaging: A review. *Food Hydrocolloids, 68*, 136–148.

Chakraborty, S. (2017). Carrageenan for encapsulation and immobilization of flavor, fragrance, probiotics, and enzymes: A review. *Journal of Carbohydrate Chemistry, 36*(1), 1–19.

Chen, K., Zhang, M., Mujumdar, A. S., & Wang, H. (2021). Quinoa protein-gum Arabic complex coacervates as a novel carrier for eugenol: Preparation, characterization, and application for minced pork preservation. *Food Hydrocolloids, 120*, 106915.

Cheng, A. W., Xie, H. X., Qi, Y., Liu, C., Guo, X., Sun, J. Y., & Liu, L. N. (2017). Effects of storage time and temperature on polyphenolic content and qualitative characteristics of freeze-dried and spray-dried bayberry powder. *LWT – Food Science and Technology, 78*, 235–240.

Cheng, M., Wang, J., Zhang, R., Kong, R., Lu, W., & Wang, X. (2019). Characterization and application of the microencapsulated carvacrol/sodium alginate films as food packaging materials. *International Journal of Biological Macromolecules, 141*, 259–267. https://doi.org/10.1016/j.ijbiomac.2019.08.240.

Choudhury, N., Meghwal, M., & Das, K. (2021). Microencapsulation: An overview of concepts, methods, properties, and applications in foods. *Food Frontiers, 2*(4), 426–442.

Comunian, T. A., & Favaro-Trindade, C. S. (2016). Microencapsulation using biopolymers as an alternative to produce food enhanced with phytosterols and omega-3 fatty acids: A review. *Food Hydrocolloids, 61*, 442–457.

Comunian, T. A., da Silva Anthero, A. G., Bezerra, E. O., Moraes, I. C. F., & Hubinger, M. D. (2020). Encapsulation of pomegranate seed oil by emulsification followed by spray drying: Evaluation of different biopolymers and their effect on particle properties. *Food Bioprocess Technology, 13*, 53–66.

Cunha, L. C. M., Monteiro, M. L. G., Costa-Lima, B. R. C., Guedes-Oliveira, J. M., Alves, V. H. M., Almeida, A. L., Tonon, R. V., Rosenthal, A., & Conte-Junior, C. A. (2018). Effect of microencapsulated extract of pitaya (*Hylocereus costaricensis*) peels on color, texture, and oxidative stability of refrigerated ground pork patties submitted to high-pressure processing. *Innovative Food Science & Emerging Technologies, 49*, 136–145.

Cunha, L. C. M., Monteiro, M. L. G., da Costa-Lima, B. R. C., Guedes-Oliveira, J. M., Rodrigues, B. L., Fortunato, A. R., Baltar, J. D., Tonon, R. V., Koutchma, T., & Conte-Junior, C. A. (2021). Effect of microencapsulated extract of pitaya (*Hylocereus costaricensis*) peel on oxidative quality parameters of refrigerated ground pork patties subjected to UV-C radiation. *Journal of Food Processing and Preservation, 45*(3), e15272.

de Moura, S. C. S. R., Schettini, G. N., Gallina, D. A., Dutra Alvim, I., & Hubinger, M. D. (2022). Microencapsulation of hibiscus bioactive and its application in yogurt. *Journal of Food Processing and Preservation, 46*(4), e16468.

De Oliveira, T. T. B., dos Reis, I. M., de Souza, M. B., da Silva Bispo, E., Maciel, L. F., Druzian, J. I., Tavares, P. P. L. G., de Oliveira Cerqueira, A., Morte, E. D. S. B., & Glória, M. B. A. (2021). Microencapsulation of Spirulina sp. LEB-18 and its incorporation in chocolate milk: Properties and functional potential. *LWT, 148*, 111674.

Dong, Y., Wei, Z., & Xue, C. (2021). Recent advances in carrageenan-based delivery systems for bioactive ingredients: A review. *Trends in Food Science & Technology, 112*, 348–361.

Đorđević, V., Balanč, B., Belščak-Cvitanović, A., Lević, S., Trifković, K., Kalušević, A., & Nedović, V. (2015). Trends in encapsulation technologies for delivery of food bioactive compounds. *Food Engineering Reviews, 7*, 452–490.

Farias-Cervantes, V. S., Chávez-Rodríguez, A., García-Salcedo, P. A., García-López, P. M., Casas-Solís, J., & Andrade-González, I. (2018). Antimicrobial effect and in vitro release of anthocyanins from berries and Roselle obtained via microencapsulation by spray drying. *Journal of Food Processing Preservation, 42*(10), e13713.

Ferreira, S., Nicoletti, V. R., & Dragosavac, M. (2023). Novel methods to induce complex coacervation using dual fluid nozzle and metal membranes: Part II–Use of metal membrane technology to induce complex coacervation. *Food and Bioproducts Processing, 139*, 46–60.

Fredes, C., Becerra, C., Parada, J., & Robert, P. (2018). The microencapsulation of maqui (*Aristotelia chilensis* (Mol.) Stuntz) juice by spray-drying and freeze-drying produces powders with similar anthocyanin stability and bioaccessibility. *Molecules, 23*(5), 1227.

Gómez, B., Barba, F. J., Domínguez, R., Putnik, P., Kovačević, D. B., Pateiro, M., Toldrá, F., & Lorenzo, J. M. (2018). Microencapsulation of antioxidant compounds through innovative technologies and its specific application in meat processing. *Trends in Food Science & Technology, 82*, 135–147. https://doi.org/10.1016/j.tifs.2018.09.019.

Gómez-Contreras, P., Figueroa-Lopez, K. J., Hernández-Fernández, J., Cortés Rodríguez, M., & Ortega-Toro, R. (2021). Effect of different essential oils on the properties of edible coatings based on yam (*Dioscorea rotundata* L.) starch and its application in strawberry (*Fragaria vesca* L.) preservation. *Applied Sciences, 11*(22), 11057.

Gonçalves, N. D., Pena, F. de L., Sartoratto, A., Derlamelina, C., Duarte, M. C. T., Antunes, A. E. C., & Prata, A. S. (2017). Encapsulated thyme (*Thymus vulgaris*) essential oil is used as a natural preservative in bakery products. *Food Research International, 96*, 154–160. https://doi.org/10.1016/j.foodres.2017.03.009.

Halahlah, A., Piironen, V., Mikkonen, K. S., & Ho, T. M. (2022). Polysaccharides as wall materials in spray-dried microencapsulation of bioactive compounds: Physicochemical properties and characterization. *Critical Reviews in Food Science and Nutrition*, 1–33 .

How, Y. H., Lai, K. W., Pui, L. P., & In, L. L. A. (2022). Co-extrusion and extrusion microencapsulation: Effect on microencapsulation efficiency, survivability through gastrointestinal digestion and storage. *Journal of Food Process Engineering, 45*(3), e13985.

Huang, K., Yuan, Y., & Baojun, X. (2021). A critical review on the microencapsulation of bioactive compounds and their application. *Food Reviews International, 1–41.*

Huq, T., Riedl, B., Bouchard, J., Salmieri, S., & Lacroix, M. (2014). Microencapsulation of nisin in alginate-cellulose nanocrystal (CNC) microbeads for prolonged efficacy against Listeria monocytogenes. *Cellulose, 21*(6), 4309–4321.

Iñiguez-Moreno, M., Ragazzo-Sánchez, J. A., & Calderón-Santoyo, M. (2021). An extensive review of natural polymers used as coatings for postharvest shelf-life extension: Trends and challenges. *Polymers, 13*(19), 3271.

Jafari, S. M., Esfanjani, A. F., Katouzian, I., & Assadpour, E. (2017). Release, characterization, and safety of nanoencapsulated food ingredients. *Nanoencapsulation of Food Bioactive Ingredients*, 401–453.

Jafari, S. M., Ghanbari, V., Dehnad, D., & Ganje, M. (2018). Neural networks modeling of *Aspergillus flavus* growth in tomato paste containing microencapsulated olive leaf extract. *Journal of Food Safety, 38*(1), e12396.

Jafari, S., Jafari, S. M., Ebrahimi, M., Kijpatanasilp, I., & Assatarakul, K. (2022). A decade overview and prospect of spray drying encapsulation of bioactives from fruit products: Characterization, food application and in vitro gastrointestinal digestion. *Food Hydrocolloids, 123*, 108068.

Jeyakumari, A., Zynudheen, A., & Parvathy, U. (2016). Microencapsulation of bioactive food ingredients and controlled release-A review. *Food Processing & Technology.*

Kandasamy, S., & Naveen, R. (2022). A review of the encapsulation of bioactive components using spray-drying and freeze-drying techniques. *Journal of Food Process Engineering, 45*(8), e14059.

Kavas, N., Kavas, G., Kinik, Ö., Ates, M., & Kaplan, M., Şatir, G. (2021). Symbiotic microencapsulation to enhance Bifidobacterium longum and Lactobacillus paracasei survival in goat cheese. *Food Science and Technology, 42.*

Ko, S., & Gunasekaran, S. (2014). Controlled release of food ingredients. *Nano-and Microencapsulation for Foods*, 325–343.

Kulka, K., & Sionkowska, A. (2023). Chitosan-based materials in cosmetic applications: A review. *Molecules, 28*(4), 1817.

Lavelli, V., Sri Harsha, P. S. C., & Spigno, G. (2016). Modeling the stability of maltodextrin-encapsulated grape skin phenolics used as a new ingredient in apple puree. *Food Chemistry, 209*, 323–331. https://doi.org/10.1016/j.foodchem.2016.04.047.

Lazou, A., & Krokida, M. K. (2017). Extrusion for microencapsulation. In A. S. Mujumdar (Ed.), *Advances in Drying Science & Technology: Thermal and nonthermal encapsulation methods* (1st ed., Vol. 1, pp. 137–171). CRC Press.

Lescano, C. H., Sanjinez-Argandoña, E. J., de Arruda, E. J., Kassuya, C. A. L., & Moraes, I. C. F. (2014). Acrocomia aculeata (Jacq.) Lodd. oil microencapsulation by complex coacervation: Preservation of bioactive compounds. *Journal of Encapsulation and Adsorption Sciences, 4*(04), 105.

Leyva-Porras, C., Saavedra-Leos, M. Z., López-Martinez, L. A., Espinosa-Solis, V., Terán-Figueroa, Y., Toxqui-Terán, A., & Compeán-Martínez, I. (2021). Strawberry juice powders:

Effect of spray-drying conditions on the microencapsulation of bioactive components and physicochemical properties. *Molecules, 26*(18), 5466. https://doi.org/10.3390/molecules26185466.

Lima, E. M. F., Madalão, M. C. M., dos Santos, W. C., Bernardes, P. C., Saraiva, S. H., & Silva, P. I. (2019). Spray-dried microcapsules of anthocyanin-rich extracts from *Euterpe edulis* M. as an alternative for maintaining color and bioactive compounds in dairy beverages. *Journal of Food Science and Technology, 56*, 4147–4157.

McClements, D. J. (2014). Nanoparticle-and microparticle-based delivery systems: Encapsulation, protection, and release of active compounds. CRC Press.

Mehta, N., Kumar, P., Verma, A. K., Umaraw, P., Kumar, Y., Malav, O. P., & Lorenzo, J. M. (2022). Microencapsulation as a noble technique for the application of bioactive compounds in the food industry: A comprehensive review. *Applied Sciences, 12*(3), 1424.

Moghanjougi, Z. M., Bari, M. R., Khaledabad, M. A., Almasi, H., & Amiri, S. (2020). Bio-preservation of white brined cheese (Feta) by using probiotic bacteria immobilized in bacterial cellulose: Optimization by response surface method and characterization. *LWT, 117*, 108603.

Mohammadalinejhad, S., & Kurek, M. A. (2021). Microencapsulation of anthocyanins—Critical review of techniques and wall materials. *Applied Sciences, 11*(9), 3936.

Mohite, D., & Waghmare, R. (2020). Encapsulation techniques for delivery of bioactive compounds in milk and dairy products-A review. *Journal of Dairy Research & Technology, 3*, 1–9.

Montoya-Yepes, D. F., Jiménez-Rodríguez, A. A., Aldana-Porras, A. E., Velásquez-Holguin, L. F., Méndez-Arteaga, J. J., & Murillo-Arango, W. (2023). Starches in the encapsulation of plant active ingredients: state of the art and research trends. *Polymer Bulletin*, 1–29.

Nezamdoost-Sani, N., Khaledabad, M. A., Amiri, S., & Khaneghah, A. M. (2023). Alginate and derivatives hydrogels in encapsulation of probiotic bacteria: An updated review. *Food Bioscience*, 102433.

Ng, S.-K., Jessie, L.-Y. L., Tan, C.-P., Long, K., & Nyam, K.-L. (2013). Effect of accelerated storage on microencapsulated kenaf seed oil. *Journal of the American Oil Chemists' Society, 90*, 1023–1029.

Nogueira, G. F., Oliveira, R. A. de, Velasco, J. I., & Fakhouri, F. M. (2020). Methods of incorporating plant-derived bioactive compounds into films made with agro-based polymers for application as food packaging: A brief review. *Polymers, 12*(11), 2518. https://doi.org/10.3390/polym12112518.

Onwulata, C. I. (2013). Microencapsulation and functional bioactive foods. *Journal of Food Processing and Preservation, 37*(5), 510–532.

Parente, J. F., Sousa, V. I., Marques, J. F., Forte, M. A., & Tavares, C. J. (2022). Biodegradable polymers for microencapsulation systems. *Advances in Polymer Technology, 2022*, 1–43.

Patrignani, F., Siroli, L., Serrazanetti, D. I., Braschi, G., Betoret, E., Reinheimer, J. A., & Lanciotti, R. (2017). Microencapsulation of functional strains by high-pressure homogenization for potential use in fermented milk. *Food Research International, 97*, 250–257.

Peng, X., Umer, M., Pervez, M. N., Hasan, K. F., Habib, M. A., Islam, M. S., Lin, L., Xiong, X., Naddeo, V., & Cai, Y. (2023). Biopolymers-based microencapsulation technology for sustainable textiles development: A short review. *Case Studies in Chemical and Environmental Engineering*, 100349.

Pérez-Soto, E., Cenobio-Galindo, A. D. J., Espino-Manzano, S. O., Franco-Fernández, M. J., Ludeña-Urquizo, F. E., Jiménez-Alvarado, R., Zepeda-Velázquez, A. P., & Campos-Montiel,

R. G. (2021). The addition of microencapsulated or nanoemulsified bioactive compounds influences the antioxidant and antimicrobial activities of fresh cheese. *Molecules, 26*(8), 2170.

Poshadri, A., & Aparna, K. (2010). Microencapsulation technology: A review. *Journal of Research ANGRAU, 38*(1), 86–102.

Reis, A. S. D., Diedrich, C., Moura, C. D., Pereira, D., Almeida, J. D. F., Silva, L. D. D., Plata-Oviedo, M. S. V., Tavares, R. A. W., & Carpes, S. T. (2017). Physico-chemical characteristics of microencapsulated propolis co-product extract and its effect on storage stability of burger meat during storage at −15?C. *LWT – Food Science and Technology, 76*, 306–313.

Rezvankhah, A., Emam-Djomeh, Z., & Askari, G. (2020). Encapsulation and delivery of bioactive compounds using spray and freeze-drying techniques: A review. *Drying Technology, 38*(1–2), 235–258.

Rezvankhah, A., Emam-Djomeh, Z., Safari, M., Salami, M., & Askari, G. (2022). Investigating the effects of maltodextrin, gum Arabic, and whey protein concentrate on the microencapsulation efficiency and oxidation stability of hemp seed oil. *Journal of Food Processing and Preservation, 46*(6), e16554.

Ribeiro, A., Caleja, C., Barros, L., Santos-Buelga, C., Barreiro, M. F., & Ferreira, I. C. (2016). Rosemary extracts in functional foods: Extraction, chemical characterization and incorporation of free and microencapsulated forms in cottage cheese. *Food & Function, 7*(5), 2185–2196.

Rocchetti, G., Becchi, P. P., Lucini, L., Cittadini, A., Munekata, P. E. S., Pateiro, M., Domínguez, R., & Lorenzo, J. M. (2022). Elderberry (*Sambucus nigra* L.) encapsulated extracts as meat extenders against lipid and protein oxidation during the shelf-life of beef burgers. *Antioxidants*.

Saberi, M., Saremnezhad, S., Soltani, M., & Faraji, A. (2023). Functional stirred yogurt manufactured using co-microencapsulated or free forms of grape pomace and flaxseed oil as bioactive ingredients: Physicochemical, antioxidant, rheological, microstructural, and sensory properties. *Food Science & Nutrition*, 1–13.

Sapper, M., Palou, L., Pérez-Gago, M. B., & Chiralt, A. (2019). Antifungal starch–gellan edible coatings with thyme essential oil for the postharvest preservation of apple and persimmon. *Coatings, 9*(5), 333.

Savaiano, D. A., & Hutkins, R. W. (2021). Yogurt, cultured fermented milk, and health: A systematic review. *Nutrition Reviews, 79*(5), 599–614.

Shamil, A., Nautiyal, A., & Omar, I. (2023). Microencapsulated bioactive compounds for fortification of dairy products: A review. *The Pharma Innovation Journal, 12*(4), 1855–1865.

Shao, P., Yu, J., Chen, H., & Gao, H. (2021). Development of microcapsule bioactive paper loaded with cinnamon essential oil to improve the quality of edible fungi. *Food Packaging and Shelf Life, 27*, 100617. https://doi.org/10.1016/j.fpsl.2020.100617.

Silveira, M. P., Almeida, F. L. C., Andreola, K., Alvim, I. D., & Prata, A. S. (2023). Influence of composition on the internal diffusion mechanism of pectin–starch gel beads. *Journal of Applied Polymer Science, 140*(9), e53570.

Tang, Y., Wang, X., Yu, J., Song, L., & Lin, S. (2020). Fish skin gelatin-based emulsion as a delivery system to protect lipophilic bioactive compounds during in vitro and in vivo digestion: The case of benzyl isothiocyanate. *LWT, 134*, 110145.

Thakur, N., Sharma, R., & Thakur, A. (2022). Optimization of lyophilized microencapsulated phenolic extract concentration for the enrichment of yogurt and effect on chemical parameters, bioactive compounds, antioxidant activity, and sensory quality under storage. *South African Journal of Botany, 151*, 413–422.

Tomsone, L., Galoburda, R., Kruma, Z., Durrieu, V., & Cinkmanis, I. (2020). Microencapsulation of horseradish (*Armoracia rusticana* L.) juice using spray-drying. *Foods, 9*(9), 1332.

Wang, B., Akanbi, T. O., Agyei, D., Holland, B. J., & Barrow, C. J. (2018). Coacervation technique as an encapsulation and delivery tool for hydrophobic biofunctional compounds. In A. Grumezescu & A. M. Holban (Eds.), *Role of Materials Science in Food Bioengineering* (pp. 235–261). Elsevier.

Wang, Y. F., Jia, J. X., Tian, Y. Q., Shu, X., Ren, X. J., Guan, Y., & Yan, Z. Y. (2018). Antifungal effects of clove oil microcapsule on meat products. *LWT – Food Science and Technology, 89*, 604–609.

Wu, Q., & Zhou, J. (2021). The application of polyphenols in food preservation. *Advances in Food Nutrition Research, 98*, 35–99.

Xu, L., Cheng, J. R., Liu, X. M., & Zhu, M. J. (2019). Effect of microencapsulated process on the stability of mulberry polyphenol and oxidation property of dried minced pork slices during heat processing and storage. *LWT – Food Science and Technology, 100*, 62–68.

Yamashita, C., Chung, M. M. S., dos Santos, C., Mayer, C. R. M., Moraes, I. C. F., & Branco, I. G. (2017). Microencapsulation of an anthocyanin-rich blackberry (Rubus spp.) by-product extract by freeze-drying. *LWT – Food Science and Technology, 84*, 256–262.

Yan, C., Kim, S.-R., Ruiz, D. R., & Farmer, J. R. (2022). Microencapsulation for Food Applications: A Review. *ACS Applied Bio Materials, 5*(12), 5497–5512.

Yan, S., Zhang, S., Zhu, H., Qi, B., & Li, Y. (2022). Recent advances in protein-based multilayer emulsions: Fabrication, characterization, and applications: A Review. *Food Reviews International, 1–18.*

Yang, Y., Aghbashlo, M., Gupta, V. K., Amiri, H., Pan, J., Tabatabaei, M., & Rajaei, A. (2023). Chitosan nanocarriers containing essential oils as a green strategy to improve the functional properties of chitosan: A review. *International Journal of Biological Macromolecules*, 123954.

Yoplac, I., Vargas, L., Robert, P., & Hidalgo, A. (2021). Characterization and antimicrobial activity of microencapsulated citral with dextrin by spray drying. *Heliyon, 7*(4), e06737.

Zhang, D., Ivane, N. M., Haruna, S. A., Zekrumah, M., Elysé, F. K. R., Tahir, H. E., Wang, G., Wang, C., & Zou, X. (2022). Recent trends in the micro-encapsulation of plant-derived compounds and their specific application in meat as antioxidants and antimicrobials. *Meat Science*, 108842.

Zhang, L., Yu, D., Regenstein, J. M., Xia, W., & Dong, J. (2021). A comprehensive review of natural bioactive films with controlled release characteristics and their applications in foods and pharmaceuticals. *Trends in Food Science & Technology, 112*, 690–707.

Zhang, R., Belwal, T., Li, L., Lin, X., Xu, Y., & Luo, Z. (2020). Recent advances in polysaccharides stabilized emulsions for encapsulation and delivery of bioactive food ingredients: A review. *Carbohydrate Polymers, 242*, 116388.

CHAPTER 9

Biological Activities and Food Applications of Citrulline From Watermelon (*Citrullus lanatus*)

KARLA SEGURA-ALMARAZ,[1]
ADRIANA CAROLINA FLORES-GALLEGOS,[1]
SANDRA CECILIA ESPARZA GONZÁLEZ,[2]
JUAN ALBERTO ASCACIO-VALDÉS,[1]
SONIA YESENIA SILVA-BELMARES,[1] and RAÚL RODRIGUEZ-HERRERA[1]

[1]*Department of Food Research, School of Chemical Sciences, Autonomous University of Coahuila, Blvd. V. Carranza e Ing. Jose Cardenas, Saltillo, Coahuila, Mexico*

[2]*School of Odontology, Autonomous University of Coahuila, Dra Cuquita Cepeda de Dávila, Saltillo, Coahuila, Mexico*

ABSTRACT

Citrulline is considered a non-essential amino acid for the human body. However, because of the different performance benefits that it could offer to athletes who consume it as a dietary supplement, the interest in this compound has increased. Among these benefits are the increase in energy synthesis, muscle development, increased blood flow, delayed fatigue, and prevention of catabolism. Citrulline is synthesized and sold as citrulline malate; however, it is found naturally and abundantly in fruits from the Cucurbitaceae family, and greater proportion in the watermelon (*Citrullus lanatus*) shell. On the other hand, in the last decade, demand for the consumption of functional foods has increased; thereby, the food industry has developed new foods with high nutritional value, which provide health benefits, such as snacks,

Sustainable Agricultural and Food Systems Engineering: New Research, Practices, and Developments.
Jose Juan Buenrostro-Figueroa, Julio César Tafolla-Arellano, Leonardo Sepúlveda Torre, Porteen Kannan, A. K. Haghi (Eds.)

energy-protein bars, yogurt, cereals, gums, truffles, etc. The objectives of this document are to describe the health benefits of the consumption of natural citrulline and its application for the development of functional foods. Among the scientific advances to be highlighted are that citrulline is reported to be of great ergogenic help, which also provides benefits for disease treatment, and that supplements with citrulline significantly increase health benefits. Among the effects reported are a decrease in insulin resistance, body weight, arterial stiffness, blood pressure, muscle mass, vasodilation, blood flow, and the elimination of excess ammonia. In conclusion, the food industry is constantly at the forefront of developing functional food from a natural source (watermelon). Not only will it take advantage of the nutrients from watermelon shells, but it will also minimize environmental problems by using agro-industrial waste.

9.1 INTRODUCTION

Fruit consumption provides nutrients such as carbohydrates, protein, fiber, vitamins, and minerals (Bhol, Lanka, & Bosco, 2016). However, during its consumption, some plant structures such as shells and seeds are discarded, which may contain bioactive compounds (Nagulb & Tantawy, 2019). In most cases, fruit shells are considered agro-industrial waste and are thrown into the environment, although they are sometimes used as livestock feed (Bhol, Lanka, & Bosco, 2016). However, fruit shell extracts contain phytochemicals that can be used to treat or prevent diseases (Nagulb & Tantawy, 2019). These phytochemicals can also be incorporated into foods to add value. Currently, an important market has developed for this type of food, which is marketed as juices, bread, candies, yogurts, custards, truffles, etc.

Citrulline (Table 9.1) is a compound present in fruit pulp and shells from the Cucurbitaceae family, mainly in watermelon (*Citrullus lanatus*) (Saulskaya, Fofonova, Sudorgina, & Komarova, 2011). This compound may have different benefits for human health when incorporated into foods. This molecule has not received much importance and was classified only as a non-essential amino acid for the human body, where its importance lay only in being an intermediary in the urea cycle (Ahmadi, Babalar, Sarcheshmeh, Morshedloo, & Shokrpour, 2020). However, as knowledge about this molecule accumulated, it was found to have other metabolic contributions that are beneficial to human health. Recently, it has been associated with antioxidant and anti-cancer effects (Joshi, Joshi, Silwal, Noonan, Rodriguez, & Penalosa, 2019). While in plants, citrulline is involved in nutrient transport, signaling, and the interaction process with pathogens and environmental stress. In human beings, this compound promises to be a good ergogenic aid, improving physical conditions, sports performance, delaying fatigue, and improving the synthesis of nitric oxide (NO), which helps to eliminate toxic compounds and decrease oxidative damage (Kiyici, Eroğlu, Kishali, & Burmaoglu, 2017; Valenzuela, Morales, Emanuele, Pareja-Galeano, & Lucia, 2019). However, it is necessary to determine the maximum amount of citrulline that can be added to each food to avoid an excess in l-arginine production, since excess l-arginine consumption has been associated with gastrointestinal side effects, such as nausea and diarrhea, because of the rapid increase in NO production in the gastrointestinal tract (Ridwan, Abdul Razak, Adenan, & Md. Saad, 2019). The objectives of this document are to discuss the health benefits of citrulline and the possibilities of its application in the development of functional foods.

TABLE 9.1 Physical Characteristics and Biological Description of L-Citrulline According to the US National Library of Medicine

Chemical Formula	**C6H13N3O3**
Molar Mass	175.19/mol
Chemical Structure	$H_2N–C–N–CH_2–CH_2–CH_2–CH–COO^-$ ‖ | | O H $^+NH_3$
Biological Description	H–Cit–OH
Name IUPAC	DL-Citrulline
	2-amino-5-(carbamoylamino)pentanoic acid.

9.1.1 CITRULLINE SYNTHESIS

L-Citrulline was isolated about 100 years ago from *Citrullus lanatus*, where its name came from. It has also been identified in other species from the Cucurbitaceae family (Saulskaya, Fofonova, Sudorgina, & Komarova, 2011), including melon, cucumbers, and pumpkin (Caruso, Rossi, Gahn, & Caruso, 2017). This amino acid can be synthesized by two pathways, from the amino acids L-glutamine (Figure 9.1) and L-arginine. The pathway from L-glutamine is:

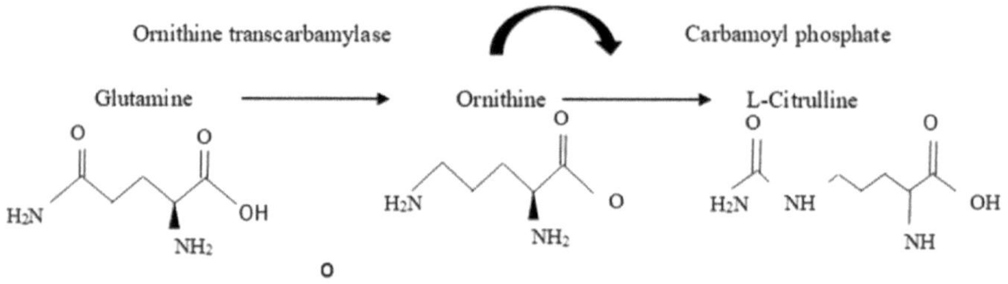

FIGURE 9.1 Citrulline synthesis in mitochondria during the urea cycle.

The second route of citrulline synthesis is carried out by using L-arginine (McNeal, Meininger, Reddy, Wilborn, & Wu, 2016), as a precursor, which is catalyzed by nitric oxide synthase and co-producer of nitric oxide (NO) synthesis (Ahmadi, Babalar, Sarcheshmeh, Morshedloo, & Shokrpour, 2020; Caruso, Rossi, Gahn, & Caruso, 2017):

$$\text{L-arginine} + 3\ \text{NADPH} + H^+ + 4\ O_2 \rightarrow 2\ \text{L-citrulline} + 2\ \text{NO} + 4\ H_2O + 3\ \text{NADP}^+$$

It is not yet known at the cellular level, which the best nitric oxide precursor, arginine or citrulline (Ahmadi, Babalar, Sarcheshmeh, Morshedloo, & Shokrpour, 2020), but the best alternative could be citrulline, as it is more efficient for increasing the availability of arginine, obtaining a higher yield (Umang, Inka, Yang, Xiaoying, & Marini, 2017). However, the human body is capable of synthesizing citrulline from other amino acids (from glutamine, glutamate, and proline catabolism in the small intestine), making direct consumption unnecessary (McNeal, Meininger, Reddy, Wilborn, & Wu, 2016).

Endogenous NO synthesis occurs through the L-arginine-NO pathway, whereby L-arginine is converted to NO and L-citrulline by nitric oxide synthase (NOS) enzymes (Gonzalez & Trexler, 2020).

9.2 METABOLIC PATHWAY OF CITRULLINE IN THE HUMAN BODY

Citrulline is important during the urea cycle (Figure 9.2) because it converts ammonia into urea, a process that takes place in the liver. Therefore, people with problems excreting ammonia into the blood have received citrulline to assist them in this elimination (Caruso, Rossi, Gahn, & Caruso, 2017). In addition, the combination of L-citrulline with malate (an amino acid) can act as a mediator in the tricarboxylic acid cycle, thus increasing the rate of adenosine triphosphate (ATP) production, muscle contractile force (Glenn, Gray, Wethington, Stone, Stewart, & Moyen, 2017), improving mitochondrial efficiency, and calcium handling (Gonzalez & Trexler, 2020). However, excessive amounts of citrulline may cause the known disorder called citrullination, a chemical process that occurs when there are high amounts of citrulline, and where a process of post-translation hydrolysis of arginine residues occurs to create citrulline and ammonia residues, with catalysis by peptidyl arginine deiminases (PAD), causing extensive protein modification (Caruso, Rossi, Gahn, & Caruso, 2017). Citrullination is a promising therapeutic use for autoimmune, neurodegenerative diseases, and certain cancers (Mondal & Thompson, 2021).

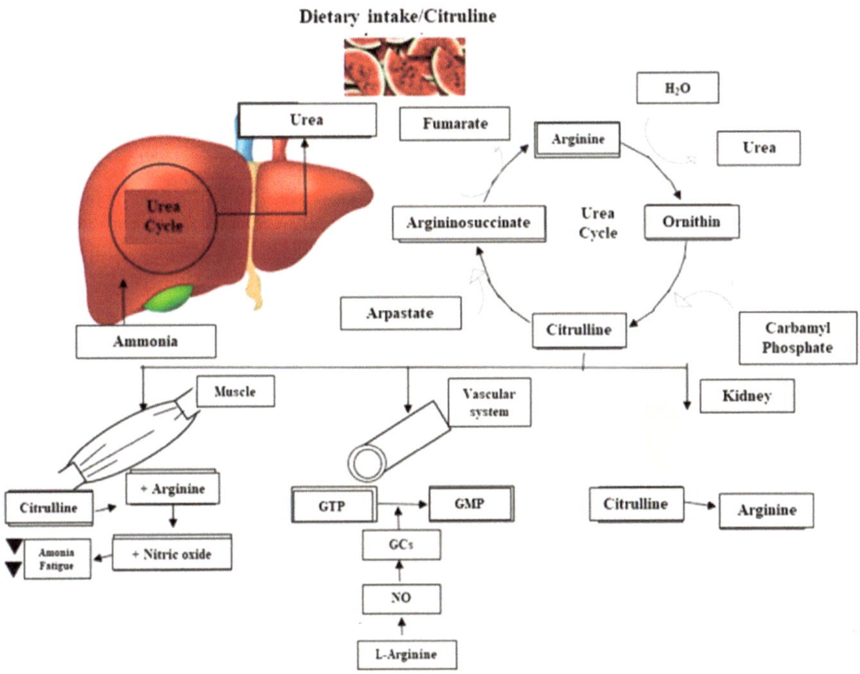

FIGURE 9.2 Citrulline in the human body.

9.3 NITRIC OXIDE (NO) AND CITRULLINE RATIO

Citrulline increases the availability of arginine and consequently increases the synthesis of nitric oxide (NO), which is a gas synthesized from L-arginine and oxygen by the endothelial nitric oxide synthetase (eNOS) enzyme. This gas is secreted by the endothelium and activates the soluble guanylate cyclase (SGC), increases the production of cyclic guanosine monophosphate (cGMP) and promotes vasodilation of blood vessels (Azizi et al., 2020). This process aids blood circulation and effectively improves mitochondrial function (Suzuki et al., 2019), especially during physical exercise (Glenn, Gray, Wethington, Stone, Stewart, & Moyen, 2017).

In conjunction with calmodulin, the enzyme nitric oxide synthase (NOS) is activated, and in the presence of other cofactors such as oxygen, NADPH, FAD, FMN, and biopterin, the substrate L-Arginine is converted to NO and L-Citrulline, activated by the peroxisome proliferator-γ co-activator-1α (PGC-1α), which acts to regulate mitochondrial function. NO has a series of physiological functions among which the following stand out: increase the effectiveness of muscle contraction, control of oxygen consumption, relaxation of the capillaries, and increase of exercise tolerance (therefore, it is an intercellular messenger and forms reactive nitrogen species). In addition, this gas has certain benefits of dietary nitrite (NO_3^-) supplements and their beneficial effects on sports performance. In contrast, other authors report no improvement in endurance performance following NO_3^- supplementation in highly trained athletes (Suzuki et al., 2019). Nitric oxide deficiency has been implicated in several health problems, including mitochondrial disease, and this deficiency can be better remedied with citrulline supplementation than with arginine (Caruso, Rossi, Gahn, & Caruso, 2017).

9.4 DISEASE PREVENTION AND TREATMENT

Citrulline has been used as therapy for mitochondrial diseases, and it is important to note that this compound is well-tolerated, unlike arginine supplementation (Caruso, Rossi, Gahn, & Caruso, 2017), despite the fact that citrulline is a precursor of L-arginine (itself a precursor of nitric oxide) (Glenn, Gray, Wethington, Stone, Stewart, & Moyen, 2017; Azizi et al., 2020). In addition, the effectiveness and benefits of L-citrulline administration have been proven as a therapeutic use for multiple diseases. However, its major application has been in athletes to increase their performance during continuous physical activity. These diseases are triggered by overweight and obesity. Therefore, in order to reduce weight, several hypocaloric diets have been proposed, which reduce fat mass while preserving lean mass, as the restitution of muscle mass is very complicated. Leucine and citrulline are two amino acids that help preserve muscle function during restrictive diets.

Recently, interest in non-essential plant-based amino acids (NPAA) has intensified, as they have been associated with health benefits, including antioxidant, cardioprotective, and anti-inflammatory activities (Azizi et al., 2020) (Table 9.2), and their incorporation into dietary supplements and pharmacology has been investigated (Joshi, Joshi, Silwal, Noonan, Rodriguez, & Penalosa, 2019). In addition, citrulline has the capacity to increase the availability of arginine in the plasma, and later the synthesis of nitric oxide, performing in the process of vasodilation, blood flow, respiration of the mitochondria, and oxidation of glucose (Caruso, Rossi, Gahn, & Caruso, 2017). It has been suggested to be involved in physiological processes where nitrogen transport occurs (Joshi, Joshi, Silwal, Noonan, Rodriguez, & Penalosa, 2019). Citrulline is recognized as an amino acid with great contributions to cardio-metabolic health. Although research continues to confirm the potential of this substance, not only as ergogenic on vascular function and physical performance but also in the treatment of various diseases.

TABLE 9.2 Function and Effect of Citrulline by System

Systems	Function	Effect	References
Metabolism	• Increased glucose results in muscle vasodilation. • Cellular activity increases. • Lipolysis is increased.	• Decreases insulin resistance. • Energy synthesis is increased. • Low body weight.	Ogurtsova et al. (2017); Darabi, Darand, Yari, Hedayati, Faghihi, & Hekmatdoost (2019)
Musculoskeletal	• Activates muscle synthesis. • Muscle blood flow increases significantly.	• Muscle development. • It anticipates catabolism.	Kiyici, Eroğlu, Kishali, & Burmaoglu (2017); Ridwan, Abdul Razak, Adenan, & Md Saad (2019)
Immunological	• It reduces the number of white blood cells circulating.	• The white blood cells look midway, in the face of endothelial damage.	Kuropteva, Baider, Nagler, Bogatyrenko, & Belaiac (2019)
Cardiovascular	• Synthesis of NO. • Arginine synthesis.	• Vasodilation. • It lowers arterial stiffness. • It lowers blood pressure. • Prevents myocardial damage and endothelial dysfunction. • Improve blood flow to the male member (helping to develop erections).	Hess, Baker, Gyenes, Tsuyuki, Newman, & Le Melledo (2017); Hotta, Shiota, Kataoka, Motonari, Maeda, Morita, & Kimura (2013); Morita, Hayashi, Ochiai, Maeda, Yamaguchi, Ina, & Kuzuya (2014); Ridwan, Abdul Razak, Adenan, & Md Saad (2019)
Liver	• In the metabolism of urea.	• In the blood, it removes ammonia excess.	Briseño-Bass, Chávez-Pérez, & López-Zendejas (2018)

9.4.1 CITRULLINE AND CENTRAL NERVOUS SYSTEM

The brain is made up of, among other things, astrocytes, which are cells responsible for the detoxification of ammonia by the amidation of glutamate to form glutamine. The presence of citrulline in the central nervous system (CNS) has been reported for more than 50 years, and it was thought that the origin of citrulline was only hepatic; however, it was discovered over the years that this amino acid can be synthesized at the brain level. In the central nervous system (CNS), specifically within the astrocytes, there is a key mechanism for the regulation of excitatory neurotransmission in a normal way. This is evidence that citrulline is present in the central nervous system (CNS). In addition, the presence of citrulline has been reported in neurons, microglia, and oligodendrocytes, and it is suggested that this amino acid is involved in the physiology and pathophysiology of the CNS (Pérez-Neri, & Diéguez-Campa, 2018).

9.4.2 CITRULLINE AND CARDIOVASCULAR DISEASE

Cardiovascular diseases are the result of heart and blood vessel disorders, due to tobacco use, lack of physical exercise, and poor diet (World Health Organization, 2020). Currently, this type of disease is increasing at an alarming rate. According to the global forecast for 2030, it is estimated that 23.6 million people will die from these diseases (Mirenayat, Moradi, Mohammadi, & Rouhani, 2018). The first risk factor for cardiovascular disease is hypertension, which can trigger myocardial infarction, vascular accidents, and even death. For the treatment of hypertension, the use of L-citrulline alone or in combination with L-arginine has been suggested to increase plasma nitric oxide. However, in other qualitative and quantitative studies where L-citrulline is supplemented, overall, no significant effect on aortic systolic and aortic diastolic blood pressure was found. In order to promote the effectiveness of citrulline and obtain its benefits, physical activity is important, and further research and clinical trials are needed to confirm these effects (Mirenayat, Moradi, Mohammadi, & Rouhani, 2018). For proper endothelial health, proper NO production is needed. When this production decreases, low-density lipoprotein (LDL) oxidation increases and plaques form in the arteries, which trigger the development of cardiovascular disease (Hess, Baker, Gyenes, Tsuyuki, Newman, & Le Melledo, 2017).

On the other hand, it has been observed that citrulline, added to nutritional supplements, can help the blood flow in the male reproductive system, benefiting the impulse of erections (Davies, 2015; Hotta et al., 2013). The control of erectile dysfunction and subjective improvement in sexual relations are some of the benefits that have been observed when administering this amino acid, proposing itself as a more natural alternative for those who suffer from mild or moderate dysfunction (Miyake et al., 2014; Morita et al., 2014).

9.4.3 CITRULLINE AND DIABETES MELLITUS

Diabetes is a disease that occurs when the pancreas is not producing enough insulin, or when the body cannot effectively use the insulin that it is producing. There are 3 types of diabetes: type 1, type 2, and gestational diabetes, where the main problem is hyperglycemia (increased blood glucose), which occurs when the disease is not controlled. This metabolic disease is one of the leading causes of death in most countries. In 2015, the International Diabetes Federation mentioned that 8.8% of adults between 20 and 79 years of age suffered from this disease, and they state that by 2040 these data will increase to 10.4%, which is alarming since diabetes brings with it both micro- and macrovascular conditions when patients are in chronic hyperglycemia (Briseño-Bass, Chávez-Pérez, & López-Zendejas, 2018). Hyperglycemia produces some compounds that cause inflammation, and chronic stimulating cytokines such as TNF-α are secreted, and overexpressed in fatty and muscular tissue, giving resistance to insulin. It has been mentioned that a decrease in NO affects both directly and indirectly the metabolism of skeletal muscle, and insulin resistance will develop. Some reports mention that citrulline can significantly decrease the cytokines responsible for inflammation, and also increase anti-inflammatory cytokines (Darabi et al., 2019), this is the main reason why citrulline supplementation is believed to bring benefits in diabetes (Azizi et al., 2020), by increasing insulin signaling, thus improving oxidative stress, glucose tolerance, and lipid metabolism to prevent mitochondrial deterioration (Azizi et al., 2020; Darabi et al., 2019).

A meta-analysis on diabetes mentions that citrulline and watermelon extract supplementation have an antioxidant and anti-inflammatory effect. In some analyzed articles, where different metabolic variables related to diabetes were evaluated, it was found that supplementation offered various benefits in patients with diabetes, including reduction of inflammation, improvement of glycemic status with hypolipidemic, and antioxidant effect (Azizi et al., 2020). On the other hand, diabetes has been associated with the risk of suffering cardiovascular diseases (CVD), such as hypercholesterolemia, hypertension, and even activating inflammation, and it has been mentioned that with very low levels of citrulline, production of NO by the endothelium will decrease in patients with diabetes mellitus (apparently physically healthy) (Hess et al., 2017). However, more studies are needed to establish the role of citrulline in patients with diabetes, as well as to determine optimal doses of citrulline for this condition.

9.4.4 CITRULLINE AND LIVER DISEASES

The disease known as non-alcoholic fatty liver (NAFLD) is characterized by an accumulation of 5–10% of lipids in the liver tissue, without chronic alcohol consumption, viral infection, or other specific liver disease conditions. It has been reported that 20% of patients with NASH may have progressive liver fibrosis, leading to liver cirrhosis and hepatocellular carcinoma (Darabi et al., 2019). This disease is associated with obesity, insulin resistance, hypertension, dyslipidemia, and metabolic syndrome. NASH begins with steatosis, where a liver lesion is observed, and a series of inflammatory reactions and lobular bulging begins, which, if not treated in a timely manner, can lead to cirrhosis (an irreversible disease) (Briseño-Bass, Chávez-Pérez, & López-Zendejas, 2018).

The rate of inflammation, liver elasticity, and changes in echogenicity are some of the effects sought to be tested using pancreatic enzymes in patients with NASH disease. Thus, in one study, 2 g/day was administered to newly diagnosed patients with NASH, without any previous treatment, and significant changes in activity (TNF-α) (a marker of chronic inflammation) were observed. In addition to a relevant decrease in hepatic steatosis. However, more evidence is lacking to determine the maximum doses of the amino acid that can be supplemented for patients (Darabi et al., 2019).

9.4.5 CITRULLINE AND EXERCISE PERFORMANCE

Recently, people's physical activity has increased, as it has been shown that regular exercise contributes to maintaining good physical health and preventing disease (World Health Organization, 2020). In addition, the sports culture aims to develop athletes with greater mental and physical endurance, which is why the demand for ergogenic supplements has increased. Secondly, these supplements are intended to improve emotional stress (Ridwan, Abdul Razak, Adenan, & Md Saad, 2019).

The development of nutritional supplements that help improve individuals' exercise performance, delay fatigue, or enable early recovery must be within the legal framework (Kiyici, Eroğlu, Kishali, & Burmaoglu, 2017). Since during physical exercise, there are hormones that accelerate fatigue, such as cortisol, lactate, ammonia, etc., supplements are intended to reduce the levels of these compounds (e.g., lactate) (Ridwan, Abdul Razak, Adenan, & Md Saad, 2019). Lactate is an indicator of physical exhaustion, since during exercise it increases the amounts of hydrogen ions, due to the accumulation of lactic acid. Citrulline activates muscle fatigue by decreasing ammonia, promoting NO production, and increasing blood oxygenation (Joshi & Fernie, 2017). It has also been reported that L-citrulline from watermelon is necessary to suppress the activation of phosphofructokinase (PFK) and the enzyme of the glycolytic system, inhibiting the re-synthesis of adenosine triphosphate (ATP) (Ridwan, Abdul Razak, Adenan, & Md Saad, 2019). Citrulline supplementation increases the concentration of L-arginine more than when pure L-arginine is administered. However, the excess of L-arginine generates a rapid production of nitric oxide (NO), which can cause gastrointestinal discomfort such as nausea and diarrhea (Kuropteva, Baider, Nagler, Bogatyrenko, & Belaiac, 2019).

Branched-chain amino acids (BCAAs) have been reported to help relieve post-training fatigue (Chen, Wu, Chen, Chou, & Chang, 2016), and this is another application of citrulline in sports practice, as it decreases muscle pain (Caruso, Rossi, Gahn, & Caruso, 2017). Arginine and NO levels increase thanks to citrulline; therefore, more blood flows through the muscle, which increases the transfer of oxygen and nutrients, providing more muscle energy, and ammonia is eliminated from the body (Kiyici, Eroğlu, Kishali, & Burmaoglu, 2017). In addition, glucose uptake, contractility, blood flow, and muscle repair are increased, thus benefiting sports activity (Glenn, Gray, Wethington, Stone, Stewart, & Moyen, 2017). On the other hand, it has been reported that women may have a greater response to citrulline malate (CM) supplementation because women produce more estrogen, which regulates NO production (Glenn, Gray, Wethington, Stone, Stewart, & Moyen, 2017).

Citrulline malate (CM) is the combination of citrulline with malic acid (malate). This mixture is part of certain chemical reactions useful for energy (ATP) production; therefore, it helps those people who, in addition to improving their muscle composition, seek to replenish strength (more energy) in their day-to-day activity (Gonzalez, Spitz, Ghigiarelli, Sell, & Mangine, 2017). CM supplementation has been proposed to improve sports performance since it reduces muscle fatigue. CM is an intermediate in the tricarboxylic acid (TCA) cycle, and when consumed, it improves muscle energy production and therefore physical performance. The administration of 6 g for 15 days increases up to 20% recovery of creatine phosphate post-training, controlling 40% of the sensation of fatigue or muscular pain, and increasing 34% production of adenosine triphosphate (ATP) (Kiyici, Eroğlu, Kishali, & Burmaoglu, 2017).

In watermelon pulp juice, there are bioactive metabolites in addition to citrulline, which have a synergistic interaction with other amino acids, carotenoids, micronutrients, and antioxidant enzymes, and can be used during the production of ATP (energy), which can be used during exercise, and after training can speed up the recovery process. This increases the production of NO, which in turn increases blood flow, causing vasodilation and increased blood perfusion, thus eliminating more toxic metabolites, such as lactate and ammonia (Ridwan, Abdul Razak, Adenan, & Md Saad, 2019). Oral consumption of citrulline and arginine mixture has been shown to increase the concentration of arginine in plasma more effectively and efficiently than if either of these compounds were supplied separately. Administering a citrulline and arginine mixture (1:1) at a dose of 2.4 g/day for one week improved the perception of "leg muscle pain" and "ease of pedaling" during an ergometry test (Suzuki et al., 2019).

9.5 SOURCES AND FOOD APPLICATIONS OF CITRULLINE

Citrulline is a free amino acid found in plants of the Cucurbitaceae family, among which watermelon stands out (Azizi et al., 2020; Gonzalez, Spitz, Ghigiarelli, Sell, & Mangine, 2017). It has also been identified, although in smaller quantities, in pumpkin, melon, cucumber, and other agricultural commodities such as almond, cocoa (dark chocolate), chickpea, garlic, onion, nuts, peanuts, mushrooms, in aquaculture products such as salmon and even in certain fermented foods and beverages, such as soybean, wine (Joshi & Fernie, 2017), meat, fish, and vegetables (Agarwal, Didelija, Yuan, Wang, & Marini, 2017).

Citrulline is considered a drug-nutrient for the human organism, since it may be able to transform itself into other compounds (Agarwal, Didelija, Yuan, Wang, & Marini, 2017). In Table 9.3, different supplements available in the market are shown.

TABLE 9.3 Different Citrulline Products

Product Name	Attributes	Price, Company Name (City, State, and Country Where it is Located)
Amazing Formulas L-Citrulina 750 mg 90 tablets	• Excellent cardiovascular and circulatory support. • Helps reduce fatigue and promotes endurance. • Promotes sexual health. • Increases physical endurance.	MXN 1,139.00 Amazing nutrition Jersy City, Nueva York, U.S.A.
Nutricost L-Citrulline 600 mg, 195 scoops.	• Increases physical endurance.	US $27.99 Simaro Nutricost, Vineyard, Utah. U.S.A.
L-Citulline 120 mg Tablets	• Support the extra force. • Aids in the function of the immune system. • Supports protein metabolism. • GMP certified, vegan/vegetarian.	US $18.35 Now sports U.S.A.

The recommended doses for each objective have not been established yet. However, studies are continuing. According to previously applied clinical trials, tentative amounts of citrulline ranging from 1 to 10 grams per day have been proposed (Agarwal, Didelija, Yuan, Wang, & Marini, 2017). The Mexican Supplements Guide proposed recommended doses for certain objectives: for cardiovascular health 6 to 10 g/day, for sexual health 1.5 g/day, for increasing muscle mass, and sports exercise performance 6 to 10 g/day. Information about the time of day, when most absorption of the citrulline supplement is performed, is still lacking. For athletes, it is recommended that the intake is made 1 or 2 hours before physical exercise, and if the dose exceeds 1–1.5 grams, it should be divided until the recommended daily grams are completed.

9.5.1 WATERMELON AS SOURCE OF CITRULLINE

Watermelon (*Citrullus lanatus* Thunb. Matsum and Nakai) is the fruit with the highest quantity of citrulline and is among the most important vegetable crops worldwide, with its annual production reaching approximately 118 million tons. This fruit is consumed for its great taste and nutritional content. It is primarily composed of sugar, fiber, vitamin C, and minerals such as calcium, magnesium, potassium, sodium, zinc, etc. (Agarwal, Didelija, Yuan, Wang, & Marini,

2017), lycopene, and amino acids (citrulline and arginine) (Romdhane et al., 2017; Grumet, Katzir, & Garcia-Mas, 2017). It is estimated that watermelon contains 0.7 and 3.6 mg/g fresh weight (Gonzalez & Trexler, 2020) arginine and citrulline, respectively, and 40% more lycopene than raw tomato (Azizi et al., 2020), other watermelon compounds are lutein and sucrose. Watermelon fruit presents great diversity in terms of shape, size, color, texture, flavor, and nutritional composition (Grumet, Katzir, & Garcia-Mas, 2017). However, its cropping can be affected by many environmental conditions such as temperature, heavy metals, soil salinity, drought, and pests or soil-borne pathogens (Garcia-Lozano et al., 2019), coupled with the fact that it is composed of 95% water, which accelerates its deterioration (Grumet, Katzir, & Garcia-Mas, 2017; Bellary, Indiramma, Prakash, Baskaran, & Rastogi, 2016). In addition, at least 18 polyphenolic compounds have been reported in the watermelon's shell (Gonzalez, Spitz, Ghigiarelli, Sell, & Mangine, 2017). These compounds have anti-mutagenic properties that are directly related to cancer (Zamora-Ros et al., 2018), or they promote cardiovascular health during the treatment of diabetes, etc. (Durán, Villegas, & Nieves, 2017).

Watermelon contains large amounts of L-citrulline (Kuropteva, Baider, Nagler, Bogatyrenko, & Belaiac, 2019). In watermelon, L-citrulline is found in seeds, leaves, xylem and phloem sap, fruit pulp, and shell, although most of this amino acid is observed in the shell on a dry weight basis (Joshi & Fernie, 2017). Several authors mentioned that the amount of citrulline in watermelon increases progressively as the fruit approaches its maturity. The citrulline content then decreases after ripening, and this does not depend on the watermelon cultivar used (Joshi & Fernie, 2017), while the red-fleshed watermelon has 16 mg/100 g. It has also been reported that citrulline levels increase in the plant under abiotic stress conditions (Joshi, Joshi, Silwal, Noonan, Rodriguez, & Penalosa, 2019). Watermelon chemicals have been reported to improve exercise performance when orally supplied (Ridwan, Abdul Razak, Adenan, & Md Saad, 2019).

9.5.2 FOOD APPLICATIONS OF CITRULLINE

Watermelon citrulline has been reported as an ergogenic compound (Kuropteva, Baider, Nagler, Bogatyrenko, & Belaiac, 2019) since it helps to improve performance during exercise, and it is in greater amounts in the watermelon shell. At an industrial level, citrulline is produced using microbial strains. However, this amino acid is produced naturally and in large quantities in watermelon. From one ton of watermelon pulp and rind, 19 kg of citrulline was extracted, which makes this process economically viable (Joshi & Fernie, 2017). In addition, watermelon rind is characterized by its high dietary fiber content, a compound used for treatment and prevention of chronic diseases (Naguib & Tantawy, 2019), and bioactive compounds, which is why it has been proposed for incorporation in products mainly of bakery, as a substitute for flours (Durán, Villegas, & Nieves, 2017), including in these foods also pectins (Petkowicz, Vriesmann, & Williams, 2017). In addition, flour from watermelon shells has a low energy value and could help to reduce fatigue and tiredness after physical effort and improve the cardiovascular system (Durán, Villegas, & Nieves, 2017).

It has been reported that the watermelon shell contains large amounts of total phenolic compounds, which are powerful antioxidants. This is why its incorporation into foods could be a very promising application, generating foods rich in natural dietary fiber, with a better final product, in terms of its organoleptic characteristics (color, flavor, texture, aroma), with higher nutritional quality and low energy inputs (Naknaen, Itthisoponkul, Sondee, & Angsombat, 2016). In some countries, the lack of organization in the dates of sowing watermelon leads to harvest concentration in a short time of the year, which causes that much fruit is not harvested because it is not affordable, and watermelon is lost (Gobierno de México, 2017). Also, in most cases, only the fruit pulp is utilized, and the shell or bark is only used in some cases as livestock feed and mostly only thrown into the environment as waste, causing serious environmental problems of pests and pollution. Therefore, the use of watermelon shells during food development is a great alternative.

Some companies aim to develop new foods which include improvements in their nutritional quality. According to this, in most cases, plant extracts are incorporated into foods, since these compounds have powerful antioxidant activity (Bassani et al., 2016) and anti-cancer (Gomez-Cadena et al., 2015) effects, so their incorporation provides added value with health benefits. However, the recommended doses of its bioactive compounds in food to provide the desired functionality are still unknown. Therefore, it is important to establish the appropriate food formulations to meet the desired functionality, which leads to the design of foods with different proportions of their active ingredients.

9.5.2.1 HEALTHY EATING

Consumers' food choices have changed over the years. Nowadays, the choice of food is based on the dietary preferences of each individual, as they look for low-calorie foods that may provide an extra health benefit, so the food industry has created a trend towards healthy food consumption (Altamimi, 2019).

The Codex Alimentarius defines food as a substance prepared for human consumption from the time it is manufactured, prepared, or treated as food and mentions that a healthy diet protects against malnutrition and a range of chronic and degenerative diseases (World Health Organization, 2018). Over the years, the lifestyle has accelerated, and habits have been modified, including food, increasing consumption of industrialized products, and high-calorie fast food, but in addition, without a great nutritional contribution, because its composition includes saturated fats, free sugars, sodium, etc. (World Health Organization, 2018). The consumption of this type of food is greater in some regions where "snack" is consumed, since it is food for daily consumption due to its practicality and can be consumed at any time of the day (Chacón Orduz, Muñoz Rincón, & Quiñónez Mosquera, 2017). In addition, if vegetables and fruits are included daily, the risk of chronic diseases will be reduced (Naguib & Tantawy, 2019).

The food industry has developed functional foods, which are classified as substances that have a basic nutritional value, but also provide an extra health benefit. The European Food Safety Authority requires that if any food is declared functional, it must have an accepted scientific basis and justification. Therefore, the evaluation of food for the identification of the bioactive compound is important. However, many foods claimed to be functional do not have studies that validate their functionality and safety. In addition, there is no institution that regulates and determines the appropriate dosage, efficacy, and bioavailability of the active ingredient of each functional food, deliberately allowing the sale of foods with these characteristics. These foods are known as dietary supplements, and excessive consumption of some of them is associated with hepatocellular lesions and high levels of transaminases, about certain untested products (Ettel, Gonzalez, Gera, Eze, Sigal, & Park, 2017).

9.5.2.2 INCORPORATION OF FUNCTIONAL FOODS AND SUPPLEMENTS

The selection of foods by consumers has changed over the years. They generally choose them based on their dietary preferences, the number of calories, and the benefits that they provide for health. Therefore, the food industry maintains a trend in the development of healthy foods (Altamimi, 2019).

Accordingly, citrulline isolated from agro-industrial residues of plants has the potential to formulate functional foods or food supplements. Consequently, countries producing *C. lanatus* such as Mexico could take advantage of plant material for its preparation, since *C. lanatus* is a promising species because it is the most studied to recover this precious amino acid.

The citrulline content that developed foods or supplements should contain is still unknown. For that reason, food formulations must be established to fulfill the desired function. Therefore, it is proposed to prepare the product with the functional doses of synthetic citrulline established in previous studies. According to clinical trials to evaluate the biological activities of citrulline, the amino acid has a dose effect of 1 to 10 grams per day (Agarwal, Didelija, Yuan, Wang, & Marini, 2017). In Table 9.4, the recommended daily doses of citrulline to improve some health conditions are shown.

TABLE 9.4 Recommended Daily Doses to Improve Some Health Conditions

Property	Doses (g/day)	References
Cardiovascular health	6 to 10	Suzuki et al. (2019); Romdhane et al. (2017); Bouillanne et al. (2018)
Improve erectile function. Sexual health.	1.5	Nagulb & Tantawy (2019); Hotta et al. (2013); Chen, Wu, Chen, Chou, & Chang (2016)
Increase muscle mass and improve athletic performance.	1–8	Joshi, Joshi, Silwal, Noonan, Rodriguez, & Penalosa (2019); Davies (2015); Hotta et al. (2013); Ogurtsova et al. (2017); Darabi et al. (2019); Bellary, Indiramma, Prakash, Baskaran, & Rastogi (2016)
Hypoglycemic treatment in diabetes mellitus II	1	Kuropteva, Baider, Nagler, Bogatyrenko, & Belaiac (2019); Figueroa, Wong, Jaime, & Gonzales (2017)

Consumption of the amino acid is considered safe but should not be consumed for more than six months without medical supervision (Mohan, Wu, Shin, & Fung, 2012).

Information on the incorporation of citrulline in food is scarce. Although some researchers propose its incorporation as a substitute for flour in bakery products (Durán, Villegas, & Nieves, 2017), others incorporate *C. lanatus* husk flour. Since it has a low energy value, and due to its citrulline content, it can reduce fatigue after physical effort and improve the cardiovascular system (Durán, Villegas, & Nieves, 2017). Alternatively, the development of food supplements should include storage tests that have conditions such as a cool and dry place free of unpleasant odors and sunlight to keep it in excellent condition. Besides, it is necessary to evaluate its effectiveness for two years after the date of manufacture.

According to the Codex Alimentarius, a healthy diet protects from both malnutrition and a variety of chronic and degenerative diseases WHO (World Health Organization, 2018). Therefore, the development of a "Snack" containing citrulline from *C. lanatus* could be promising, given that this type of food has a high daily consumption due to its practicality (Chacón Orduz, Muñoz Rincón, & Quiñónez Mosquera, 2017). Since intake in the daily diet of plant derivatives reduces the risk of chronic diseases (Gobierno de México, 2017; Ettel, Gonzalez, Gera, Eze, Sigal, & Park, 2017). Also, the European Food Safety Authority declares it a functional food, a product developed with a scientific basis. As a result, evaluating food composition, its biological activities, and the reduction of disease risk is important. However, in Mexico, many food products affirm their functionality without having studies validate their function and safety. Consequently, their study represents an area of opportunity to investigate since currently, no institution regulates the adequate dose, efficacy, and bioavailability of the active ingredient of each functional food.

9.6 CONCLUSION

Citrulline is an amino acid, to which different health benefits have been attributed and has been proposed to treat diseases such as diabetes mellitus, liver disease, heart disease, CNS, and even cancer. Among the positive effects of the administration of this compound are decreased insulin resistance, body weight, arterial rigidity, and blood pressure, thus increasing energy synthesis, muscle mass, vasodilation, and blood flow, and eliminating ammonia excess, being known as a natural Viagra. Currently, citrulline is added to dietary supplements, and most of the one used for this purpose is synthesized chemically or from microbiological origin. Citrulline is used as a sport ergogenic, because of its benefits in sports performance. However, watermelon is rich in citrulline, which offers the advantage that it is of natural origin and could be used for the development of functional foods, which is undoubtedly an excellent proposal since it would be taking full advantage of natural resources that were considered agro-industrial waste (shells) and minimizing environmental pollution.

KEYWORDS

- **agro-industrial waste**
- **Citrullus lanatus**
- **codex Alimentarius**
- **dietary supplements**
- **food supplement**
- **functional food**
- **health benefits**

REFERENCES

Agarwal, U., Didelija, I. C., Yuan, Y., Wang, X., & Marini, J. C. (2017). Supplemental citrulline is more efficient than arginine in increasing systemic arginine availability in mice. *The Journal of Nutrition, 147*, 596–602. https://doi.org/10.3945/jn.116.239335.

Ahmadi, H., Babalar, M., Sarcheshmeh, M. A. A., Morshedloo, M. R., & Shokrpour, M. (2020). Effects of exogenous application of citrulline on prolonged water stress damages in hyssop (*Hyssopus officinalis* L.): Antioxidant activity, biochemical indices, and essential oils profile. *Food Chemistry, 127433*. https://doi.org/10.1016/j.foodchem.2019.127433.

Altamimi, J. Z. (2019). Awareness of the consumption of dietary supplements among students in a university in Saudi Arabia. *Journal of Nutrition and Metabolism, 2019*, 4641768. https://doi.org/10.1155/2019/4641768.

Azizi, S., Mahdavi, R., Vaghef-Mehrabany, E., Maleki, V., Karamzad, N., & Ebrahimi-Mameghani, N. (2020). Potential roles of citrulline and watermelon extract on metabolic and inflammatory variables in diabetes mellitus, current evidence, and future directions: A systematic review. *Clinical and Experimental Pharmacology and Physiology, 47*, 187–198. https://doi.org/10.1111/1440-1681.13264.

Bassani, B., Rossi, T., De Stefano, D., Pizzichini, D., Corradino, P., Macrì, N., Noonan, D. M., Albini, A., & Bruno, A. (2016). Potential chemopreventive activities of a polyphenol-rich purified extract from olive mill wastewater on colon cancer cells. *Journal of Functional Foods, 27*, 236–248. https://doi.org/10.1016/j.jff.2016.09.002.

Bellary, A. N., Indiramma, A. R., Prakash, M., Baskaran, R., & Rastogi, N. K. (2016). Anthocyanin-infused watermelon rind and its stability during storage. *Innovative Food Science & Emerging Technologies, 33*, 554–562. https://doi.org/10.1016/j.ifset.2015.11.024.

Bhol, S., Lanka, D., & Bosco, S. J. D. (2016). Quality characteristics and antioxidant properties of breads incorporated with pomegranate whole fruit bagasse. *Journal of Food Science and Technology, 53*, 1717–1721. https://doi.org/10.1007/s13197-016-2200-1.

Bouillanne, O., Melchior, J. C., Faure, C., Paul, M., Canouï-Poitrine, F., Boirie, Y., Chevenne, D., Forasassi, C., Guery, E., Herbaud, S., Le Corvoisier, P., Neveux, N., Nivet-Antoine, V., Astier, A., Raynaud-Simon, A., Walrand, S., Cynober, L., & Aussel, C. (2018). Impact of 3-week citrulline supplementation on postprandial protein metabolism in malnourished older patients: The Ciproage randomized controlled trial. *Clinical Nutrition, 38*, 564–574. https://doi.org/10.1016/j.clnu.2018.03.005.

Briseño-Bass, P., Chávez-Pérez, R., & López-Zendejas, M. (2018). Prevalence and relation of hepatic steatosis with lipid and hepatic profile in check-up patients. *Journal of Gastroenterology of Mexico, 84*, 290–295. https://doi.org/10.1016/j.rgmxen.2018.07.007.

Caruso, A., Rossi, M., Gahn, C., & Caruso, F. (2017). A structural and computational study of citrulline in biochemical reactions. *Structural Chemistry, 28*, 1581–1589. https://doi.org/10.1007/s11224-017-0988-1.

Chacón Orduz, G., Muñoz Rincón, A., & Quiñónez Mosquera, G. A. (2017). Description of the healthy snack market in Villavicencio, Meta. *Libre Empresa, 14*, 33–45. https://doi.org/10.18041/1657-2815/libre.2017v14n2.269.

Chen, I. F., Wu, H. J., Chen, C. Y., Chou, K. M., & Chang, C. K. (2016). Branched-chain amino acids, arginine, citrulline alleviate central fatigue after 3 simulated matches in taekwondo athletes: A randomized controlled trial. *Journal of the International Society of Sports Nutrition, 13*, 28. https://doi.org/10.1186/s12970-016-0140-4.

Chen, I.-F., Wu, H.-J., Chen, C. Y., Chou, K. M., & Chang, C. K. (2016). Branched-chain amino acids, arginine, citrulline alleviate central fatigue after 3 simulated matches in taekwondo athletes: A randomized controlled trial. *Journal of the International Society of Sports Nutrition, 13*, 28. https://doi.org/10.1186/s12970-016-0140-4.

Darabi, Z., Darand, M., Yari, Z., Hedayati, M., Faghihi, A., & Hekmatdoost, A. (2019). Inflammatory markers response to citrulline supplementation in patients with non-alcoholic fatty liver disease: A randomized, double-blind, placebo-controlled clinical trial. *BMC Research Notes, 12*, 89. https://doi.org/10.1186/s13104-019-4131-4.

Davies, K. P. (2015). Development and therapeutic applications of nitric oxide-releasing materials to treat erectile dysfunction. *Future Science OA, 1*, FSO53. https://doi.org/10.4155/fso.15.53.

Durán, R., Villegas, M. E., & Nieves, I. (2017). Caracterización y extracción de citrulina de la corteza de la sandía (*Citrullus lanatus* "thunb") consumida en Valledupar. *Temas Agrarios, 22*, 60–67. https://doi.org/10.21897/rta.v22i2.1066.

Ettel, M., Gonzalez, G. A., Gera, S., Eze, O., Sigal, S., & Park, J. S. (2017). Frequency and pathological characteristics of drug-induced liver injury in a tertiary medical center. *Human Pathology, 68*, 92–98. https://doi.org/10.1016/j.humpath.2017.08.016.

Figueroa, A., Wong, A., Jaime, S. J., & Gonzales, J. U. (2017). Influence of L-citrulline and watermelon supplementation on vascular function and exercise performance. *Current Opinion in Clinical Nutrition and Metabolic Care, 20*, 92–98. https://doi.org/10.1097/MCO.0000000000000356.

Garcia-Lozano, M., Dutta, S. K., Natarajan, P., Tomason, Y. R., Lopez, C., Katam, R., Levi, A., Nimmakayala, P., & Reddy, U. K. (2019). Transcriptome changes in reciprocal grafts involving watermelon and bottle gourd reveal molecular mechanisms involved in the increase of the fruit size, rind toughness, and soluble solids. *Plant Molecular Biology, 102*, 213–223. https://doi.org/10.1007/s11103-019-00909-1.

Glenn, J. M., Gray, M., Wethington, L. N., Stone, M. S., Stewart, R. W., & Moyen, N. E. (2017). Acute citrulline malate supplementation improves upper- and lower-body submaximal weightlifting exercise performance in resistance-trained females. *European Journal of Nutrition, 56*, 775–784. https://doi.org/10.1007/s00394-015-1116-2.

Gobierno de México. (2017). Recovered on 2019 de October 12, from: https://www.gob.mx/senasica/galerias/produccion-de-sandia-en-mexico (accessed on 30 May 2024).

Gomez-Cadena, A., Martinez-Usatorre, A., Urueña, C., Prieto, K., Barreto, A., Romero, P., & Fiorentino, S. (2015). Immune system activation through immunogenic cell death and tumor recruitment of dendritic cells is required for the anti-tumor activity of a plant-derived polyphenol-rich fraction. *Journal for ImmunoTherapy of Cancer, 3*, 301–301. https://doi.org/10.1186/s40425-015-0075-y.

Gonzalez, A. M., & Trexler, E. T. (2020). Effects of citrulline supplementation on exercise performance in humans. *Journal of Strength and Conditioning Research, 34*, 1480–1495. https://doi.org/10.1519/JSC.0000000000002283.

Gonzalez, A. M., Spitz, R. W., Ghigiarelli, J. J., Sell, K. M., & Mangine, G. T. (2017). Acute effect of citrulline malate supplementation on upper-body resistance exercise performance in recreationally resistance-trained men. *Journal of Strength and Conditioning Research, 32*, 3088–3094. https://doi.org/10.1519/JSC.0000000000002167.

Grumet, R., Katzir, N., & Garcia-Mas, J. (2017). Genetics and genomics of Cucurbitaceae, plant genetics, and genomics: crops and models. *Springer International Publishing, 20*, 199. https://doi.org/10.1007/978-3-319-90365-7.

Hess, S., Baker, G., Gyenes, G., Tsuyuki, R., Newman, S., & Le Melledo, J. M. (2017). Decreased serum L-arginine and L-citrulline levels in major depression. *Psychopharmacology, 234*, 3241–3247. https://doi.org/10.1007/s00213-017-4695-5.

Hotta, Y., Shiota, A., Kataoka, T., Motonari, M., Maeda, Y., Morita, M., & Kimura, K. (2013). Oral L-citrulline supplementation improves erectile function and penile structure in castrated rats. *International Journal of Urology, 21*, 608–612. https://doi.org/10.1111/iju.12369.

Joshi, V., & Fernie, A. R. (2017). Citrulline metabolism in plants. *Amino Acids, 49*, 1543–1559. https://doi.org/10.1007/s00726-017-2433-9.

Joshi, V., Joshi, M., Silwal, D., Noonan, K., Rodriguez, S., & Penalosa, A. (2019). Systematized biosynthesis and catabolism regulate citrulline accumulation in watermelon. *Phytochemistry, 162*, 129–140. https://doi.org/10.1016/j.phytochem.2019.02.011.

Kiyici, F., Eroğlu, H., Kishali, N. F., & Burmaoglu, G. (2017). The effect of citrulline/malate on blood lactate levels in intensive exercise. *Biochemical Genetics, 55*, 387–394. https://doi.org/10.1007/s10528-017-9806-7.

Kuropteva, Z. V., Baider, L. M., Nagler, L. G., Bogatyrenko, T. N., & Belaiac, O. L. (2019). L-Arginine and nitric oxide synthesis in the cells with inducible NO synthase. *Russian Chemical Bulletin, 68*, 174–180. https://doi.org/10.1007/s11172-019-2375-0.

McNeal, C. J., Meininger, C. J., Reddy, D., Wilborn, C. D., & Wu, G. (2016). Safety and effectiveness of arginine in adults. *The Journal of Nutrition, 146*, 2587S–2593S. https://doi.org/10.3945/jn.115.228863.

Mirenayat, M. S., Moradi, S., Mohammadi, H., & Rouhani, M. H. (2018). Effect of L-citrulline supplementation on blood pressure: A systematic review and meta-analysis of clinical trials. *Current Hypertension Reports, 20*, 98. https://doi.org/10.1007/s11906-018-0898-8.

Miyake, M., Kirisako, T., Kokubo, T., Miura, Y., Morishita, K., Okamura, H., & Tsuda, A. (2014). Randomized controlled trial of the effects of L-ornithine on stress markers and sleep quality in healthy workers. *Nutrition Journal, 13*, 1–8. https://doi.org/10.1186/1475-2891-13-53.

Mohan, S., Wu, C. C., Shin, S., & Fung, H. L. (2012). Continuous exposure to L-arginine induces oxidative stress and physiological tolerance in cultured human endothelial cells. *Amino Acids, 44*, 1179–1188. https://doi.org/10.1007/s00726-012-1424-0.

Mondal, S., & Thompson, P. R. (2021). Chemical biology of protein citrullination by the protein A arginine deiminases. *Current Opinion in Chemical Biology, 63*, 19–27. https://doi.org/10.1016/j.cbpa.2021.01.007.

Morita, M., Hayashi, T., Ochiai, M., Maeda, M., Yamaguchi, T., Ina, K., & Kuzuya, M. (2014). Oral supplementation with a combination of L-citrulline and L-arginine rapidly increases plasma L-arginine concentration and enhances NO bioavailability. *Biochemical and Biophysical Research Communications, 454*, 53–57. https://doi.org/10.1016/j.bbrc.2014.10.027.

Naguib, D. M., & Tantawy, A. A. (2019). Anticancer effect of some fruits peels aqueous extracts. *Oriental Pharmacy and Experimental Medicine, 19*, 415–420. https://doi.org/10.1007/s13596-019-00372-6.

Nagulb, D. M., & Tantawy, A. A. (2019). Anticancer effect of some fruits peels aqueous extracts. *Oriental Pharmacy and Experimental Medicine, 19*, 415–420. https://doi.org/10.1007/s13596-019-00391-5.

Naknaen, P., Itthisoponkul, T., Sondee, A., & Angsombat, N. (2016). Utilization of watermelon rind waste as a potential source of dietary fiber to improve health-promoting properties and reduce glycemic index for cookie making. *Food Science and Biotechnology, 25*, 415–424. https://doi.org/10.1007/s10068-016-0061-5.

Ogurtsova, K., da Rocha Fernandes, J. D., Huang, Y., Linnenkamp, U., Guariguata, L., Cho, N. H., Cavan, D., Shaw, J. E., & Makaroff, L. E. (2017). IDF Diabetes Atlas: Global estimates for the prevalence of diabetes for 2015 and 2040. *Diabetes Research and Clinical Practice, 128*, 40–50. https://doi.org/10.1016/j.diabres.2017.03.024.

Pérez-Neri, I., & Diéguez-Campa, C. E. (2018). Relevancia de la citrulina para el sistema nervioso central. *Archivos de Neurociencia, 23*, 50–54. https://doi.org/10.31157/an.v23i2.234.

Petkowicz, C., Vriesmann, L., & Williams, P. (2017). Pectins from food waste: Extraction, characterization, and properties of watermelon rind pectin. *Food Hydrocolloids, 65*, 57–67. https://doi.org/10.1016/j.foodhyd.2016.10.016.

Ridwan, R., Abdul Razak, H. R., Adenan, M. I., & Md Saad, W. M. (2019). Supplementation of 100% flesh watermelon [*Citrullus lanatus* (Thunb.) Matsum. and Nakai] juice improves swimming performance in rats. *Preventive Nutrition and Food Science, 24*, 41–48. https://doi.org/10.3746/pnf.2019.24.1.41.

Romdhane, M. B., Haddar, A., Ghazala, I., Ben Jeddou, K., Helbert, C. B., & Ellouz-Chaabouni, S. (2017). Optimization of polysaccharides extraction from watermelon rinds: Structure, functional and biological activities. *Food Chemistry, 216*, 355–364. https://doi.org/10.1016/j.foodchem.2016.08.056.

Saulskaya, N. B., Fofonova, N. V., Sudorgina, P. V., & Komarova, A. S. (2011). Danger-associated sound signals activate the nitrergic system of the medial part of the nucleus accumbens. *Neuroscience and Behavioral Physiology, 41*, 452–458. https://doi.org/10.1007/s11055-011-9416-4.

Suzuki, I., Sakuraba, K., Horiike, T., Kishi, T., Yabe, J., Suzuki, T., Morita, M., Nishimura, A., & Suzuki, Y. (2019). A combination of oral L-citrulline and L-arginine improved 10-min full-power cycling test performance in male collegiate soccer players: A randomized crossover trial. *European Journal of Applied Physiology, 119*, 1075–1084. https://doi.org/10.1007/s00421-019-04101-y.

Umang, A., Inka, C. D., Yang, Y., Xiaoying, W., & Marini, J. C. (2017). Supplemental citrulline is more efficient than arginine in increasing systemic arginine availability in mice. *The Journal of Nutrition, 147*, 596–602. https://doi.org/10.3945/jn.116.242990.

Valenzuela, P. L., Morales, J. S., Emanuele, E., Pareja-Galeano, H., & Lucia, A. (2019). Supplements with purported effects on muscle mass and strength. *European Journal of Nutrition, 58*, 2983–3008. https://doi.org/10.1007/s00394-018-1875-5.

World Health Organization. (2018). Healthy eating. Retrieved from: https://www.who.int/es/news-room/fact-sheets/detail/healthy-diet (accessed on 30 May 2024).

World Health Organization. (2020). Cardiovascular diseases. Retrieved from: https://www.who.int/cardiovascular_diseases/es/ (accessed on 30 May 2024).

World Health Organization. (2020b). Global strategy on diet, physical activity, and health. Retrieved from: https://www.who.int/dietphysicalactivity/pa/es/ (accessed on 30 May 2024).

Zamora-Ros, R., Cayssials, V., Jenab, M., Rothwell, J. A., Fedirko, V., Aleksandrova, K., Tjønneland, A., Kyrø, C., Overvad, K., Boutron-Ruault, M. C., Carbonnel, F., Mahamat-Saleh, Y., Kaaks, R., Kühn, T., Boeing, H., Trichopoulou, A., Valanou, E., Vasilopoulou, E., Masala, G., & Scalbert, A. (2018). Dietary intake of total polyphenol and polyphenol classes and the risk of colorectal cancer in the European Prospective Investigation into Cancer and Nutrition (EPIC) cohort. *European Journal of Epidemiology, 33*, 1063–1075. https://doi.org/10.1007/s10654-018-0428-6.

CHAPTER 10

Perspective to Develop Hypoglycemic Nutraceuticals From Polyphenolic Extracts of Apiaceae Using Sustainable Techniques

ANDREA ELIZABETH CORTES MORÁN,[1] LUIS ENRIQUE COBOS PUC,[1] ANNA ILINÁ,[1] ELDA PATRICIA SEGURA CENICEROS,[1] MARÍA DEL CARMEN RODRÍGUEZ SALAZAR,[1] JUAN ALBERTO ASCACIO VALDÉS,[1] LAURA MARÍA SOLÍS SALAS,[2] and SONIA YESENIA SILVA BELMARES[1]

[1]*Department of Food Research, Faculty of Chemistry, Autonomous University of Coahuila, Blvd. Venustiano Carranza S/N, Saltillo, Coahuila, Mexico*

[2]*Department of Basic Sciences, Antonio Narro Agrarian University, Calz. Antonio Narro, Saltillo, Coahuila, Mexico*

ABSTRACT

Diabetes mellitus has been a public health problem for years and continues to increase rapidly. Many factors influence the development of the disease, such as genetic information, diet, physical activity, and sociocultural aspects. Therefore, it is necessary to implement sustainable management and prevention systems based on products of natural origin to avoid collateral damage from the prolonged use of antidiabetic drugs and develop them under environmentally friendly processes. Plants have been used since ancient times for their properties against diseases such as diabetes, mainly due to their chemical composition. Plants cultured at an industrial level are a sustainable solution to develop nutraceuticals compared to wild plants. Apiaceae plants

Sustainable Agricultural and Food Systems Engineering: New Research, Practices, and Developments.
Jose Juan Buenrostro-Figueroa, Julio César Tafolla-Arellano, Leonardo Sepúlveda Torre, Porteen Kannan, A. K. Haghi (Eds.)

are abundantly produced and consumed for their nutritional properties. However, it is known that their extracts have hypoglycemic activity. Consequently, they represent a source to develop hypoglycemic nutraceuticals through sustainable, less expensive, and effective technologies. This review describes the scientific findings on the properties of Apiaceae plants and the mechanisms of hypoglycemic action.

10.1 INTRODUCTION

Diabetes mellitus (DM) is a health problem worldwide (Arroba & Aguilar-Diosdado, 2022). By 2035, the diabetic population will be approximately more than 400 million people (Bi, Lim, & Henry, 2017). Elevated plasma glucose levels characterize this chronic metabolic disorder. Mainly, type 1, type 2 diabetes, and gestational diabetes are known. In type 1 diabetes, autoimmune damage occurs in pancreatic β cells, which decreases insulin production. Type 2 diabetes has several causes such as the internal resistance of cells to insulin (Lovic et al., 2019), inefficient production of insulin from the pancreas (Laakso, 2019), and faulty regulation of glucagon, which is an insulin antagonist, causing elevated blood glucose levels (Haedersdal, Lun, Knop, & Vilsboll, 2018).

As a result of hyperglycemia, the body's homeostasis is affected since there is an imbalance in the metabolism of carbohydrates, fats, and proteins (Bi, Lim, & Henry, 2017). However, type 2 diabetes is the most prevalent among the population. It is usually associated with secondary factors that can trigger the disease, such as high blood pressure, the elevation of parameters such as cholesterol and triglycerides, body weight, and abnormal uric acid values (Ceriello & Prattichizzo, 2021), as shown in Figure 10.1. The change phase from normoglycemia to hyperglycemia is known as IFG (impaired fasting glucose) or IGT (impaired glucose tolerance), which are precedents for diabetes (Lovic et al., 2019). Generally, the pharmacological treatments prescribed for these cases are the drugs such as biguanides, sulfonylureas, thiazolidinediones, and α-glucosidase inhibitors. Unfortunately, these drugs are administered for a prolonged time causing adverse effects (Navin, Srivastava, Prakash, Kumari, & Kumar, 2021), including diarrhea, fatigue, lactic acidosis, hepatotoxicity (de Paulo Farias et al., 2021), and reduction of their effectiveness (Bi, Lim, & Henry, 2017). Furthermore, other limitations are that they are sometimes unavailable and are expensive in most cases (Navin, Srivastava, Prakash, Kumari, & Kumar, 2021).

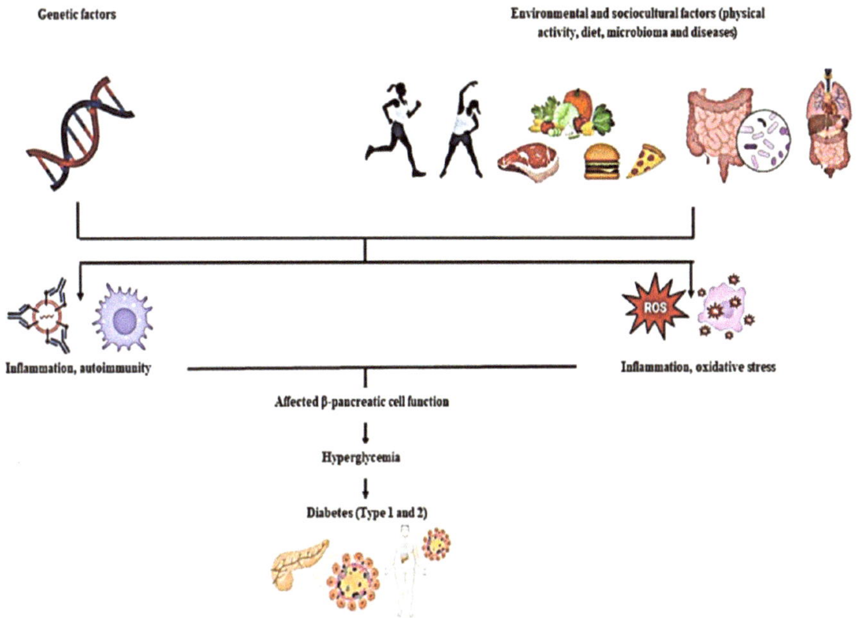

FIGURE 10.1 Factors involved in the development of hyperglycemia.

In addition to the damage to health, pharmacological treatments have a great environmental impact due to the processes generated by the pharmaceutical industries since they produce a wide range of compounds (Panizza, 2018) that contaminate effluents, harming aquatic life and humans due to their toxicity, genotoxicity, and mutagenicity (Rana, Singh, Kandar, Singh, Dobhal, & Gupta, 2017). Nevertheless, the main route of contamination is the excretion of active compounds through urine and feces, added to the incorrect disposal of waste (Rogowska & Zimmermann, 2022). For this reason, the pharmaceutical industry constantly seeks efficient and sustainable solutions to reduce or eliminate environmental impact (Li et al., 2019). Therefore, the pharmaceutical industry tries to carry out procedures based on "green synthesis." However, it is necessary to implement viable actions based on the use of methodologies with less environmental impact. Consequently, it increasingly takes into account the use of plants with medicinal properties that humanity has used since the early stages of the development of their civilizations. Since it is estimated that around 80% of the populations of developing countries use plants as primary care for different diseases. Furthermore, about 25% of prescribed drug treatments are synthesized from plants (Mishra et al., 2021).

The main interest of the pharmaceutical industry is due to the active compounds that plants contain to treat different diseases. Most of these compounds are synthesized by the secondary metabolism of plants. This metabolism does not involve vital functions such as growth and development. These metabolites are found in products derived from medicinal plants and are classified into three main groups: alkaloids, terpenes, and phenolic compounds (Grigoriadou, Krigas, Lazar, & Maloupa, 2019). Among the advantages offered by using products derived from medicinal plants are accessibility, low cost, efficacy, and lower cost compared to synthetic drugs (Chandran, Meena, Barupai, & Sharma, 2020). However, there are doubts about their use, efficacy, insecurity, and harm. Currently, plant products are marketed with no scientific basis to prove their efficacy (Lapraz, 2019). Therefore, it is essential to study their composition and biological properties to guarantee their safety. On the other hand, the ecological impact that could cause the exploitation of plants of pharmacological interest should be considered, especially if they are endemic since their survival and geographical distribution could be affected and put them in danger of extinction (Kurnaz & Kurnaz, 2021). In addition, it could affect their chemical composition due to the modification of edaphic factors.

Accordingly, strategies have been created to conserve species through in-situ, *ex-situ* practices, and greenhouse systems (Chen, Yu, Luo, Wu, Li, & Steinmetz, 2016). In addition, biotechnology has developed the production of bioactive metabolites using cell cultures and plant precursors. Nonetheless, some challenges are constantly increasing the studies of a large number of medicinal plants to treat diseases of global importance, such as diabetes. Additionally, they have given rise to a constant effort to satisfy the demand for natural resources that include integral methods that are sustainable, safe, and scientifically effective (Kurnaz & Kurnaz, 2021).

For this reason, current studies focus on studying medicinal plants from families cultivated on a large scale for food. The Apiaceae family is one of the most prominent, consisting of approximately 3,000 species in 434 genera (Sayed-Ahmad, Talou, Saad, Hijazi, & Merah, 2017). Its main species are *Anethum graveolens* (dill), *Angelica archangelica* (angelica), *Apium graveolens* (celery), *Carum carvi* (caraway), *Coriandrum sativum* (coriander), *Cuminum cyminum* (cumin), *Daucus carota* (carrot), *Foeniculum vulgare* (fennel), *Petroselinum crispum* (parsley), *Pimpinella anisum* (anise). They are recognized for their use as a culinary spice (Sousa, Cunha, & Fernandes-Ferreira, 2021) and for their chemical composition since they are rich in phytochemicals such as terpenoids, flavonoids, triterpenoids, coumarins, steroids, saponins, and polyacetylenes. These compounds are a natural capital with the potential for application against various diseases such as antitumor, analgesic, antioxidant, diuretic, anti-inflammatory, and especially anti-diabetic (Sayed-Ahmad, Talou, Saad, Hijazi, & Merah, 2017).

Apiaceae plants represent a potential source to develop hypoglycemic nutraceuticals to prevent diabetes or to complement conventional drug-based treatments (Sousa, Cunha, & Fernandes-Ferreira, 2021; Saleem, Sarkar, Ankoleikar, & Shetty, 2017) because they contain active compounds that increase pancreatic insulin secretion in the Langerhans cell islets (Saleem, Sarkar, Ankoleikar, & Shetty, 2017). Therefore, they could be included in the design of natural, economic, and sustainable products as part of the new trends. Accordingly, the review focused on identifying Apiaceae plants to propose them as a source of compounds hypoglycemic agents to develop nutraceuticals under criteria of efficacy, profitability, and sustainability.

10.2 APIACEAE SPECIES WITH POTENTIAL HYPOGLYCEMIC EFFECT

As described, the Apiaceae family has excellent potential for developing functional products to treat hyperglycemia. It is cosmopolitan and widely distributed in temperate climate geographical areas (Sayed-Ahmad, Talou, Saad, Hijazi, & Merah, 2017). The most representative species of Apiaceae are shown in Figure 10.2, while Figure 10.3 shows the phenolic compounds with the hypoglycemic effect that these plants synthesize most frequently. Some Apiaceae species with hypoglycemic properties, bioactive extracts, polyphenolic compounds that they biosynthesize, and mechanisms of action are described in subsections.

10.2.1 CORIANDRUM SATIVUM

C. sativum, better known as coriander, is one of the medicinal plants of this family that has been used in glucose control and insulin production. *C. sativum* extracts can decrease blood glucose levels. In mice with induced diabetes, a regeneration of β-pancreatic cells and inhibition of the enzyme α-glucosidase after treatment with extracts of this plant were demonstrated (Cardoso et al., 2019) because it promotes glucose oxidation (Tripathy, Sahoo, & Sahoo, 2021). They also act by controlling enzymes such as glycogen phosphatase and glucose-6-phosphatase, thus inhibiting the activity of the metabolic pathways of glycogenolysis and gluconeogenesis (Sobhani et al., 2022). The mechanism of action to increase glycogen concentration was proposed for coriander seed extracts, making it a good prospect for developing functional products or nutraceuticals with hypoglycemic properties (Sahib et al., 2013).

The main compounds responsible for the described effect are isoquercetin, caffeic acid, quercetin, hydroxybenzoic acid, and vanillic acid (Sanlier & Gencer, 2019).

10.2.2 *ANETHUM GRAVEOLENS*

A. graveolens (dill) is a plant from Asia, rich in nutrients such as calcium, magnesium, sodium, vitamin A, niacin, and essential oils (limonene, α-phellandrene, pinene) (Zeilab et al., 2020). Dill has a good response against hyperglycemia; since it reduces the homeostatic pattern of HOMA-IR, which decreases insulin resistance (Haidari et al., 2020). On the other hand, the seeds of *A. graveolens* demonstrated inhibitory activity of α-glucosidase which slows down the absorption of starch (Saleem, Sarkar, Ankoleikar, & Shetty, 2017). The antidiabetic property of this plant is mainly due to the fusion of bile acids in the intestine, the inhibitory activity of cholesterol in the intestine, and an increase in fecal excretion (Mohammed et al., 2019).

Additionally, dill and other species of Apiaceae have been documented to possess antioxidant activity closely related to antihyperglycemic activity because oxidative stress increases in diabetic patients due to non-enzymatic glycosylation, metabolic stress, and autoxidative glycosylation. Therefore, the decrease in oxidative stress decreases the production of inflammatory cytokines and reduces insulin resistance (Sabu & Kuttan, 2002). In consequence, the hypoglycemic effect of *A. graveolens* is related to the content of quercetin, flavonoids, tannins, and alkaloids (Goodarzi, Khodadadi, Tavilani, & Oshaghi, 2016) but is also associated with the content of caffeic acid, which belongs to the family of hydroxycinnamic acids (HA) (Mohammed et al., 2019).

10.2.3 *APIUM GRAVEOLENS*

A. graveolens (celery) is a vegetable native to southern Europe (Consentino et al., 2020). Seed extracts control serum glucose levels by producing insulin in the islet β cells of the pancreas (Hedayati, Naeini, Mohammadinejad, & Mohajeri, 2019). Celery has a better inhibitory capacity of the α-amylase enzyme associated with the phenolic content of extract and its antioxidant capacity (Neagu, Paun, Moroeanu, Ungureanu, & Radu, 2019).

In addition, the ethanolic extract of celery regulates blood glucose levels in diabetic rats (Mans & Aburjai, 2022). The effect is because the extract stimulates λ cells to produce insulin, increases glucose transport into cells, and increases glycogen production in pancreatic cells (Wasfi & AL-Kabi, 2019). The phenolic compounds responsible for the hypoglycemic effect are flavonoids (apigenin) and luteolin, which stimulate the NF-KB pathway in macrophages. In addition, they improve the glucose uptake response in 3T3-L1 adipocytes. Another essential compound with a hypoglycemic effect that it contains is 3-n-butylphthalide (Hedayati, Naeini, Mohammadinejad, & Mohajeri, 2019).

10.2.4 *CUMINUM CYMINUM*

C. cyminum (cumin) is a seed commonly used in Asia to make beverages, pharmaceuticals, flavorings, and foods (Karimian, Farrokhzad, & Jalili, 2021). The seeds have shown a favorable effect in treating metabolic abnormalities in diabetic rat models. Its effect is similar to the mode of action of glibenclamide. Achieving better management of oxidative stress and AGE (formation of glycation end products) (Srinivasan, 2018). The methanolic extract significantly reduces blood glucose levels and prevents the formation of glycosylated hemoglobin and creatinine. Furthermore, it improves serum insulin and insulin production in alloxan- and streptozotocin-induced diabetic rats. In addition to the hypoglycemic effect, cumin reduces serum cholesterol levels in rats (Joshi & Soulimani, 2020).

Additionally, cumin powder in diabetic animal models reduces glucose and creatinine levels and controls elevated blood urea levels by 50%. Also, it produces aldose reductase and α-glucosidase inhibition, improving glucose levels and other biochemical parameters regularly altered in diabetes, such as cholesterol and triglycerides (Sowbhagya, 2013). The crude seed extract reduces the plasma glucose level by 38.34%, similar to glibenclamide with a value of 37.73%, but it also reduces the plasma lipid profile, inflammation, and oxidative stress (Mohamed, Hamed, & Fouda, 2018). The antihyperglycemic effect of cumin is related to the content of cumin aldehyde (Karimian, Farrokhzad, & Jalili, 2021), flavonoids (Singh, Gangadharappa, & Mruthunjaya, 2017), coumarins, tannins, and essential oils such as safranal, limonene, and eugenol (Singh, Yadav, Kumar, & Narashiman, 2021).

10.2.5 *DAUCUS CAROTA*

D. carota is the world's most important and widely cultivated root plant (Ahmad et al., 2019). Generally, the orange variety is the most produced. However, in Turkey, India, and Egypt, the most consumed variety is purple

(Rasheed et al., 2022). The purple prevents problems derived from the metabolism of carbohydrates because the juice in rats with a diet high in fat and carbohydrates improves glucose tolerance and prevents obesity, hypertension, cardiac fibrosis, and inflammation (Grover & Pandey, 2020). The fermented extracts with *Aspergillus oryzae* and *Lactobacillus plantarum* have antidiabetic properties, and the intake of fermented juice improves triglyceride levels in the blood. Safe consumption as it has no toxicity to the liver and kidneys (Rasheed et al., 2022). On the other hand, dichloromethane extract can activate PPARγ in 3T3-L1 adipocytes (Kumar et al., 2020), and separate fractions of the extract control insulin-dependent glucose (El-Houri et al., 2015).

The effect is due to the content of phenolic compounds in purple/black carrots, among which anthocyanins and flavonoids stand out (Grover & Pandey, 2020). These compounds reduce the risk of diabetes by reducing ROS (reactive oxygen species) related to anti-inflammatory and antioxidant effects. Also, they interact with signal transduction pathways, such as NF-KB and AP-1, so their response is related to oxidative signals that delay glycogenolysis. Cyanidin and cyanidin-3-xylosyl galactoside can inactivate α-glucosidase, glycogen phosphorylase, and α-amylase involved in carbohydrate metabolism (Oliveira et al., 2020). While chlorogenic acid, luteolin, and polyacetylenes are strong inactivators of the α-amylase enzyme (Ahmad et al., 2019; Rasheed et al., 2022).

10.2.6 PETROSELINUM CRISPUM

P. crispum (parsley) has been used for years for its aromatic and culinary properties in the food, pharmaceutical, and cosmetic industries (Jankovic, Mitic, Arsic, & Stankov-Jovanovic, 2021). Extracts of leaves and aerial parts of *P. crispum* with different solvents have hypoglycemic properties. The aqueous extract at 2 g/kg helps in glycemic control. Furthermore, it increases lipid peroxidation and decreases glutathione levels, and has hepatoprotective properties (Agyare, Appiah, Boakye, & Apenteng, 2017). The aqueous extract of *P. crispum* decreases body weight, may be related to the diuretic effect, and reduces elevated plasma glucose levels (Abou Khalil et al., 2016).

In addition, the extract has antioxidant activity in diabetic rat models, which explains the reduction in glucose levels, since parsley improves oxidative damage caused by hyperglycemia and not through cell regeneration in the pancreas (Sarwar et al., 2016). As well, the extract has an antioxidant activity of $EC_{50} \leq 10$ μg/mL, which is related to 37.48 ± 0.33% inhibition of α-amylase according to the Matthews correlation coefficient (MCC) analysis (Bashkin et al., 2021). The aqueous extracts reduce the levels of transaminases such as ALP and ALT and have a cardioprotective effect in rats with induced diabetes (Punosevac et al., 2021).

The hypoglycemic effect and the improvement of parameters associated with metabolic syndrome could be due to the content of flavonoids such as apigenin, apigenin glycosides (Jankovic, Mitic, Arsic, & Stankov-Jovanovic, 2021), hydroxytyrosol, quercetin, ferulic acid, cinnamic acid (Slighoua et al., 2020) and luteolin (Seczyk, Swieca, & Gawlik-Dziki, 2015). As well as coumarins such as umbelliferone, esculetin and psoralen (Li, Yao, & Li, 2017).

10.2.7 *FOENICULUM VULGARE*

F. vulgare (fennel) is commonly used in foods and beverages for its aroma (Sharopov et al., 2017). The essential oil has hypoglycemic and antioxidant properties in diabetic rats (Naquibuddin, Md., Hamiduddin, & Reyaz, 2020), while the methanolic extract reduces total cholesterol, triglycerides, and AST and ALT transaminases (Mehra, Tamta, & Nand, 2021). The compounds contained in the aqueous extracts of leaves and rhizomes participate in the inhibition of enzymes involved in carbohydrate metabolism. But the rhizome extract has better α-glucosidase and α-amylase inhibitory capacity with IC_{50} of 165.90 ± 1.2 and 194.30 ± 4.8 μg/mL, respectively. The inhibitory capacity was slightly better than acarbose (311.20 ± 1.38 μg/mL). However, both showed dose-dependent α-amylase inhibitory behavior (Sayah et al., 2021).

Quinic acid, chlorogenic acid, caffeic acid, 3,4-di-O-caffeoylquinic, p-coumaric acid, rutin, luteolin, apigenin-7-o-glucoside (Sayah et al., 2021; Kalleli et al., 2019), kaempferol, and esculetin (Malin et al., 2022). These compounds have been related to the hypoglycemic effect through various mechanisms of action. Caffeic acid acts against hyperglycemia through its antioxidant potential to promote insulin production (Sayah et al., 2021).

10.2.8 *CENTELLA ASIATICA*

C. asiatica (Gotu kola) is an aromatic plant native to India, China, and South Africa. It is used in traditional medicine for its antihyperglycemic effect

(Oyenihi, Langa, Mukaratirwa, & Masola, 2019). The ethanolic fraction contains compounds that inhibit the α-glucosidase of the yeast *Saccharomyces cerevisiae* with IC_{50} of 73.12 μg/mL (Dewi & Maryani, 2015). The methanolic extract of *C. asiatica* controls the increase in body weight of diabetic rats by 69% compared to the control group. Also, it decreases triglyceride levels in the blood (Oyenihi, Chegou, Oguntibeju, & Masola, 2017).

The ethanolic extract can control blood glucose levels in rats with streptozotocin-induced diabetes for 30 days. Then the glycemic index is restored to typical values (Sasikala & Naidu, 2019). It stimulates the production of insulin and renews the β-pancreatic cells. The percentage of regeneration after treatment is difficult to determine, but it increases the antioxidant capacity in tissues such as the liver and kidney. In addition, it reduces proinflammatory cytokines to prevent diseases derived from diabetes (Oyenihi, Ahiante, Oyenihi, & Masola, 2020). The leaf extract at 250, 500, and 100 mg/kg reduced blood glucose in diabetic rats after 3 hours of administration, presenting glucose percentage reductions of 32.6%, 38.8%, and 29.9%, respectively (Prakash, Jaiswal, & Srivastava, 2017).

The antihyperglycemic activity of *C. asiatica* is related to its content of quercetin, kaempferol (Dewi & Maryani, 2015), glycosides (Sachdeva, Roy, & Bharadvaja, 2020), rutin, flavanones, and tannins (Cu, Fu, J, & Oj, 2020).

10.2.9 CARUM CARVI

C. carvi (caraway) is an aromatic plant used to treat gastrointestinal problems and is particularly interesting for its anti-diabetic properties (Hajlaoui et al., 2021). Caraway aqueous extract is hypoglycemic in streptozotocin-induced diabetic rats (Javed, Hanif, Rehman, Hanif, & Tung, 2019). The oil has an antioxidant effect that could lead to an antihyperglycemic effect (Mahboubi, 2019). The oily part of the seeds also has an antioxidant effect, particularly the carveol since it inhibits the enzymes involved in the hydrolysis of carbohydrates (Ahmed, Khan, Kury, & Shah, 2020), and also controls blood glucose levels with an effect dose-dependent and time compared to the diabetes-induced group. As well, it significantly decreases the activities of transaminases (ALT, AST) due to hepatoprotective effects.

The hypoglycemic effect is due to D-limonene and carveol (Joshi & Soulimani, 2020). However, phenolic compounds such as caffeic acid, flavonoids (quercetin and kaempferol), coumarins, and tannins could participate in the effect (Mahboubi, 2019).

10.2.10 *ANGELICA KEISKEI*

A. keiskei (Ashitaba) is a perennial plant native to Japan, used in cooking and traditional medicine (Ohkura et al., 2018). The plant has antioxidant properties and an antihyperlipidemic effect. Therefore, its consumption for a long time reduces blood glucose levels since it inhibits the formation of free radicals. It also reduces liver weight and the expression of catalase and glutathione reductase, indicating a protective action against reactive oxygen species in rats fed a high-fat diet (Kil, Pham, Seo, & Jafari, 2017). The ethanolic extract of the leaves has an antidiabetic effect and functional properties in diabetic rats since they improve glucose utilization and regulate insulin production (Zhang, Jin, Luo, Wu, & Wang, 2018).

The ethanolic extract rich in flavonoids at 125, 250, and 500 μg/mL reduces glucose levels in the hyperglycemic zebrafish model. These organisms have glucose regulatory mechanisms similar to those of mammals. The effect of the extract was dose-dependent (Tu et al., 2022). Glucose levels decreased by 57.7% after treatment with 500 μg/mL. The effect was similar to the group treated with pioglitazone hydrochloride as a positive control. The hypoglycemic effect of this plant is related to its content of flavonoids such as luteolin, apigenin, and their glycosides (Tu et al., 2022). In addition, it contains chalcones and coumarins with this bioactivity (Salehi et al., 2021).

10.2.11 *HERACLEUM PERSICUM*

H. persicum is a perennial plant native to Iran. It has been used as a flavoring agent in pickles (Changxing et al., 2019). The root n-hexane extract has antioxidant activity and participates in the inhibition of α-glucosidase simultaneously. The IC_{50} was 13.5 μg/mL and more effective than acarbose (IC_{50} = 15.2 μg/mL) (Dehghan, Sarrafi, Salehi, & Ebrahimi, 2017).

The methanolic extract of *Heracleum dissectum* at 125, 250, and 500 μg/mL administered orally to mice showed a slight hypoglycemic effect compared to metformin. The group treated at 125 and 250 μg/mL of extract decreased glucose levels by 19.7–29.1%, respectively. Thus, the extract regulates glucose tolerance in mice. In addition, it slightly reduces the weight of the mice after three weeks of treatment compared with their initial weight and group of diabetic mice (Zhang et al., 2017). The hypoglycemic activity of this species is due to coumarins such as psoralen, bergapten, fraxetin (Dehghan, Sarrafi, Salehi, & Ebrahimi, 2017), scopoletin, alkaloids, β-sitosterol, and essential oils (Changxing et al., 2019).

FIGURE 10.2 Representative species of Apiaceae. (A) *Anethum graveolens*; (B) *Petroselinum crispum*; (C) *Coriandrum sativum*; (D) *Daucus carota*; (E) *Cuminum cyminum*; (F) *Pimpinella anisum*; and (G) *Apium graveolens*.

10.3 SUSTAINABLE TRENDS IN NUTRACEUTICAL PRODUCTION

A nutraceutical is defined as a food or ingredient with properties that benefit the health of those who consume it. They are composed of non-essential nutrients for the human body, such as vitamins and minerals. But they may contain polyphenols, polyunsaturated fatty acids, carotenoids, alkaloids, aliphatic acids, and alkaloids (Dima, Assadpour, Dima, & Jafari, 2020), prebiotics, and probiotics (Lee, 2017). These products are classified into two categories:

- Established nutraceuticals; and
- Potential nutraceuticals.

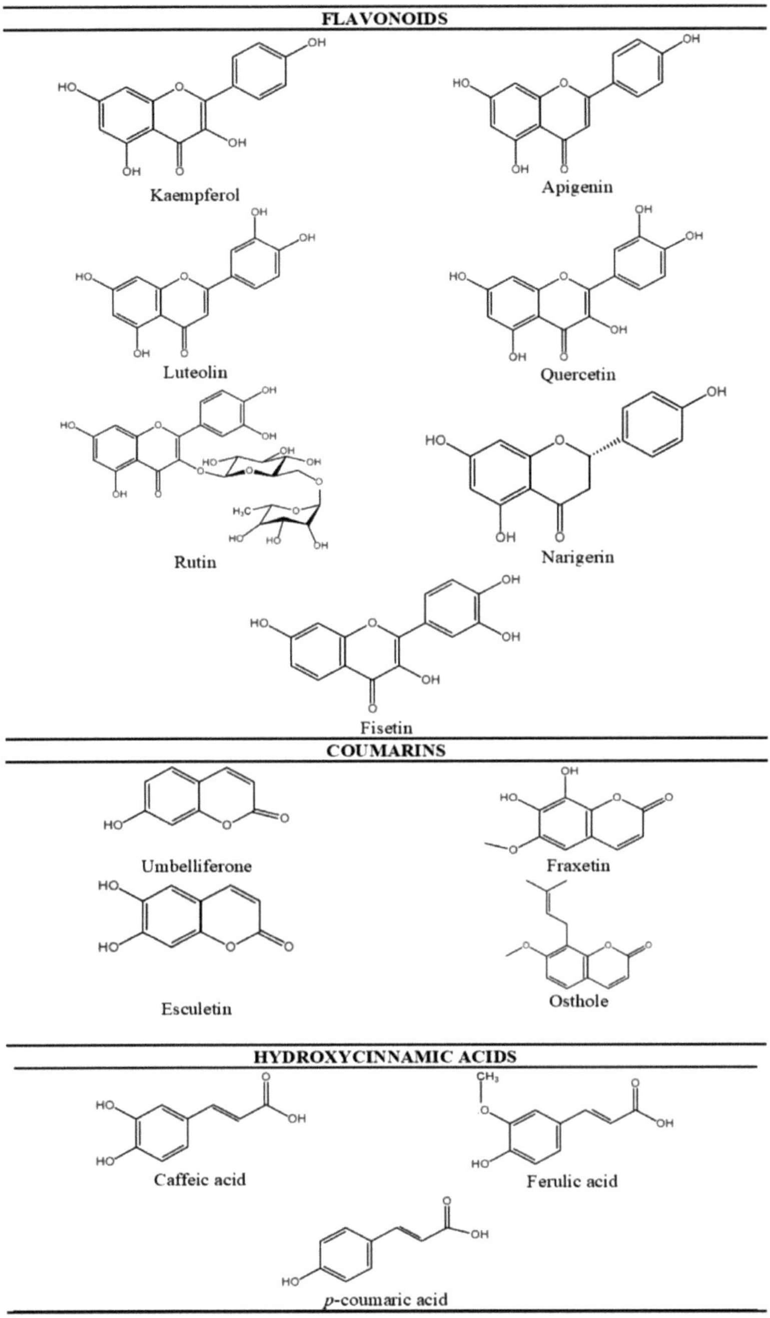

FIGURE 10.3 Main bioactive compounds with hypoglycemic activity.

Established nutraceuticals correspond to ingredients such as vitamins and minerals because they are widely known to be incorporated into food. Potential sources of nutraceuticals correspond to plants that synthesize bioactive metabolites such as polyphenols, whose incorporation into food is becoming increasingly important. Consequently, Apiaceae plants represent a potential source to develop new nutraceutical products. Accordingly, much research is focused on validating their properties, effective doses, mechanisms of action, interactions with food or drugs, and safety. But above all, evaluating their stability in a food matrix to guarantee their effectiveness.

Although nutraceuticals cannot replace synthetic drugs, they are used as auxiliaries to prevent or treat different diseases (Chandra, Saklani, Kumar, Kim, & Coutinho, 2022). Their consumption and production present advantages, but the manufacture of nutraceuticals still has challenges. Hence, in recent years studies have focused on improving the bioavailability of various nutraceutical agents as plant extracts by using a variety of food matrices to facilitate their administration (Dima, Assadpour, Dima, & Jafari, 2020). On the other hand, the exact mechanisms of action and effective doses have been clarified through tests (Durazzo, Lucarini, & Santini, 2020) *in vivo* and *in vitro* (Dima, Assadpour, Dima, & Jafari, 2020) to standardize the concentrations incorporated in nutraceuticals. Such as the studies of Apiaceae in which it is described that the polyphenolic composition of its extracts is responsible for the hypoglycemic effect and that they act by inhibiting enzymes that participate in the metabolism of carbohydrates or by reducing oxidative stress. The most frequently studied Apiaceae extracts are the aqueous, ethanolic, and juices, since these solvents have a high degree of safety for consumption. Additionally, there are different presentations of nutraceuticals, generally developed by incorporating the extract with excipients to increase its bioavailability (Dima, Assadpour, Dima, & Jafari, 2020). Because bioavailability is a critical point in the formulation of nutraceuticals since, after absorption, the compounds are less available. Before absorption, their bioactivity is affected by factors such as pH, polymer composition, gastrointestinal enzyme activity, and peristaltic movements. However, systems can be designed to facilitate better bioavailability by considering compound integrity, improving solubility, rate of release, increased intestinal permeability, and regulation of metabolic functions (Ting, Jiang, Ho, & Huang, 2014).

Consequently, the food industry focuses more and more on the manufacture of functional products based on integrated systems and new technologies that ensure the functionality of everything involved in their production due to the great demand of consumers for new sources of food that benefit health. So now it has new concepts to develop nutraceuticals.

It is well known that many crops are produced in large quantities to satisfy the world's diet. Apiaceae plants like carrot, parsley, coriander, dill, cumin, celery, fennel, gotu kola, caraway, ashitaba, and golpar are examples. But during the harvest, losses occur that often become waste. Therefore, Apiaceae residues could be used to recover bioactive compounds and incorporate them as nutraceuticals. This extraction could be done from the biorefinery point of view to obtain valuable products that can be marketed since this system uses "green chemistry" to implement safe and environmentally friendly processes. Ethanol is the most used solvent to extract active compounds because it achieves high yields. However, sometimes the food industry uses solvents such as hexane and acetone, but they represent a risk because they are toxic, flammable, and volatile (Dias et al., 2022). Water is considered a solvent for these purposes due to its lower cost and selectivity. Furthermore, it requires less aggressive reaction conditions than the synthesis of bioactive compounds (Benincasa, Santoro, Nardi, Cassano, & Sindona, 2019). Over time, extraction techniques have also changed. Emerging technologies are used (microwave-assisted extraction, ultrasound-assisted extraction, ohmic heating, and pulsed electric fields). These technologies have helped to better exploit bioactive molecule extraction in conjunction with intelligent solvents such as deep eutectic solvents (Dwamena, 2019). Natural deep eutectic solvents have low toxicity. It consists of a hydrogen bond acceptor and a hydrogen bond donor at a certain molarity that will result in a drop in the melting point of the solvent (Dwamena, 2019). Therefore, the techniques could be used to recover active compounds from Apiaceae wastes. Concerning this, the techniques for obtaining and applying metabolites in the production of nutraceuticals have been improving. Consequently, the selection of the extraction technique depends on the bioavailability of the plant, the type of compounds of interest, and the costs. Enzymes are an option as a pretreatment in the production of nutraceuticals. However, they represent a very high cost in the manufacturing processes. Thus, solid-state fermentation is used to recover active compounds from plant waste, and during fermentation, bioactive compounds are produced by microbial metabolic pathways. To perform is chosen the plant used as a substrate as well as pretreatments are established to favor the availability of the compounds retained in the tissues of the plant (Cano et al., 2021).

The type of microorganism to be used can come from a pure culture, an identified mixed culture, or a consortium of microorganisms, which will take the requirements for the survival of the medium in which it is found (Sadh, Duhan, & Duhan, 2018). Then, the interaction between the substrate and the microorganism will allow the synthesis of bioactive compounds to manufacture nutraceuticals. The extraction of bioactive compounds is carried out according to the physicochemical characteristics of the substrate and its cellular structures that preserve them. However, sometimes enzymatic hydrolysis is required to improve extraction yield. Consequently, solid-state fermentation is a technique that can be industrially exploited for its utility in the pretreatment of the substrate of interest or to achieve the biosynthesis of biologically active molecules. Additionally, this technique has the advantage of using little water, which reduces the cost of industrial processing and produces less wastewater. But above all, a large number of by-products can be used to innovate value-added products (Cano et al., 2021). However, its expansion at the industrial level represents a challenge due to the purification of the desired product and the generation of biomass (Singhania, Patel, Soccol, & Pandey, 2009). Furthermore, it presents a disadvantage in control of humidity, temperature, pH, and oxygen as critical factors constantly monitored. Also, fermentations that last for long periods need an efficient mixing system. Otherwise, the development of microorganisms such as fungi will need more time to grow (Cano et al., 2021).

In addition, metabolic engineering can be employed to recover active compounds from Apiaceae. Because GRAS microorganisms, generally safe, could be used to carry out the fermentation process and synthesize enantiomers of compounds (Yuan & Alper, 2019). *Bacillus subtilis* is an example of a GRAS microorganism since its genetic information is known, which makes it a potential microorganism to produce compounds in nutraceuticals from Apiaceae residues (Liu, Guan, Li, Shin, Du, & Chen, 2017).

Liposome formation is a frequently used nanotechnological technique to preserve molecules of interest. Its function is to increase the life of the compound and its bioavailability. Liposomes are a great tool for the food, pharmaceutical, and cosmetic industries since they improve the organoleptic characteristics of foods and nutraceuticals. Consequently, they could be used to manufacture nutraceutical products (Khorasani, Danaei, & Mozafari, 2018). Liposome encapsulation allows a controlled release of lipophilic and hydrophobic molecules into a food system. This technique could be used to preserve polyphenolic compounds hypoglycemic isolated from Apiaceae. This process generally occurs in three stages: the material to be encapsulated must be considered around the wall, monitoring, and control of no leakage, and finally, unwanted materials must be removed (Subramani

& Ganapathyswamy, 2020). An additional application of these systems is through phytosomes, which produce drugs generated from dispersions of plant extracts in phospholipid matrices, widely applied to low solubility compounds and bioavailability (Demirci, Caglar, Cakir, & Gülseren, 2017). As already mentioned, this method can be used in the area of nutraceutical design and formulation. Nanoliposomes integrate a great variety of molecules, for example, bioactive compounds, proteins, low molecular weight compounds, and nucleic acids, thus allowing them to be released for an extended period in the target region, promoting better solubility and absorption. This mechanism also serves as protection, as healthy cells are not affected by the adverse effects or toxicity of the compound (Subramani & Ganapathyswamy, 2020). Nanoliposome production is simple and safe, as they can come from sources such as eggs, milk, and soy, which allows them to be accepted and permitted by regulatory authorities since they are food-grade products. For these reasons, liposomes and nanoliposomes have been of great interest for use in the food industry, particularly the dairy industry (Khorasani, Danaei, & Mozafari, 2018).

Accordingly, plants of the Apiaceae family represent a source of compounds with a hypoglycemic effect that can be recovered using described techniques, which are in continuous development. Since the agro-industrial production of various species produces waste that represents a source of raw material to develop effective nutraceuticals for trade.

10.4 CONCLUSION

Apiaceae plants and waste represent a good prospect for innovating products that prevent hyperglycemia and diseases associated with metabolic syndrome that seriously impact the world population. Since they are produced in large quantities as food, waste is generated during the production chain. Therefore, their use opens a sustainable way to recover bioactive compounds and incorporate them as nutraceuticals in food. These plants have in common the synthesis of polyphenolic compounds with hypoglycemic properties such as flavonoids, coumarins, and hydroxycinnamic acids, among which kaempferol, apigenin, luteolin, quercetin, rutin, naringenin, fisetin, umbelliferone, fraxetin, esculetin, osthol, stand out. Caffeic acid, ferulic acid, and p-coumaric acid. Until now, only those related to the inhibition of enzymes that participate in the metabolism of carbohydrates or the reduction of oxidative stress due to their antioxidant properties have been described as mechanisms of action. However, it is necessary to deepen the study of other mechanisms of action and overcome the challenges involved in nutraceutical manufacturing. On the other hand, the demand for natural products to improve the health of diabetic patients is constantly increasing. Therefore, the scientific community and the private sector want to satisfy the needs. But coupled with this, they have acquired the commitment to protect the environment through the design of products under comprehensive schemes to develop efficient, effective, sustainable, and economically profitable processes.

KEYWORDS

- **alkaline phosphatase**
- **Apiaceae**
- **diabetes mellitus**
- **generally recognized as safe**
- **hypoglycemic effect**
- **natural sources**
- **Petroselinum crispum**
- **phenolic compounds**
- **reactive oxygen species**

REFERENCES

Abou Khalil, N. S., Abou-Elhamd, A. S., Wasfy, S. I. A., El Mileegy, I. M. H., Hamed, M. Y., & Ageely, H. M. (2016). Antidiabetic and antioxidant impact of desert date (*Balanites aegyptiaca*) and parsley (*Petroselinum sativum*) aqueous extracts: Lessons from experimental rats. *Journal of Diabetes Research, 2016*, 2–10.

Agyare, C., Appiah, T., Boakye, Y. D., & Apenteng, J. A. (2017). *Petroselinum crispum*: A review. In *Medicinal Spices and Vegetables from Africa: Therapeutic Potential Against Metabolic, Inflammatory, Infectious and Systemic Diseases* (pp. 527–547).

Ahmad, T., Cawood, M., Iqbal, Q., Ariño, A., Batool, A., Tariq, R. M. S., Azam, M., & Akhtar, S. (2019). Phytochemicals in *Daucus carota* and their health benefits—Review article. *Foods*, *8*, 1–22.

Ahmed, M. S., Khan, A., Kury, L. T. A., & Shah, F. A. (2020). Computational and pharmacological evaluation of carveol for antidiabetic potential. *Frontiers in Pharmacology*, *22*, 1–9.

Arroba, A. I., & Aguilar-Diosdado, M. (2022). Special issue "The prevention, treatment, and complications of diabetes mellitus." *Journal of Clinical Medicine*, *11*, 18–20.

Bashkin, A., Ghanim, M., Abu-Farich, B., Rayan, M., Miari, R., Srouji, S., Rayan, A., & Falah, M. (2021). Forty-one plant extracts screened for dual antidiabetic and antioxidant functions: Evaluating the types of correlation between α-amylase inhibition and free radical scavenging. *Molecules*, *26*, 1–26.

Benincasa, C., Santoro, I., Nardi, M., Cassano, A., & Sindona, G. (2019). Eco-friendly extraction and characterization of nutraceuticals from olive leaves. *Molecules*, *24*, 1–14.

Bi, X., Lim, J., & Henry, J. (2017). Spices in the management of diabetes mellitus. *Food Chemistry*, *217*, 281–293.

Cano y Postigo, L. O., Jacobo-Velázquez, D. A., Guajardo-Flores, D., & García-Amezquita, L. E., & García-Cayuela, T. (2021). Solid-state fermentation for enhancing the nutraceutical content of agrifood by-products: Recent advances and its industrial feasibility. *Food Bioscience*, *41*, 1–10.

Cardoso, A. L., Frederico, É. H., Guimarães, C. A., Moura-Fernandes, M. C., Guedes-Aguiar, E. O., Silva, A. L. D., & Bernardo-Filho, M. (2019). Effects of *Coriandrum sativum* L. in association with physical exercise in alloxan-induced type 1 diabetes mellitus in rats. *Applied Sciences*, *9*, 1–11.

Ceriello, A., & Prattichizzo, F. (2021). Variability of risk factors and diabetes complications. *Cardiovascular Diabetology*, *20*, 1–11.

Chandra, S., Saklani, S., Kumar, P., Kim, B., & Coutinho, H. D. M. (2022). Nutraceuticals: Pharmacologically active potent dietary supplements. *Biomed Research International*, *2022*, 1–10.

Chandran, H., Meena, M., Barupai, T., & Sharma, K. (2020). Plant tissue culture as a perpetual source for production of industrially important bioactive compounds. *Biotechnology Reports*, *26*, 1–10.

Changxing, L., Dongfang, G., Lixue, Z., Saeed, M., Alagawany, M., Farag, M. R., Chenling, M., & Jianhua, L. (2019). *Heracleum persicum*: Chemical composition, biological activities and potential uses in poultry nutrition. *World's Poultry Science Journal*, *75*, 207–217.

Chen, S. L., Yu, H., Luo, H. M., Wu, Q., Li, C. F., & Steinmetz, A. (2016). Conservation and sustainable use of medicinal plants: Problems, progress, and prospects. *Chinese Medicine*, *11*, 1–10.

Consentino, B. B., Virga, G., La Placa, G. G., Sabatino, L., Rouphael, Y., Ntatsi, G., Iapichino, G., La Bella, S., Mauro, R. P., D'Anna, F., & Tuttolomondo, T., De Pasquale, C. (2020). Celery (*Apium graveolens* L.) performances as subjected to different sources of protein hydrolysates. *Applied Sciences*, *9*, 1–13.

Cu, O. N., Fu, I., J, A., & Oj, P. H. W. (2020). Nutrient and phytochemical composition of *Centella asiatica* leaves. *Medicinal & Aromatic Plants*, *9*, 1–8.

de Paulo Farias, D., Fernandes de Araújo, F., Neri-Numa, I. A., & Maria Pastore, G. (2021). Antidiabetic potential of dietary polyphenols: A mechanistic review. *Food Research International*, *145*, 1–16.

Dehghan, H., Sarrafi, Y., Salehi, P., & Ebrahimi, S. N. (2017). α-Glucosidase inhibitory and antioxidant activity of furanocoumarins from *Heracleum persicum*. *Medicinal Chemistry Research*, *26*, 849–855.

Demirci, M., Caglar, M. Y., Cakir, B., & Gülseren, I. (2017). Encapsulation by nanoliposomes. In *Nanoencapsulation Technology for the Food and Nutraceutical Industry* (pp. 74–113).

Dewi, R. T., & Maryani, F. (2015). Antioxidant and α-glucosidase inhibitory compounds of *Centella asiatica. Procedia Chemistry*, *17*, 147–152.

Dias, M., Roamaní-Pérez, M., Romaní, A., de la Cruz, A., Pastrana, L., Fuciños, P., & Amado, I. R. (2022). Recent technological advances in phenolic compounds recovery and applications: Source of nutraceuticals for the management of diabetes. *Applied Sciences*, *12*, 1–26.

Dima, C., Assadpour, E., Dima, S., & Jafari, S. M. (2020). Bioavailability of nutraceuticals: Role of the food matrix, processing conditions, the gastrointestinal tract, and nanodelivery systems. *Comprehensive Reviews in Food Science and Food Safety*, *19*, 954–994.

Durazzo, A., Lucarini, M., & Santini, A. (2020). Nutraceuticals in human health. *Foods*, *9*, 18–20.

Dwamena, A. K. (2019). Recent advances in hydrophobic deep eutectic solvents for extraction. *Separations*, *6*, 1–15.

El-Houri, R. B., Kotowska, D., Christensen, K. B., Bhattacharya, S., Oksbjerg, N., Wolber, G., Kristiansen, K., & Christensen, L. P. (2015). Polyacetylenes from carrots (*Daucus carota*) improve glucose uptake in vitro in adipocytes and myotubes. *Food & Function*, *6*, 2135–2144.

Goodarzi, M. T., Khodadadi, I., Tavilani, H., & Oshaghi, E. A. (2016). The role of *Anethum graveolens* L. (dill) in the management of diabetes. *Journal of Tropical Medicine*, *2016*, 2–12.

Grigoriadou, K., Krigas, N., Lazar, D., & Maloupa, E. (2019). Sustainable use of mediterranean medicinal-aromatic plants. In *Feed Additives: Aromatic Plants and Herbs in Animal Nutrition and Health* (pp. 57–74).

Grover, K., & Pandey, P. (2020). Characterization of black carrot (*Daucus carota L.*) polyphenols; role in health promotion and disease prevention: An overview. *Journal of Pharmacognosy and Phytochemistry*, *9*, 2784–2792.

Haedersdal, S., Lun, A., Knop, F. K., & Vilsboll, T. (2018). The Role of Glucagon in the Pathophysiology and Treatment of Type 2 Diabetes. *Mayo Clinic Proceedings*, *93*, 217–239.

Haidari, F., Zakerkish, M., Borazjani, F., Ahmadi Angali, K., & Amoochi Foroushani, G. (2020). The effects of *Anethum graveolens* (dill) powder supplementation on clinical and metabolic status in patients with type 2 diabetes. *Trials*, *21*, 1–11.

Hajlaoui, H., Arraouadi, S., Noumi, E., Aouadi, K., Adnan, M., Khna, M. A., Kadri, A., & Snoussi, M. (2021). Antimicrobial, antioxidant, anti-acetylcholinesterase, antidiabetic, and pharmacokinetic properties of *Carum carvi* L. and *Coriandrum sativum* L. essential oils alone and in combination. *Molecules*, *26*, 1–18.

Hedayati, N., Naeini, M. B., Mohammadinejad, A., & Mohajeri, S. A. (2019). Beneficial effects of celery (*Apium graveolens*) on metabolic syndrome: A review of the existing evidences. *Phytotherapy Research*, *33*, 3040–3053.

Jankovic, S., Mitic, M., Arsic, B., & Stankov-Jovanovic, V. (2021). The kinetic and thermodynamic studies of solid-liquid extraction of apigenin-glycosides from parsley (*Petroselinum crispum*). *Separation Science and Technology*, *56*, 2253–2265.

Javed, R., Hanif, M. A., Rehman, R., Hanif, M., & Tung, B. T. (2019). Caraway. In *Medicinal Plants of South Asia: Novel Sources for Drug Discovery* (pp. 87–100).

Joshi, R. K., & Soulimani, R. (2020). Ethno-medicinal and phytochemical potential of *Carum carvi* Linn. and *Cuminum cyminum*: A review. *International Journal of Pharmacognosy and Life Science*, *1*, 33–37.

Kalleli, F., Rebey, I. B., Wannes, W. A., Boughalleb, F., Hammami, M., Tounsi, M. S., & M'hamsi, M. (2019). Chemical composition and antioxidant potential of essential oil and methanol extract from Tunisian and French fennel (*Foeniculum vulgare* Mill.) seeds. *Journal of Food Biochemistry*, *43*, 1–14.

Karimian, J., Farrokhzad, A., & Jalili, C. (2021). The effect of cumin (*Cuminum cyminum* L.) supplementation on glycemic indices: A systematic review and meta-analysis of randomized controlled trials. *Phytotherapy Research*, *35*, 4127–4135.

Khorasani, S., Danaei, M., & Mozafari, M. R. (2018). Nanoliposome technology for the food and nutraceutical industries. *Trends in Food Science & Technology*, *79*, 106–112.

Kil, Y. S., Pham, S. T., Seo, E. K., & Jafari, M. (2017). *Angelica keiskei*, an emerging medicinal herb with various bioactive constituents and biological activities. *Archives of Pharmacal Research*, *40*, 655–675.

Kumar, S., Schachi, K., Prasad, N. K., Dubey, N. K., & Dubey, U. (2020). Anti-diabetic, haematinic and anti-cholesterolmic effects of carrot (*Daucus carota* Linn.) juice metabolites to cure alloxan monohydrate induced type-1 diabetes in albino rats. *Journal of Diabetes, Metabolic Disorders & Control*, *7*, 37–40.

Kurnaz, M. L., & Kurnaz, I. A. (2021). Commercialization of medicinal bioeconomy resources and sustainability. *Sustainable Chemistry and Pharmacy*, *22*, 1–7.

Laakso, M. (2019). Biomarkers for type 2 diabetes. *Molecular Metabolism*, *27*, S139–S146.

Lapraz, J. C. (2019). Introduction to the usage of medicinal plants. In *The Theory of Endobiogeny* (pp. 255–266).

Lee, S. (2017). Strategic design of delivery systems for nutraceuticals. In *Nanotechnology Applications in Food: Flavor, Stability, Nutrition and Safety* (pp. 65–86).

Li, H., Yao, Y., & Li, L. (2017). Coumarins as potential antidiabetic agents. *Journal of Pharmacy and Pharmacology*, *69*, 1253–1264.

Li, Y., Taggart, M. A., McKenzie, C., Zhang, Z., Lu, Y., Pap, S., & Gibb, S. (2019). Utilizing low-cost natural waste for the removal of pharmaceuticals from water: Mechanisms, isotherms and kinetics at low concentrations. *Journal of Cleaner Production*, *227*, 88–227.

Liu, L., Guan, N., Li, J., Shin, H., Du, G., & Chen, J. (2017). Development of GRAS strains for nutraceutical production using systems and synthetic biology approaches: Advances and prospects. *Critical Reviews in Biotechnology*, *37*, 139–150.

Lovic, D., Piperidou, A., Zografou, I., Grassos, H., Pittaras, A., & Manolis, A. (2019). The Growing Epidemic of Diabetes Mellitus. *Current Vascular Pharmacology*, *18*, 104–109.

Mahboubi, M. (2019). Caraway as important medicinal plants in management of diseases. *Natural Products and Bioprospecting*, *9*, 1–11.

Malin, V., Garofulic, I. E., Repajic, M., Zoric, Z., Pedisic, S., Sternisa, M., Mozina, S. S., & Dragovic-Uzelac, V. (2022). Phenolic characterization and bioactivity of fennel seed (*Foeniculum vulgare* Mill.) extracts isolated by microwave-assisted and conventional extraction. *Processes*, *20*, 1–16.

Mans, K., & Aburjai, T. (2022). Accessing the hypoglycemic effects of seed extract from celery (*Apium graveolens*) in alloxan-induced diabetic rats. *Journal of Pharmaceutical Research International*, *26*, 1–11.

Mehra, N., Tamta, G., & Nand, V. (2021). A review on nutritional value, phytochemical and pharmacological attributes of *Foeniculum vulgare* Mill. *Journal of Pharmacognosy and Phytochemistry*, *10*, 1255–1263.

Mishra, M., Sharma, M., Dubey, R., Kumari, P., Ranjan, V., & Pandey, J. (2021). Green synthesis interventions of pharmaceutical industries for sustainable development. *Current Research in Green and Sustainable Chemistry*, *4*, 1–7.

Mohamed, D. A., Hamed, I. M., & Fouda, K. A. (2018). Antioxidant and anti-diabetic effects of cumin seeds crude ethanol extract. *Journal of Biological Sciences*, *5*, 251–259.

Mohammed, F. A., Razvi, S. S., Abdul, W. M., Mohammed, K., Hakeem, K. R., Banaganapalli, B., Shaik, N. A., & Elkady, A. I. (2019). Protective role of medicinal herb *Anethum graveolens* (dill) against various human diseases and metabolic disorders. In *Plant Human Health* (Vol. 3, pp. 181–194).

Naquibuddin, Md., Hamiduddin, & Reyaz, Z. (2020). Badiyan (*Foeniculum Vulgare* Mill.): an important drug of Unani system of medicine. *European Journal of Pharmaceutical and Medical Research*, *7*, 307–312.

Navin, S., Srivastava, A., Prakash, D., Kumari, R., & Kumar, R. (2021). Management of Type-2 Diabetes Mellitus through Indian Medicinal Plants. *Journal of Diabetes & Metabolism*, *12*, 1–4.

Neagu, E., Paun, G., Moroeanu, V., Ungureanu, O., & Radu, G. L. (2019). Antioxidant and antidiabetic properties of polyphenolic-rich extracts of *Apium graveolens* and *Agropyron repens*. *Farmacia*, *64*, 909–913.

Ohkura, N., Atsumi, G., Ohnishi, K., Baba, K., & Taniguchi, M. (2018). Possible antithrombotic effects of *Angelica keiskei* (Ashitaba). *Pharmazie*, *73*, 315–317.

Oliveira, H., Fernandes, A., Brás, N. F., Mateus, N., de Freitas, V., & Fernandes, I. (2020). Anthocyanins as antidiabetic agents in vitro and in silico approaches of preventive and therapeutic effects. *Molecules*, *25*, 1–30.

Oyenihi, A. B., Ahiante, B. O., Oyenihi, O. R., & Masola, B. (2020). *Centella asiatica*: Its potential for the treatment of diabetes. In *Diabetes: Oxidative Stress and Dietary Antioxidants* (pp. 213–222).

Oyenihi, A. B., Chegou, N. N., Oguntibeju, O. O., & Masola, B. (2017). *Centella asiatica* enhances hepatic antioxidant status and regulates hepatic inflammatory cytokines in type 2 diabetic rats. *Pharmaceutical Biology*, *55*, 1671–1678.

Oyenihi, A. B., Langa, S. O. P., Mukaratirwa, S., & Masola, B. (2019). Effects of *Centella asiatica* on skeletal muscle structure and key enzymes of glucose and glycogen metabolism in type 2 diabetic rats. *Biomedicine & Pharmacotherapy*, *112*, 1–8.

Panizza, M. (2018). Fine chemical industry, pulp and paper industry, petrochemical industry and pharmaceutical industry. In *Electrochemical Water and Wastewater Treatment* (pp. 335–364).

Prakash, V., Jaiswal, N., & Srivastava, M. (2017). A review on medicinal properties of *Centella asiatica*. *Asian Journal of Pharmaceutical and Clinical Research*, *10*, 69–74.

Punosevac, M., Radovic, J., Lekovic, A., & Kundakovic-Vasovic, T. (2021). A review of botanical characteristics, chemical composition, pharmacological activity and use of parsley. *Archives of Pharmacy (Belgrade)*, *71*, 177–196.

Rana, S., Singh, P., Kandar, V., Singh, R., Dobhal, R., & Gupta, S. (2017). A review on characterization and bioremediation of pharmaceutical industries' wastewater: an Indian perspective. *Applied Water Science*, *7*, 1–12.

Rasheed, H., Shehzad, M., Rabail, R., Kowalczewski, P. L., Kidon, M., Jezowski, P., Ranjha, M. M. A. N., Rakha, A., Din, A., & Aadil, R. M. (2022). Delving into the nutraceutical benefits of purple carrot against metabolic syndrome and cancer: A review. *Applied Sciences*, *12*, 1–15.

Rogowska, J., & Zimmermann, (2022). Household pharmaceutical waste disposal as a global problem—A review. *International Journal of Environmental Research and Public Health, 19*, 1–29.

Sabu, M. C., & Kuttan, R. (2002). Anti-diabetic activity of medicinal plants and its relationship with their antioxidant property. *Journal of Ethnopharmacology, 81*, 155–160.

Sachdeva, V., Roy, A., & Bharadvaja, N. (2020). Current prospects of nutraceuticals: A review. *Current Pharmaceutical Biotechnology, 21*, 884–896.

Sadh, P. K., Duhan, S., & Duhan, J. S. (2018). Agro-industrial wastes and their utilization using solid state fermentation: A review. *Bioresource and Bioprocessing, 5*, 1–15.

Sahib, N. G., Anwar, F., Gilani, A. H., Hamid, A. A., Saari, N., & Alkharfy, K. M. N. (2013). Coriander (*Coriandrum sativum* L.): A potential source of high-value components for functional foods and nutraceuticals – A review. *Phytotherapy Research, 27*, 1439–1456.

Saleem, F., Sarkar, D., Ankoleikar, C., & Shetty, K. (2017). Phenolic bioactives and associated antioxidant and anti-hyperglycemic functions of select species of *Apiaceae* family targeting for type 2 diabetes relevant nutraceuticals. *Industrial Crops and Products, 107*, 518–525.

Salehi, B., Quispe, C., Chamkhi, I., Omari, N. E., Balahib, A., Sharifi-Rad, J., Bouyahya, A., Akram, M., Iqbal, M., Docea, A. O., Caruntu, C., Leyva-Gómez, G., Dey, A., Martorell, M., Calina, D., López, V., & Les, F. (2021). Pharmacological properties of chalcones: A review of preclinical including molecular mechanisms and clinical evidence. *Frontiers in Pharmacology, 11*, 1–21.

Sanlier, N., & Gencer, F. (2019). Role of spices in the treatment of diabetes mellitus: A minireview. *Trends in Food Science and Technology, 99*, 441–449.

Sarwar, S., Ayyub, M. A., Rezgui, M., Nisar, S., & Jilani, M. I. (2016). Parsley: A review of habitat, phytochemistry, ethnopharmacology and biological activities. *International Journal of Chemical and Biochemical Sciences, 9*, 49–55.

Sasikala, S., & Naidu, M. D. (2019). Evaluation of protective effect of *Centella asiatica* leaves on pancreas function in diabetic rats. *International Journal of Herbal Medicine, 7*, 55–60.

Sayah, K., El Omari, N., Kharbach, M., Bouyahya, A., Kamal, R., Marmouzi, I., Cherrah, Y., & Faouzi, M. E. A. (2021). Comparative study of leaf and rootstock aqueous extracts of *Foeniculum vulgare* on chemical profile and in vitro antioxidant and antihyperglycemic activities. *Advances in Pharmacological and Pharmaceutical Sciences, 2020*, 1–9.

Sayed-Ahmad, B., Talou, T., Saad, Z., Hijazi, A., & Merah, O. (2017). The *Apiaceae*: Ethnomedicinal family as source for industrial uses. *Industrial Crops and Products, 109*, 661–671.

Seczyk, L., Swieca, M., & Gawlik-Dziki, U. (2015). Changes of antioxidant potential of pasta fortified with parsley (*Petroselinum crispum* Mill.) leaves in the light of protein-phenolics interactions. *Acta Scientiarum Polonorum Technologia Alimentaria, 14*, 29–36.

Sharopov, F., Valiev, A., Satyal, P., Gulmurodov, I., Yusufi, S., Setzer, W. N., & Wink, M. (2017). Cytotoxicity of the essential oil of fennel (*Foeniculum vulgare*) from Tajikistan. *Foods, 6*, 1–11.

Singh, N., Yadav, S. S., Kumar, S., & Narashiman, B. (2021). A review on traditional uses, phytochemistry, pharmacology, and clinical research of dietary spice *Cuminum cyminum L. Journal of Biological Sciences, 35*, 1–24.

Singh, R. P., Gangadharappa, H. V., & Mruthunjaya, K. (2017). *Cuminum cyminum* – A popular spice: An updated review. *Journal of Biological Sciences, 3*, 292–301.

Singhania, R. R., Patel, A. K., Soccol, C. R., & Pandey, A. (2009). Recent advances in solid-state fermentation. *Biochemical Engineering Journal*, *44*, 13–18.

Slighoua, M., Mahdi, I., Amrati, E., Di Cristo, F., Amaghnouje, A., Grafof, A., Boucetta, N., Bari, A., & Bousta, D. (2020). Assessment of in vivo estrogenic and anti-inflammatory activities of the hydro-ethanolic extract and polyphenolic fraction of parsley (*Petroselinum sativum* Hoffm.). *Journal of Ethnopharmacology*, *265*, 1–8.

Sobhani, Z., Mohtashami, L., Amiri, M. S., Ramezani, M., Emami, S. A., & Simal-Gandara, J. (2022). Ethnobotanical and phytochemical aspects of the edible herb *Coriandrum sativum* L. *Journal of Food Science*, *87*, 1386–1422.

Sousa, R. M. O., Cunha, A. C., & Fernandes-Ferreira, M. (2021). Phytochemistry: The potential of *Apiaceae* species as sources of singular phytochemicals and plant-based pesticides. *Phytochemistry*, *187*, 1–32.

Sowbhagya, H. B. (2013). Chemistry, technology, and nutraceutical functions of cumin (*Cuminum cyminum* L): An overview. *Critical Reviews in Food Science and Nutrition*, *53*, 1–10.

Srinivasan, K. (2018). Cumin (*Cuminum cyminum*) and black cumin (*Nigella sativa*) seeds: Traditional uses, chemical constituents, and nutraceutical effects. *Food Quality and Safety*, *2*, 1–16.

Subramani, T., & Ganapathyswamy, H. (2020). An overview of liposomal nano-encapsulation techniques and its applications in food and nutraceutical. *Journal of Food Science and Technology*, *10*, 3545–3555.

Ting, Y., Jiang, Y., Ho, C. T., & Huang, Q. (2014). Common delivery systems for enhancing in vivo bioavailability and biological efficacy of nutraceuticals. *Journal of Functional Foods*, *7*, 112–128.

Tripathy, B., Sahoo, N., & Sahoo, S. K. B. (2021). Trends in diabetes care with special emphasis to medicinal plants: Advancement and treatment. *Biocatalysis and Agricultural Biotechnology*, *33*, 1–11.

Tu, L., Wang, R., Fang, Z., Sun, M., Sun, X., Wu, J., Dang, Y., & Liu, J. (2022). Assessment of the hypoglycemic and hypolipidemic activity of flavonoid-rich extract from *Angelica keiskei*. *Processes*, *27*, 1–13.

Wasfi, R. M. A. H., & AL-Kabi, J. S. A. J. (2019). Studying the hypoglycemic activity of celery herb extract (*Apium graveolens*) in blood glucose level of laboratory rats (*Sprague Dawley*). *Journal of Pure and Applied Microbiology*, *13*, 2389–2395.

Yuan, S. F., & Alper, H. S. (2019). Metabolic engineering of microbial cell factories for production of nutraceuticals. *Microbial Cell Factories*, *18*, 1–11.

Zeilab Sendijani, R., Abedian Kenari, A., Smiley, A. H., & Esmaeili, M. (2020). The effect of extract from dill (*Anethum graveolens*) on the growth performance, body composition, immune system, and antioxidant system of rainbow trout. *North American Journal of Aquaculture*, *82*, 119–131.

Zhang, H., Su, Y., Wang, X., Mi, J., Huo, Y., Wang, Z., Liu, Y., & Gao, Y. (2017). Antidiabetic activity and chemical constituents of the aerial parts of *Heracleum dissectum* Ledeb. *Food Chemistry*, *214*, 572–579.

Zhang, W., Jin, Q., Luo, J., Wu, J., & Wang, Z. (2018). Phytonutrient and anti-diabetic functional properties of flavonoid-rich ethanol extract from *Angelica keiskei* leaves. *Journal of Food Science and Technology*, *55*, 4406–4412.

CHAPTER 11

Use of Waste From the Food Industry to Obtain Biopolymers and Its Application to Control Dust Emissions into the Environment

LUIS CARLOS MATA GÓMEZ,[1] ERNESTO OYERVIDES-MUÑOZ,[1] CLAUDIA MAGDALENA LÓPEZ BADILLO,[1] and IVÁN DE JESÚS ZAPATA GONZÁLEZ[2]

[1]*Facultad de Ciencias Químicas, Universidad Autónoma de Coahuila, Saltillo, Mexico*

[2]*Centro de Investigación en Química Aplicada, Saltillo, Mexico*

ABSTRACT

There are currently major problems due to the increased amount of food waste generated due to inappropriate management. Residual materials are commonly rich in natural compounds that can be exploited directly or by obtaining specific compounds of interest due to their properties and commercial value on the market. As natural molecules with many properties, biopolymers are widely available worldwide, making them attractive materials for several industrial sectors, since they can be used for several applications, and are environmentally friendly compounds, so they must be developed in a low-cost manner. Among the applications of interest, recently there has been an increase in research on biopolymers as active ingredients for the formulation of dust suppressants. This is due to their ability to retain water and the mechanical properties that biopolymers can provide to dust particles. The main objective is to reduce the emission of particulate matter to the environment and to save up to 90% of the water used to irrigate unpaved

Sustainable Agricultural and Food Systems Engineering: New Research, Practices, and Developments.
Jose Juan Buenrostro-Figueroa, Julio César Tafolla-Arellano, Leonardo Sepúlveda Torre, Porteen Kannan, A. K. Haghi (Eds.)

roads in the agricultural industry. For this reason, it is necessary to research in detail the chemical composition, potential sources, and treatments for the various food residues and their derivative compounds in order to enhance their performance as dust suppressants.

11.1 INTRODUCTION

There are currently major problems around the world with the increasing generation of organic waste from the food industry. According to recent data published by FAO in an international journal, 1.3 billion tons of waste from food processing are generated (Blakeney, 2019; CNN, 2020). During food processing and post-harvest in agriculture, the most waste is produced (Hansen & Cheong, 2007; Malik & Grohmann, 2011), although considerable amounts of waste are also produced during marketing activities. Most of the residues represent problems for the industries that produce them; therefore, these have to be disposed of in landfills or government confinement sites or used to prepare compost and applied as fertilizer for new agricultural processes. In recent years, the development of high-value-added products from organic waste has gained interest from researchers and industries looking for alternatives to develop sustainable and global processes and an interest in the valorization of by-products into high-value commercial products (Theagarajan et al., 2019).

The amounts of waste generated from the food industry commonly depend on the geographic area, with the European Union being the area with the most waste generated worldwide, followed by India in second place, and then the United States of America (Ranganathan et al., 2020). This fact shows that first-world countries are not exempt from environmental problems caused by food waste generation; therefore, the development of sustainable global processes for obtaining highly valuable compounds is required around the world.

Biopolymers are compounds found in large amounts in organic wastes and due to their intrinsic properties, which confer structural characteristics, they are also highly biodegradable and biocompatible compounds that have interested researchers and industries for their valorization and application in several industrial areas (Bayón et al., 2018; Theagarajan et al., 2019a).

These aforementioned properties make biopolymers very useful and attractive for industries such as medicine, food industry, biosensors, plastics, ceramics, textiles, water waste treatment, cosmetics, pharmaceuticals, agriculture, mining, and others (Sánchez-Vázquez et al., 2013).

Several industries have big problems with dust emissions as environmental pollutants derived from their operational activities; the agriculture and mining industries generate large amounts of particulate matter which is emitted as an air pollutant by the passage of freight transport on unpaved roads and causes health damage in people; also, dust emissions cause truck engine failures mainly in air systems, which means an increase in maintenance costs. Biopolymers have been studied by many researchers who use them as the main ingredient to formulate dust suppressants, due to their ability to retain moisture and mechanical properties that were described in the previous paragraph; also, the wide availability around the world in natural and residual sources is an important factor to create sustainable technologies to control dust emissions.

Obtaining biopolymers from agro-industrial wastes can be achieved by direct extraction from biomass using thermal and/or chemical pre-treatments to increase the yield and purity of the interest compounds. Another way to obtain high-value compounds from industrial wastes is the fermentation of their biomass used to supplement the culture medium. This is a process that has been investigated by many authors to increase the yields for biopolymer obtaining.

The high potential of application for biopolymers around the world has been increasing their market value. In Mexico, the market value is expected to be positive. Sales estimated for 2023 will be 4.5 MOP (millions of pesos), while in 2028 they are estimated to be 8.7 MOP, which means almost double in a five-year range, showing a CAGR of 18.8%.

11.2 BIOPOLYMERS AND THEIR MAIN OBTAINING SOURCES

Biopolymers are compounds formed by polymerase chain reactions catalyzed by enzymes, carried out in cells as complex metabolic processes under natural conditions. Biodegradable polymers have been used in several industrial areas, such as the medical field (a trending application that has increased remarkably). Natural biopolymers such as cellulose, starch, pectin, chitosan, and their derivatives are mainly obtained from agricultural wastes.

On the other hand, collagen is an abundant biopolymer formed by proteins and obtained from mammals. Cattle and pig skins, and their bones obtained from slaughterhouses, are the main sources of collagen.

11.2.1 POLYSACCHARIDES

Polysaccharides are natural biopolymers widely available in plants, animals, and microorganisms. Cellulose and starch have been the polysaccharides most used for materials applications in the ceramics industry, but the interest is growing in more complex carbohydrates such as xanthan, chitin, chitosan, curdlan, pullulan, and hyaluronic acid. These polymers contain more than one type of carbohydrate unit, which makes them more complex polysaccharides with branched structures.

Cellulose and starch are composed of hundreds or thousands of D-glucopyranoside repeating units. The first one is a homopolysaccharide of glucose with β-1,4-glycosidic bonds, and its main function is as a structural polymer in plants and trees. It is a very resistant polysaccharide because its biodegradation requires the hydrolytic activity of three enzymes (Sánchez-Ramírez et al., 2016).

Cellulose is another abundant biopolymer in nature that is part of the lignocellulosic biomass, which is found from 35 to 50% in lignocellulosic biomass, where hemicelluloses, lignin, and various inorganic materials are also found. Some agro-industrial wastes such as cane bagasse, corncobs, wheat straw, banana peel, soy peel, oil palm mesocarp, rice straw (Tapia-Blacido et al., 2017), peanut shell, citrus peel, straw, and corn stover are rich in lignocellulosic biomass (Sánchez-Vazquez et al., 2017). Cellulosic materials show important mechanical properties because hydrogen bonds between hydroxyl groups form networks that play an important role in structural function, as they are resistant to oxidative agents and strong alkalis compounds. Considering these properties, cellulose fibers have been used as stabilizers in thermoplastics instead of glass fiber (Sanchez-Valzquez et al., 2013).

Hemicellulose is the second largest material in lignocellulosic biomass, bonded with the cellulose part by hydrogen bonds and by covalent with lignin structure (Peng et al., 2009). Lignin is a thermoplastic polymer consisting of G units in softwood, G-S-H units in herbaceous plants, and G-S units in hardwood. This biopolymer provides mechanical rigidity to the cell wall, and protection against microbial attack, and is a property that has been used to develop new products with novel industrial applications (Huang et al., 2019).

Starch is a polymer used as a reserve that occurs in crops used for its production such as potatoes, corn, rice, and other cereals. In all these plants, starch is produced in the form of granules, which have different sizes and compositions from plant to plant. Starch is composed of two principal molecules: one linear, amylose, about 20 weight percent of the granule, and the branched, amylopectin, the remainder of the molecule.

Starch and cellulose are composed of 100 or 1,000 D-glucopyranoside repeating units (Lenz, 1993), which are shown in Figure 11.1.

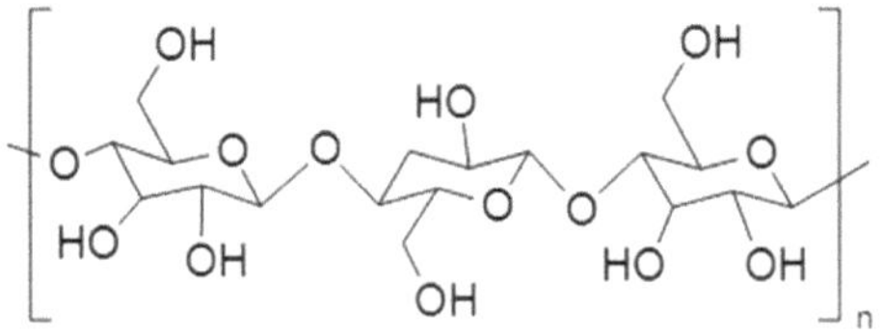

FIGURE 11.1 Starch and cellulose molecular structure.

Gelatin is another biopolymer used in food, pharmaceutical, and photographic areas (Killekar et al., 2012). Around 95% of gelatin is obtained from bovine and pigs, mainly found in bone, tendon, skin, and cartilage (derived from partial hydrolysis of fibrous collagen) of animals (Sockalingam & Abdullah, 2015).

Chitin is the second most abundant biopolymer on Earth, which consists of N-acetyl-β-glucosamine units linked by a β(1-4)-bond (Cheba et al., 2011). It is considered the main structural component in some marine species such as zooplankton, corals, jellyfish, shrimps, crabs, prawns, and lobsters (Kandile et al., 2018); also, in the cell wall of fungi and the cuticles of insects.

Chitosan is obtained from the deacetylation of chitin by enzymatic or chemical means, and it is very useful for food, medicinal, pharmaceutical, textile, agricultural, cosmetic, and water treatment applications (Al-Manhel et al., 2018). It is composed of N-acetylglucosamine and glucosamine, and its physical, mechanical, and biological properties are determined by its deacetylation degree, which is usually between 70 and 95% in its structure. It is represented in Figure 11.2 (Ababneh & Hammed, 2021).

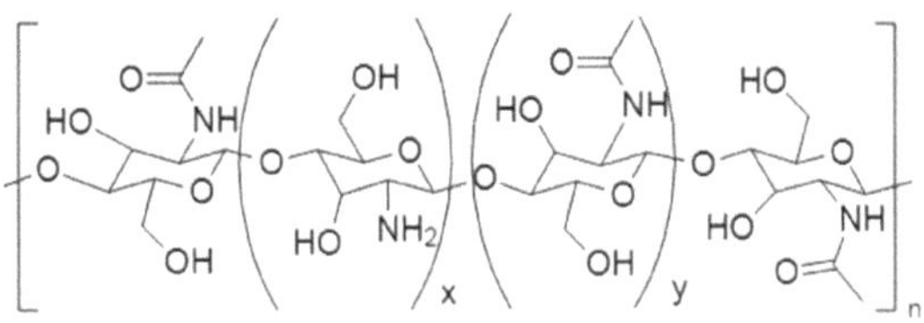

FIGURE 11.2 Chemical structure of chitosan.

11.2.2 BIOPOLYMERS OBTAINED BY MICROBIAL FERMENTATION

Xanthan gum is another important biopolymer that is estimated to reach 30,000 tons per year around the world (Stredansky & Conti, 1999). It is a polysaccharide produced by bacteria from the *Xanthomonas genus* and is useful in the food, pharmaceutical, and petrochemical industries. Studies have been developed to produce xanthan gum by microbial means using agro-industrial wastes such as cheese whey, malt bagasse, and citrus peels as sources for this biopolymer (Mesomo et al., 2009).

Polyhydroxyalkanoates (PHA) is another biopolymer that is produced by microorganisms as systems for energy storage and intracellular carbon. This biopolymer can be produced using agro-industrial wastes as carbon and nitrogen sources (Salgaonkar & Bragança, 2017), and it is produced by many bacterial species (Koller et al., 2017).

The PHA biosynthesis capacity substantially improves the survival of bacteria under numerous stress factors, such as heat shock, UV radiation, acidic pH, and osmotic pressure (Obruca et al., 2018). Poly-3-hydroxybutyrate (P3HB), poly(3-hydroxybutyrate-co-3-hydroxyhexanoate) (PHBHHx), poly(3-hydroxybutyrate-co-3-hydroxyoctanoate) (PHBO), and P(3HB-co-3HV) are the polymeric esters of the group of PHAs. For the synthesis of PHAs by bacterial fermentation, microbes such as *Alcaligenes*, *Azotobacter*, *Agrobacterium*, *Azospirilum*, *Aeromonas*, recombinant *E. coli*, *Pseudomonas* sp., *Cupriavius*, *Rhizobium*, *Rhodobacter*, *Bacillus*, and *Sphaerotilus* are used (Sanchez-Vazquez et al., 2013; Obruca et al., 2018).

11.2.2.1 OBTAINING BIOPOLYMERS FROM RESIDUAL SOURCES

Derived from the food industry activities, very large amounts of organic residues are produced, which are considered pollutants of the environment with no added value (Ravindran & Jaiswal, 2016). Biotechnology is focusing on developing new innovative and sustainable technologies, which have opened new market opportunities using biomass from organic wastes as raw materials to obtain high-value added and useful compounds for many industries (Nile et al., 2017).

11.2.2.2 STARCH: SOURCES OF OBTAINING AND APPLICATIONS

Starch from potato waste is a biopolymer that is natural, cheap, and biodegradable (Sarka & Dvoracek, 2017a, b; Ly et al., 2019). However, starch from potato waste has limited industrial applications due to low solubility in cold water, susceptibility to retrogradation, and high viscosity when it cools in solution. Therefore, studies have been conducted to increase solubility and reduce viscosity; grafting starch with co-polymers is a useful method to improve the aforementioned properties and achieve the agglomeration of particulate matter to function as an effective dust suppressant (Okonkwo & Ofoegbu, 2008; Sol et al., 2018; Liu et al., 2017a, b; Hu & Meng, 2004).

11.2.2.3 CELLULOSE: SOURCES OF OBTAINING AND APPLICATIONS

The ability of nature to produce cellulose is around 10^{11}–10^{12} tons per year, making cellulose the most abundant natural biopolymer on earth. Therefore, it is the most promising biopolymer to be used as a raw material in various industries for developing new products. Additionally, environmental regulations from federal government authorities emphasize the use of hazardous materials that pollute the environment to be replaced by materials of natural origin that are biodegradable, cheap, and renewable (Klemm et al., 2005).

11.2.2.4 STRATEGIES TO OBTAIN CELLULOSE INVOLVE TWO METHODS

The first is the degenerative process carried out to obtain low molecular weight chemical entities, for example, pyrolysis, carbonization, hydrothermal liquefaction, gasification, oxidative degradation, hydrolysis, fermentation, etc., and the second is the obtaining of high molecular weight chemical entities having intact biopolymer backbone.

Different sources have been studied in recent years; one example is the banana stem, which is a well-known source of cellulose, although it is discarded as agricultural waste. So, the researchers developed a method to use the Cavendish banana pseudostem to prepare NaCMC (Sodium carboxymethyl cellulose). The banana pseudostem was obtained after the fruit harvest. The pseudostem was dried completely and then pulverized to pass through a 20-mesh sieve. The powder was obtained using 8% sodium hydroxide (NaOH) at 100?C for 3.5 hours, followed by bleaching with 5% NaOCl at 30?C for 3 hours. Cellulose alkalinization was performed using 5–30% NaOH at 25?C for 1 hour and then sodium monochloroacetate ($ClCH_2COONa$, NaMCA) was added to 3–7 g/5 g of cellulose suspension and the temperature was maintained at 55?C for 3 hours. The CMC (Carboxymethyl cellulose) with the highest degree of substitution (DS), viscosity, purity, and crystallinity, i.e., 0.75, 4,033 cps, 98.23, and 38.33%, respectively, was obtained using NaOH at 15% and etherification using 1.2 g (w/w) of NaMCA, respectively (Adinugraha et al., 2005).

Another source used as a cellulose source is described in a method for the isolation of sago pulp (57% w/w) from sago waste to obtain CMC by etherification using NaOH and NaMCA. The reaction parameters, i.e., temperature, concentration, and reaction time, were optimized. The product thus obtained has a high DS (degree of substitution) of 0.821. The sago pulp starting material and the CMC synthesized product were characterized using

Fourier-transform infrared (FTIR) spectra. Furthermore, digital photographs of sago pulp and sago debris showed the expected rough woody structure, while CMC synthesized from the pulp showed a smooth surface morphology (Pushpamalar et al., 2006).

An earlier research paper also describes the feasible, time-efficient, and environmentally safe alternative synthesis of NaCMC from cotton stalks, an agricultural waste, using microwave heating. Response surface methodology (RSM) was applied to optimize CMC preparation. In addition, optimization of the reaction product involved pretreatment and preparation processes. The optimal pretreatment condition involved the addition of 12% alkali to the cotton stem for 6 min at 200 W microwave powers maintaining a 1:9 cotton stem to solvent ratio, yielding 87.52% cellulose. The best conditions for preparation are 195.5 W microwave power for 197 min. Under these conditions, CMC was produced with DS 0.77, viscosity of 498.0 cps, and purity of 92%. FTIR spectroscopy indicated that the product had high DS characteristics. Scanning electron microscopy (SEM) showed that the length of the particles ranged from approximately 12 to 100 μm and were almost club-shaped. However, the product met the quality standards of GB/T1904-2005 (Zhang et al., 2011).

A research study also shows the feasibility of obtaining cellulose from sugarcane bagasse, involving chemical procedures including acid hydrolysis and alkaline treatment. The extracted and purified cellulose was further treated with alkali and then etherified to obtain a cost-effective coal-water slurry (CWS) additive. The DS and intrinsic viscosity of the prepared sodium carboxymethyl cellulose (SCMC1) were determined. FTIR and thermogravimetry (TGA) analyses were also performed. Another SCMC2 sample was synthesized from microcrystalline cellulose and a commercial SCMC3 sample was used as a reference. The data thus generated showed that SCMC1 had an SD of 0.857, which was 32.7% and 44.7% higher than SCMC2 and SCMC3, respectively. SCMC1 was also shown to have a higher intrinsic viscosity, indicating that it had a higher molecular mass. The SCMC samples thus synthesized were further used as additives to prepare CWS; however, their performance was evaluated based on rheological behavior and static stability. The data revealed that due to the higher DS and higher molecular mass of SCMC1, CWS with SCMC1 showed lower apparent viscosity and higher static stability than others. SCMC1 could generate stronger steric repulsive forces and electrostatic repulsive forces between carbon particles through adsorption when added to CWS (Ge et al., 2013).

Maize husks and maize cobs, which have been used on a small scale as fuel for direct combustion in cooking and heating, were reported to be used as chemical feedstock for large-scale production of CMC, which is a more modern concept. The cellulose was extracted from the corn biomass and the etherification reaction was carried out by treating the cellulose with NaOH and monochloroacetic acid. The reaction conditions were optimized against the NaOH concentration, the monochloroacetic acid concentration, the temperature, and the reaction time. The DS was also determined. The produced CMC was characterized using FTIR spectra and X-ray diffractograms. High-purity food-grade CMC was successfully produced and used as an additive for the pharmaceutical and food industries (Mondal et al., 2015; Singh et al., 2013).

11.2.3 BIOPOLYMERS USED AS DUST SUPPRESSANTS

Currently, there are a variety of technologies used as dust suppressors that have made it possible to partially control dust emissions. The most used up to now are mainly chemical, and physical technologies are also available that are not so favorable due to their high price and low efficiency to control the erosion of particulate material.

Dust suppressants have been classified according to their chemical characteristics into three types: organic compounds, biopolymers, and chemicals, and inorganic compounds.

Biopolymers are natural compounds mainly extracted from plants, which have binding properties, and some have highly hygroscopic properties. They have recently been investigated for various industrial applications. However, interest has grown in their application as dust suppression agents due to their natural properties to retain moisture and act in a friendly way with the environment. Also, through polymerization with other compounds, different properties are obtained that improve the desired effects to contain the particulate material on the work surface.

Dust suppressants have been categorized based on their chemical characteristics into three types: organic compounds, biopolymers and chemicals, and inorganic compounds.

A research work using biopolymers is the application of guar gum as raw material, and it was chemically modified to increase the number of hydrophilic groups in order to improve its moisture retention and agglomeration properties (Zhang et al., 2018). Other researchers used carboxymethyl cellulose and polyvinyl alcohol as dust suppressant agents, employing monomers and surfactants to improve the chemical properties of the material (Xiao et al., 2015). Other polymers investigated as a dust-suppressing agents are sodium alginate, where it has been polymerized with dextrin and glutamine to increase moisture retention and thermal stability (Yan et al., 2021). Research work also focused on copolymerization where the base polymer was hydroxymethyl cellulose with acrylamide and acrylic acid, obtaining a highly fluid dust suppressant agent that is resistant to wind and rain (Ma et al., 2017). For their part, other researchers (Liu et al., 2018) developed a highly efficient and environmentally friendly dust suppressant using sodium lignosulfonate as a polymer, a residual material from the paper industry, polymerized with acrylic acid to improve its properties. Bentonite is another material that has been investigated for its water absorption properties; researchers (Bao et al., 2019) polymerized it with itaconic acid and acrylic acid to improve its dust-suppressing properties to retain water as well as resistance to erosion during heavy machinery traffic, demonstrating its ability to reduce water consumption by up to 90%. Another highly hygroscopic and environmentally friendly polymer is sodium alginate, which when polymerized with surfactants (Yan et al., 2020) produces a dust suppressant with good water retention, high humidification, and excellent dust agglomeration.

Dust suppressants are classified into three types based on their chemical characteristics: organic compounds, biopolymers and chemicals, and inorganic compounds.

To control road dust pollution, the authors prepared a composite liquid dust suppressant (CLDS) composed of starch/organo-bentonite. Potato starch was dissolved with a NaOH solution at room temperature and then cetyltrimethylammonium bromide-modified sodium bentonite (CTAB-bentonite) was incorporated prior to the polymerization reaction. The CLDS was characterized by XRD, FTIR, TG, and SEM. Its properties, dust suppression performance, dust removal effects, and economic feasibility were also evaluated. The results suggested that the CLDS featured a porous three-dimensional network structure with numerous layers and folds, and the residual CLDS could form a solidified protective layer together with the powder through high viscosity after solidification, which effectively enhanced the absorption of CLDS in dust, retarded its evaporation from water, and inhibited dust diffusion in windy weather. CLDS also had good thermal stability and biodegradability due to the addition of CTAB-bentonite. About 2.5 wt.% CLDS can dramatically improve the dust suppression effect and overcome the problem of dust re-suspension to save water consumption and bring economic benefits (Zhu et al., 2021).

Dust suppressants have been classified according to their chemical characteristics into three types: organic compounds, biopolymers, and chemicals, and inorganic compounds. Another dust suppressant made with starch was carried out with mechanical activation-assisted solid-phase reaction (MASPR) and conventional liquid phase (LP) methods to synthesize different superabsorbent polymers (SAP), defined as SAP-MA and SAP-LP, respectively. The rheological properties, crystal structure, functional group changes, and dust suppression performance of SAPs prepared by these two methods were compared, and the mechanism of dust suppression of SAPs through the adsorption experiment between them was discussed. The results showed that the SAPs were successfully prepared by both methods. Compared with SAP-LP, SAP-MA with lower molecular weight, higher grafting rate, and better flowability and water absorption showed excellent suppression performance. This improvement could be attributed to the fact that SAP-MA showed lower crystallinity and better film forming, anti-evaporation, anti-consolidation, and MA-induced permeability. Furthermore, the effective chemical adsorption between the SAPs and the dust particles had a stable consolidation effect. This environmentally friendly method for the preparation of starch-based superabsorbent polymer for road dust suppressants can provide new insights into cassava starch valorization and large-scale production of dust suppressants (Liang et al., 2021).

As observed in the literature, recent investigations focused on the use of biopolymers as dust suppression agents have presented good results in controlling the erosion of particulate material in areas where dust represents a high risk to the safety and health of operating personnel, mainly due to the ability of biopolymers to retain high moisture. Given their diverse characteristics, it is very important to evaluate different biopolymers, as well as their performance in different types of soils, climatic and operational conditions, and concentrations. Therefore, extensive research studies are required to achieve the highest efficiency through simple application processes that improve the operating conditions where they are applied.

11.3 CONCLUSION

Large amounts of industrial waste derived from the activities of the food and agriculture sector are generated daily, accumulating in large deposits that represent highly contaminated sites. Their transfer also represents extra costs for entrepreneurs' operations. With the intention of reducing the amount of waste that accumulates, technological strategies have been proposed to process this waste with two important objectives: to reduce its environmental impact and to obtain bioactive compounds of high industrial interest and greater commercial value. Therefore, biopolymers of natural origin, obtained from industrial waste, represent an alternative as sustainable technologies that can meet various industrial needs, whether new or existing, adequately replacing them in the market of interest. Obtaining biopolymers from residual materials is shown as an opportunity for the development of sustainable technologies used to control particulate matter emitted by the industrial sectors that need it most.

11.4 FUTURE PROSPECTS

Currently, there are a large number of procedures that allow us to obtain biopolymers from residual materials through sustainable and highly replicable methods. However, an opportunity and challenge for the future is the processing of these food residues on a larger scale (pilot-industrial), in order to obtain the necessary amounts of biopolymers that allow us to meet the demands of said materials for the formulation of dust suppressors.

On the other hand, the most recently investigated biopolymers as dust suppressants have been evaluated by their water retention capacity, but on very few occasions by their ability to agglomerate particulate matter. Therefore, it is important to develop methodological strategies at the laboratory level that allow us to have a clearer perspective of the capacity of the organic materials obtained to be used as dust suppressants.

Another area of opportunity is the application of chemical modifications to polymeric materials to increase their performance as biodegradable, high-efficiency dust suppressants in the industry that demands them.

KEYWORDS

- **biopolymers**
- **dust emissions**
- **food waste**
- **organic materials**
- **polymeric materials**
- **residual materials**
- **residual sources**

REFERENCES

Ababneh, H., & Hameed, B. H. (2021). Chitosan-derived hydrothermally carbonized materials and its applications: A review of recent literature. *International Journal of Biological Macromolecules, 186*, 314–327.

Adinugraha, M. P., Marseno, D. W., & Haryadi, P. (2005). Synthesis and characterization of sodium carboxymethylcellulose from *Cavendish banana* pseudo stem (*Musa cavendishii* Lambert). *Carbohydrate Polymers, 62*, 164–169.

Al-Manhel, A. J., Al-Hilphy, A. R. S., & Niamah, A. K. (2018). Extraction of chitosan, characterization and its use for water purification. *Journal of the Saudi Society of Agricultural Sciences, 17*, 186–190.

Arasukumar, B., Prabakaran, G., Gunalan, B., & Moovendhan, M. (2019). Chemical composition, structural features, surface morphology and bioactivities of chitosan derivatives from lobster (*Thenus unimaculatus*) shells. *International Journal of Biological Macromolecules, 135*, 1237–1245.

Arunmozhivarman, K., Abraham, R. J. J., Appa, V., & Parthiban, M. (2017). Extraction and molecular characterization of collagen from poultry meat processing by-product (chicken skin). *International Journal of Pure and Applied Bioscience, 5*, 1085–1091.

Bao, Q., Nie, W., Liu, C., Zhang, H., Wang, H., Jin, H., & Liu, Q. (2020). The preparation of a novel hydrogel based on crosslinked polymers for suppressing coal dust. *Journal of Cleaner Production, 249*, 119343.

Bayar, N., Friji, M., & Kammoun, R. (2018). Optimization of enzymatic extraction of pectin from *Opuntia ficus*-indica cladodes after mucilage removal. *Food Chemistry, 241*, 127–134.

Bayón, B., Berti, I. R., Gagneten, A. M., & Castro, G. R. (2018). Biopolymers from wastes to high-value products in biomedicine. In *Waste to Wealth* (pp. 1–44). Springer, Singapore.

Cheba, B. A. (2011). Chitin and chitosan: Marine biopolymers with unique properties and versatile applications. *Global Journal of Biotechnology & Biochemistry, 6*, 149–153.

CNN. (2020). Global food waste is twice as high as previously estimated, the study says. *CNN Report*. Retrieved from: https://edition.cnn.com/2020/02/20/health/global-food-waste-higher/index.html (accessed on 30 May 2024).

Englyst, H. N., Kingman, S. M., & Hudson, G. J. (1996). Measurement of resistant starch in vitro and in vivo. *British Journal of Nutrition, 75*(5), 749–755.

Ge, Y., & Li, Z. (2013). Preparation and evaluation of sodium carboxymethylcellulose from sugarcane bagasse for applications in coal–water slurry. *Journal of Macromolecular Science, Part A, 50*, 757–765.

Gou, M., Wu, H., & Saleh, A. S. M. (2019). Effects of repeated and continuous dry heat treatments on properties of sweet potato starch. *International Journal of Biological Macromolecules, 129*, 869–877.

Hansen, C. L., & Cheong, D. Y. (2007). Agricultural waste management in food processing. In *Handbook of Farm Dairy and Food Machinery*. Elsevier Inc.

Huang, J., Fu, S., & Gan, L. (Eds.). (2019). *Lignin Chemistry and Applications*. Elsevier, pp. 25–50.

Kandile, N. G., Zaky, H. T., Mohamed, M. I., Nasr, A. S., & Ali, Y. G. (2018). Extraction and characterization of chitosan from shrimp shells. *Open Journal of Organic Polymer Materials, 8*, 33–42.

Kaya, M., Baran, T., & Karaarslan, M. (2015). A new method for fast chitin extraction from shells of crab, crayfish, and shrimp. *Natural Product Research, 29*, 1477–1480.

Killekar, V. C., Koli, J. M., Sharangdhar, S. T., & Metar, S. Y. (2012). Functional properties of gelatin extracted from the skin of black kingfish (*Rachycentron canadum*). *Indian Journal of Fundamental and Applied Life Sciences, 2*, 106–116.

Klemm, D., Heublein, B., Fink, H. P., & Bohn, A. (2005). Cellulose: Fascinating biopolymer and sustainable raw material. *Angewandte Chemie International Edition, 44*, 3358–3393.

Koller, M., & Braunegg, G. (2015). Biomediated production of structurally diverse poly (hydroxyalkanoates) from surplus streams of the animal processing industry. *Polimery, 60*.

Lenz, W. Robert. (1993). Biodegradable polymers. *Advances in Polymer Science, 107*. Springer-Verlag Berlin Heidelberg.

Liang, Z., Cai, X., Hu, H., Zhang, Y., Chen, Y., & Huang, Z. (2021). Synthesis of starch-based super absorbent polymer with high agglomeration and wettability for application in road dust suppression. *International Journal of Biological Macromolecules, 183*, 982–991.

Liu, Y., Nie, W., Mu, Y., Zhang, H., Wang, H., Jin, H., & Liu, Z. (2018). A synthesis and performance evaluation of a highly efficient ecological dust depressor based on the sodium lignosulfonate–acrylic acid graft copolymer. *RSC Advances, 8*(21), 11498–11508.

Lucas-Bautista, J. A., Bautista-Baños, S., Ventura-Aguilar, R. I., & Gómez-Ramírez, M. (2019). Determinación de quitina en hongos postcosecha y de quitinasas en frutos de papaya "Maradol." *Revista Mexicana de Fitopatología, 37*, 1–7.

Lv, Y., Zhang, L., & Li, M. (2019). Physicochemical properties and digestibility of potato starch treated by ball milling with tea polyphenols. *International Journal of Biological Macromolecules, 129*, 207–213.

Ma, Y., Zhou, G., Ding, J., Li, S., & Wang, G. (2018). Preparation and characterization of an agglomeration-cementing agent for dust suppression in open pit coal mining. *Cellulose, 25*(7), 4011–4029.

Malik, A., & Grohmann, E. (Eds.). (2011). *Environmental Protection Strategies for Sustainable Development* (3rd ed.). Springer Science & Business Media.

Maran, J. P., Sivakumar, V., Thirugnanasambandham, K., & Sridhar, R. (2013). Optimization of microwave-assisted extraction of pectin from orange peel. *Carbohydrate Polymers, 97*(2), 703–709.

Medina, E., Caro, N., Abugoch, L., Gamboa, A., Díaz-Dosque, M., & Tapia, C. (2019). Chitosan thymol nanoparticles improve the antimicrobial effect and the water vapor barrier of chitosan-quinoa protein films. *Journal of Food Engineering, 240*, 191–198.

Mesomo, M., Silva, M. F., Boni, G., Padilha, F. F., Mazutti, M., Mossi, A., de Oliveira, D., Cansian, R. L., & di Luccio, M. (2009). Xanthan gum produced by *Xanthomonas campestris* from cheese whey: Production optimization and rheological characterization. *Journal of the Science of Food and Agriculture, 89*, 2440–2445.

Mondal, M. I. H., Yeasmin, M. S., & Rahman, M. S. (2015). Preparation of food-grade carboxymethyl cellulose from corn husk agro waste. *International Journal of Biological Macromolecules, 79*, 144–150.

Muñoz, I., Rodríguez, C., Gillet, D., & Moerschbacher, B. M. (2018). Life cycle assessment of chitosan production in India and Europe. *International Journal of Life Cycle Assessment, 23*, 1151–1160.

Nile, S. H., Nile, A. S., Keum, Y. S., & Sharma, K. (2017). Utilization of quercetin and quercetin glycosides from onion (*Allium cepa* L.) solid waste as an antioxidant, urease, and xanthine oxidase inhibitor. *Food Chemistry, 235*, 119–126.

Obruca, S., Sedlacek, P., Koller, M., Kucera, D., & Pernicova, I. (2018). Involvement of polyhydroxyalkanoates in stress resistance of microbial cells: Biotechnological consequences and applications. *Biotechnology Advances, 36*, 856–870.

Peng, F., Ren, J. L., Xu, F., Bian, J., Peng, P., & Sun, R. C. (2009). Comparative study of hemicelluloses obtained by graded ethanol precipitation from sugarcane bagasse. *Journal of Agricultural and Food Chemistry, 57*(14), 6305–6317.

Pushpamalar, V., Langford, S. J., Ahmad, M., & Lim, Y. Y. (2006). Optimization of reaction conditions for preparing carboxymethyl cellulose from sago waste. *Carbohydrate Polymers, 64*, 312–318.

Ranganathan, S., Dutta, S., Moses, J. A., & Anandharamakrishnan, C. (2020). Utilization of food waste streams for the production of biopolymers. *Heliyon, 6*(9), e04891.

Ravindran, R., & Jaiswal, A. K. (2016). Exploitation of food industry waste for high-value products. *Trends in Biotechnology, 34*(1), 58–69.

Salgaonkar, B. B., & Bragança, J. M. (2017). Utilization of sugarcane bagasse by *Halogeometricum borinquense* strain E3 for biosynthesis of poly (3-hydroxybutyrate-co-3-hydroxyvalerate). *Bioengineering, 4*, 50.

Sanchez-Ramirez, J., Martinez-Hernandez, J. L., Segura-Ceniceros, P., Lopez, G., Saade, H., Medina-Morales, M. A., & Ilyina, A. (2017). Cellulase immobilization on chitosan-coated magnetic nanoparticles: Application for Agave Atrovirens lignocellulosic biomass hydrolysis. *Bioprocess and Biosystems Engineering, 40*, 9–22.

Sanchez-Vazquez, S. A., Hailes, H. C., & Evans, J. R. G. (2013). Hydrophobic polymers from food waste: Resources and synthesis. *Polymer Reviews, 53*, 627–694.

Sarka, E., & Dvoracek, V. (2017a). New processing and application of waxy starch (a review). *Journal of Food Engineering, 206*, 77–87.

Sarka, E., & Dvoracek, V. (2017b). Waxy starch as a perspective raw material (a review). *Food Hydrocolloids, 69*, 402–409.

Schmitz, C., Auza, L. G., Koberidze, D., Rasche, S., Fischer, R., & Bortesi, L. (2019). Conversion of chitin to defined chitosan oligomers: Current status and prospects. *Marine Drugs, 17*, 452.

Singh, R. K., & Singh, A. K. (2013). Optimization of reaction conditions for preparing carboxymethyl cellulose from corn cobic agricultural waste. *Waste and Biomass Valorization, 4*, 129–137.

Sivaraman, I., Krishnan, M., & Radhakrishnan, K. (2019). Better management practices for sustainable small-scale shrimp farming. *Journal of Cleaner Production, 214*, 559–572.

Sockalingam, K., & Abdullah, H. Z. (2015). Extraction and characterization of gelatin biopolymer from black Tilapia (*Oreochromis mossambicus*) scales. *AIP Conference Proceedings, 1669*(1), 020053.

Srinivasan, H., Kanayairam, V., & Ravichandran, R. (2018). Chitin and chitosan preparation from shrimp shells Penaeus monodon and its human ovarian cancer cell line, PA-1. *International Journal of Biological Macromolecules, 107*, 662–667.

Stredansky, M., & Conti, E. (1999). Xanthan production by solid-state fermentation. *Process Biochemistry, 34*, 581–587.

Tapia-Blácido, D. R., Maniglia, B. C., & Martelli-Tosi, M. (2017). Biopolymers from sugarcane and soybean lignocellulosic biomass. *Sustainable Polymers from Biomass*, 227–253.

Thakur, B. R., Singh, R. K., Honda, A. K., & Rao, M. A. (1997). Chemistry and uses of pectin – A review. *Critical Reviews in Food Science and Nutrition, 37*(1), 47–73.

Theagarajan, T., Malur Narayanaswamy, L., Dutta, S., Moses, J. A., & Chinnaswamy, A. (2019). Valorization of grape pomace (cv. Muscat) for the development of functional cookies. *International Journal of Food Science & Technology, 54*(4), 1299–1305.

Tsakanika, A., Clauzet, M., & May, P. H. (2018). Envolvendo os pescadores artesanais no desenvolvimento sustentável urbano e periurbano no Brasil. *Revibec, 28*, 1–20.

Wang, J., Cheng, F., & Zhu, P. (2014). Structure and properties of urea-plasticized starch films with different urea contents. *Carbohydrate Polymers, 101*, 1109–1115.

Wu, H., Williams, G. R., Wu, J., Niu, S., Li, H., Wang, H., & Zhu, L. (2018). Regenerated chitin fibers reinforced with bacterial cellulose nanocrystals as suture biomaterials. *Carbohydrate Polymers, 180*, 304–313.

Xiao, Y. C., Zheng, C., Gan, M. K., & Lai, S. L. (2015). The preparation of CMC-PVA-AM dust suppression agent and its application appraisement. In *Advanced Materials Research* (Vol. 1120, pp. 628–634). Trans Tech Publications Ltd.

Yan, J., Nie, W., Xiu, Z., Yuan, M., Zhou, W., Bao, Q., & Yu, F. (2021). Development and characterization of a dust suppression spray agent based on an adhesive NaAlg−gln−poly/polysaccharide polymer. *Science of the Total Environment, 785*, 147192.

Yan, J., Nie, W., Zhang, H., Xiu, Z., Bao, Q., Wang, H., & Zhou, W. (2020). Synthesis and performance measurement of a modified polymer dust suppressant. *Advanced Powder Technology, 31*(2), 792–803.

Zhang, G., Zhang, L., Deng, H., & Sun, P. (2011). Preparation and characterization of sodium carboxymethyl cellulose from cotton stalk using microwave heating. *Journal of Chemical Technology & Biotechnology, 86*, 584–589.

Zhang, H., Nie, W., Wang, H., Bao, Q., Jin, H., & Liu, Y. (2018). Preparation and experimental dust suppression performance characterization of a novel guar gum-modification-based environmentally-friendly degradable dust suppressant. *Powder Technology, 339*, 314–325.

Zhu, Y., Cui, Y., Shan, Z., Dai, R., Shi, L., & Chen, H. (2021). Fabrication and characterization of a multi-functional and environmentally-friendly starch/organo-bentonite composite liquid dust suppressant. *Powder Technology, 391*, 532–543.

CHAPTER 12

Importance of Zinc (Zn) in Poultry Nutrition

A. VARUN,[1] M. NITHYA QUINTOIL,[1] and PORTEEN KANNAN[2]

[1]*Rajiv Gandhi College of Veterinary Education and Research, Puducherry, Tamil Nadu, India*

[2]*Tamil Nadu Veterinary and Animal Sciences University, Chennai, Tamil Nadu, India*

ABSTRACT

In the present-day context, new-generation birds demand timely revision of essential nutrients. In broilers, various trace minerals are required to maintain adequate body functions and general life processes. Trace minerals account for only 0.3% of the total body mineral concentration. The major sources of trace minerals for chickens are feed ingredients such as corn, soybean meal, and wheat, which can vary in their trace mineral content, potentially resulting in an imbalance or depletion of trace minerals. Therefore, mineral supplementation is necessary to meet the nutritional requirements of the bird. Zinc plays essential roles in a wide array of processes including cell proliferation and animal growth, immune development and response, reproduction, gene regulation, and defense against oxidative stress and damage. Zinc sources commonly used are ZnO (72% Zn) and $ZnSO_4$ (36% Zn). Of these, 80 to 90% is ZnO, which is less bioavailable for poultry. The poor availability of zinc in plant feed ingredients is due to the binding of zinc by phytate. Chelated minerals can be absorbed by the intestine intact as a chelate or the mineral can be released in ionic form. Also, chelated trace minerals possess high stability and therefore do not react with single ions, allowing them to be used in lower concentrations. In recent years, researchers have used

Sustainable Agricultural and Food Systems Engineering: New Research, Practices, and Developments.
Jose Juan Buenrostro-Figueroa, Julio César Tafolla-Arellano, Leonardo Sepúlveda Torre,
Porteen Kannan, A. K. Haghi (Eds.)

products from nanotechnology as additives in broiler nutrition, for example, nanosilver, nanoselenium, and zinc oxide nanoparticles.

12.1 INTRODUCTION

In the post-independence era of India, particularly during the last two decades, the revolutionary changes in size, quality, and productivity of broiler chickens are a matter of pride. It can be visualized that at market age, the broiler which attained a live weight of 1,520 grams (g) in 80 days with a feed conversion ratio of 3.5 in 1953 has improved to 2,200 g live weight at 42 days of age with an FCR of 1.6 in 2015. Similarly, India, with a poultry number of 489 million and predicted more than 65 billion egg production, ranks among the top three countries in egg production globally. On the other hand, the production of broilers is increasing at a rate of about 8% to 10% each year. Meanwhile, the increase in the production of poultry/chicken meat has grown from 0.12 million tons in 1981 to 3.4 million tons at present.

Many dietary nutrients are required to promote good health and normal growth of poultry. Studies have shown that mineral deficiencies cause many problems in poultry. Most diets are supplemented with inorganic and organic forms of trace minerals. Inorganic trace minerals such as mineral sulfates and oxides form the bulk of trace mineral supplementation. Trace mineral salts tend to dissociate in the lower pH space of the upper gastrointestinal tract, leading to the minerals being available in different ranges of nutrient and ingredient antagonisms that impair absorption (Underwood & Suttle, 1999). Organically complexed minerals are a type of mineral linked to protein/peptide/amino acids that have higher bioavailability than those inorganic salts (Brown & Zeringue, 1994).

Metals bound to amino acids are practically without electrical charge, so they do not react to changes in pH while moving toward the digestive tract. Trace minerals such as zinc are needed for broiler growth and are considered in various biosynthetic, digestive, and physiological processes. They are components of several proteins that take part in immune defense systems (Dieck et al., 2003), metabolism, and hormone secretion pathways. Going beyond a normal limit, the application of inorganic salts provides less nutrient absorption as well as smaller mineral bioavailability. In addition, current immoderate mineral intakes bring about a polluted environment due to excessive heavy mineral discharge. Because of recent concerns about the existence of potentially polluted minerals, scientists are still focusing on reducing mineral discharge with minimum negative effects on production performance (Devrim et al., 2010).

12.2 ZINC

Zinc (atomic number 30; relative atomic mass 65.38) is a metallic element belonging to group IIb and the fourth period of the periodic table. Zinc is a chalcophilic element like copper and lead and a trace constituent in most rocks. Zinc rarely occurs in its metallic state, but many minerals contain zinc as a major component from which the metal may be economically recovered. Sphalerite (ZnS) is the most important ore mineral and is the principal source of zinc production (WHO, 2001). Adriano (1986) reported an average zinc concentration of 40–90 ppm, with a minimum of 1 ppm and a maximum of 2,000 ppm in non-contaminated soils worldwide. Low levels are found in sandy soils (10–30 ppm), while high contents are found in clays (95 ppm).

12.2.1 BIOLOGICAL UPTAKE OF ZINC

Zinc is first and foremost absorbed from the small intestine. Current data from rat models indicate that zinc absorption occurs by pericellular and carrier-mediated processes, with uptake at the brush border membrane being rate-limiting (Rucker et al., 1994). At concentrations of 0.2 mmol, zinc uptake by isolated brush border membranes is saturated and occurs via an energy-independent carrier-mediated process. At 1 mmol, zinc uptake is non-saturated, suggesting passive diffusion. Intestinal perfusion studies showed that zinc absorption increased linearly between 0.1 and 1.5 mmol, after which it reached a plateau. Low-molecular-weight complexes of zinc facilitate its delivery to the luminal surface of mucosal cells, and newly absorbed zinc is then associated with proteins (e.g., metallothionein and cysteine-rich intestinal protein). The transport across the basolateral surface is known to be energy-dependent. Absorbed zinc retained by the mucosal cells can be lost by desquamation.

12.2.2 FUNCTIONS OF ZINC

Zn is a functional compound of several enzyme systems, including carbonic anhydrase, carboxypeptidase A and B, several dehydrogenases, alkaline phosphatases, ribonuclease, and DNA polymerase. Zinc has significant roles in birds probably because it is a cofactor of more than 200 enzymes. One of the most significant functions of Zn is related to its antioxidant role and its participation in the antioxidant defense system (Powell, 2000).

Zinc acts as a co-factor for the enzyme carbonic anhydrase, which is found in erythrocytes, kidney tubes, gastrointestinal mucosa, and glandular epithelium (Vergil et al., 1970). An important function of zinc is necessary for the growth of germinal and somatic cells, as well as in the maturation of spermatozoa (Underwood et al., 1969). Falchuk (1998) suggested that zinc determines both the type of mRNA transcripts synthesized and the rate of transcription itself. Bughdadi (2013) suggested that zinc supplementation to 15% feed-restricted birds could maintain a fertility level nearly similar to that of the control unrestricted birds. So, the use of zinc oxide supplementation economically could be considered as a beneficial tool for maintaining sound healthy fertility parameters.

12.2.3 DEFICIENCY OF ZINC

Supplee (1961) reported that zinc deficiency in young turkeys is indicated by feather abnormalities characterized by thinning of the web of the primary and secondary wing feathers, expanded hocks, curved or "knock-kneed" legs, and poor growth. This syndrome is easily identified at 2 to 3 weeks of age. Researchers investigated the impact of zinc deficiency on chondrocyte proliferation, differentiation, and apoptosis in the epiphyseal growth plate of juvenile chickens (Wang et al., 2002). Newly hatched broiler chickens were fed either a low-zinc (10 mg/kg diet) or a sufficient zinc (68 mg/kg diet) soy protein-based purified diet. Zinc deficiency significantly inhibited chondrocyte proliferation, promoted cell differentiation, and induced cell apoptosis in the growth plate. Therefore, it is evident that zinc deficiency has a continuous effect on the differentiation, proliferation, and apoptosis of growth plate chondrocytes.

Kopec et al. (2013) investigated how a nutrition plan containing 4% SDBC (Spray-Dried Blood Cells) with or without zinc (Zn) would influence the concentration of 2 histidine heterodipeptides and the antioxidant status of broiler blood and breast muscles. The results showed that when Zn supplements are properly mixed with Spray-Dried Blood Cells in the broiler nutrition plan, an enhancement of superoxide dismutase and glutathione peroxidase activity could be expected. Alkaladi (2013) studied the effect of zinc (Zn) deficiency or supplementation on insulin synthesis and muscular insulin signals in chickens. The study indicated that Zn deficiency leads to a decrease in hepatic glycogen, serum insulin, pancreatic cytosolic Zn, insulin receptor phosphorylation (IRP), serine/threonine kinase

(AKT), phosphoinositide-3-kinase (PI3K), and glucose transporter protein 4 (GLUT4) concentrations, and an increase in blood glucose levels, while Zn supplementation reverses these results. Therefore, it was shown that Zn deficiency adversely affects insulin synthesis and muscular insulin signals.

12.3 BIOAVAILABILITY OF TRACE MINERALS

"Bioavailability" is a term that has been defined as the degree to which consumed nutrient is absorbed in a form that can be utilized in metabolism by the normal animal (Ammerman et al., 1995), thus the mineral is usable at the tissue level as well as at the dietary level.

12.4 BIOAVAILABILITY OF ZINC (ZN)

In birds, numerous factors have a marked effect on Zn absorption. One important factor is the kind of chemical compound. Most studies on Zn bioavailability determine the relative bioavailability mainly related to Zn sulfate and also showed that the bioavailability of Zn from Zinc oxide (ZnO) was only 0.44 compared to that from the sulfates, based on studies reported by Wedekind & Baker (1990) in chicks. Only a few studies address the absolute apparent bioavailability. The absolute bioavailability was established to be about 22 for Zn oxide, 23 for Zn sulfate, and 19 for Zn acetate (Poulsen & Larsen, 1995; Poulsen & Carlson, 2001).

Ranking of criteria to assess the bioavailability of Zn from different sources was described by Sandoval et al. (1997) for broilers. They showed that the concentration of Zn in bones was closely related to bioavailability, followed by Zn content in the liver and pancreas. Mohanna & Nys (1997) showed that more than 90% of the inorganic form of dietary Zn is not bioavailable but is excreted in a commercial production condition. They also showed that sub-therapeutic level supplementation of inorganic Zn in poultry diets may cause soil phytotoxicity.

Burrell et al. (2003) indicated that broilers fed with AvailaZn (Zn pro; USA) amino acid complex showed improved zinc utilization throughout a 45-day production cycle. Furthermore, the AvailaZn amino acid complex might reduce zinc concentrations excreted into the environment, even under circumstances where dietary zinc was being added in excess of the recommended 40 mg/kg of supplemental zinc.

Hudson et al. (2005) observed that zinc from AA chelates has been reported to be more bioavailable than Zn from inorganic sources. They also reported that broilers fed diets with ZnAA complex had a higher body weight than broilers provided diets with Zn sulfate at day 17 and indicated that broiler chick performance may be improved by replacing 40 mg Zn kg^{-1} diet from $ZnSO_4$ with 40 mg Zn kg^{-1} from the zinc-amino acid complex.

Tronina et al. (2007) evaluated the bioavailability of zinc for broiler chickens from the inorganic (ZnO) and organic (Zn-glycine) sources and suggested that, as far as chickens are concerned, better bioavailability of zinc from Zn-glycine contributes to a reduction in environmental emission of this element, increased content of fat in the breast and leg muscles, increased content of zinc in the breast and leg muscle, liver, kidneys, and the thigh bone. Richards et al. (2010) showed that the bioavailability of chelated Zn is between 160% and 250% that of inorganic Zn as sulfate. The rise of bioavailability of chelated trace minerals more or less depends on the reduction of antagonistic reactions with other dietary constituents in the GIT.

Saenmahayak et al. (2010) studied that feeding organic or chelated trace minerals, either with partial replacement of inorganic trace minerals or as top-dressed supplementation, analyzed with industry levels of inorganic trace minerals (sulfates) resulted in an upgrade in high-quality paws in broilers reared up to 49 to 54 days of age. They also reported that when complexed-Zn replaced 40 ppm of $ZnSO_4$, broilers had the most improvement in skin quality, breast meat yield, and FPD severity and incidence.

Brooks et al. (2013) investigated the moderate bioavailability of Zn from Zn propionate relative to feed-grade Zn sulfate using BW gain and bone Zn as response criteria. The authors concluded that supplementation of Zn into a nutrition plan containing basal levels of Zn (22 mg of Zn/kg of diet) would augment feed intake, improve FE, weight gain, and tibia Zn concentration, and also suggested that when fed a corn-based diet with the dietary antagonist phytate, the relative bioavailability of Zn from Zn propionate was 116% of the standard feed-grade $ZnSO_4$ source evaluated.

Sahraei et al. (2013) evaluated the bioavailability and chemical properties of different zinc sources in broiler chickens. Diets included an unsupplemented corn-soybean meal basal diet or the basal diet supplemented with 100, 150, or 200 mg/kg of DM as zinc sources. Tibia and liver zinc capacity was augmented linearly by adding various levels of zinc sources except zinc oxide. The highest zinc concentration in the pancreas was observed only in the basal diet supplemented with 200 mg zinc oxide. Hence, it was concluded that Bioplex Zn at the same level was more bioavailable than others.

12.5 EFFECT OF ORGANIC ZINC ON GROWTH PERFORMANCE IN POULTRY

Zinc is a vital trace element that is needed for bone development, growth, structure, feathering, enzyme function, and appetite in all avian species (Wedekind & Baker, 1990). Zinc is generally added as a supplement to all formulated poultry nutrient plans. Currently, there are two feed-grade Zn sources typically considered by the animal feed industry: ZnO (72% Zn) and $ZnSO_4$ (36% Zn). Of the supplemental Zn fed, 80 to 90% is ZnO, which is less bioavailable for poultry than reagent-grade or feed-grade Zn sulfate (Sandoval et al., 1997; Edwards & Baker, 1999).

However, sulfates (acid salts) are highly water-soluble, allowing reactive metal ions to promote free radical formation. This reaction can lead to the breakdown of vitamins and ultimately to the degradation of fats and oils, decreasing the nutrient value of the diet. Oxides are less reactive but also less bioavailable.

12.5.1 BODY WEIGHT GAIN

Collins & Moran (1999) reported that broilers fed with inorganic Zn and Mn at levels exceeding NRC recommendations (0:0, 60:50, 120:100, or 180:150 ppm) did not influence the body weight throughout the 49 days of the experimental period. Hudson et al. (2005) observed that providing broilers with diets having added zinc from ZnAA and $ZnSO_4$ significantly improved the body weight at 17 days of age.

Osman & Ragab (2007) reported that broiler chicks fed with a diet supplemented with 0.5 g/kg Zn methionine had the heaviest live body weight gain during the periods from 21 to 28, 36 to 42, and 21 to 49 days of age. Nollet et al. (2008) carried out an experiment in broiler chickens supplementing diets with different levels of organic trace minerals (Cu, Fe, Mn, and Zn) and found that the group of birds fed with 50% and 67% bioplexed minerals had significantly better growth rates than the negative control during the starter period.

Abdallah et al. (2009) found that chicks fed a diet containing 100% organic minerals (Zn, Cu, Mn, and Fe) were significantly higher in body weight than those fed with 50% organic minerals. Ao et al. (2009) reported that dietary supplementation of zinc increased weight gain when the broiler chicks were supplemented with organic Zn. Abbas et al. (2010) conducted a study on Ross broiler chicks for 49 days by supplementing feed with inorganic and organic chelates of Zn, Mn, and Cu and reported that different sources and levels did not affect the body weight at 21, 42, and 49 days.

Vladimir et al. (2010) carried out an experiment on broiler chickens fed with dietary supplementation of inorganic Cu, Zn, Fe, Mn, and sodium selenite. They found that when analyzed, feeding trace elements in proteinated form limited to 50% Cu, 20% Fe, Zn, and Mn, and at regular levels of Se, had the same effect on body weight. Zhao et al. (2010) reported that supplementation of chelated trace minerals blended with inorganic trace minerals at 40 ppm of Zn and 60 ppm of Mn in commercial meat broilers showed improved body weight compared to those fed with the inorganic trace minerals. Gheisari et al. (2011) carried out an experiment on broiler chickens supplemented with different levels and sources of zinc, copper, and manganese. They found that neither dietary concentrations nor sources significantly altered the body weight.

Iqbal et al. (2011) reported that body weight gain of broiler chickens was not significantly influenced by supplementation of zinc from different sources (organic and inorganic) and concentrations (40 and 80 mg/kg) in a six-week (0–6 weeks) feeding trial when combined with different sources and concentrations of copper. Ellen et al. (2012) conducted a feeding trial on broilers by supplementing diets with amino acid chelates of copper, zinc, manganese, and iron and found that 2.5 times the minimum suggested level of chelated trace minerals was required to equal the body weight gain of broilers fed diets with inorganic trace mineral premix. Osama et al. (2012) carried out an experiment on Arbor Acres female broiler chicks by feeding a diet supplemented with organic zinc, manganese, and copper and found that there was a remarkable gain in body weight of birds fed with a diet with a combination of 50 mg Zn, 60 mg Mn, and 8 mg Cu.

12.5.2 FEED INTAKE

Vladimir et al. (2010) carried out an experiment on broiler chickens fed with dietary supplementation of inorganic Cu, Zn, Fe, Mn, and sodium selenite. They found that groups fed with trace elements in proteinated form restricted to 50% Cu, 20% Fe, Zn, and Mn, and at regular levels of Se had the same effect on feed intake.

Gheisari et al. (2011) carried out an experiment on broiler chickens supplemented with different levels and sources of zinc, copper, and manganese. They found that feed consumption was significantly higher when the concentration of these trace minerals was 140, 140, and 17 mg/kg from both inorganic and organic sources in the starter period. However, in the grower period, the groups fed with a diet containing 40, 40, and 7 mg/kg of these minerals from inorganic sources had higher feed intake. Iqbal et al. (2011) reported that the feed intake of broiler chickens was not significantly influenced by the supplementation of zinc from different sources (organic and inorganic) and concentrations (40 and 80 mg/kg) in a six-week feeding trial when combined with different sources and concentrations of copper.

Ellen et al. (2012) conducted a feeding trial on broilers by supplementing diets with amino acid chelates of copper, zinc, manganese, and iron. The results showed that 2.5 times the minimum inclusion level of chelated trace minerals was adequate to support normal feed intake, similar to those fed with inorganic trace minerals. Osama et al. (2012) carried out an experiment on Arbor Acres female broiler chicks by feeding them a diet supplemented with organic zinc, manganese, and copper. They found that there was a significant decrease in birds fed with a diet with a combination of 50 mg Zn, 60 mg Mn, and 8 mg Cu.

12.5.3 FEED EFFICIENCY

Abbas et al. (2010) conducted a study on Ross broiler chicks for 49 days by supplementing feed with inorganic and organic chelates of Zn, Mn, and Cu. The broilers fed diets containing 40, 40, and 7 mg/kg Zn, Mn, and Cu, respectively from organic sources had improved feed conversion ratio. Vladimir et al. (2010) carried out an experiment on broiler chickens fed with dietary supplementation of inorganic Cu, Zn, Fe, Mn, and sodium selenite and 50% of the inorganic source in bioplexed form and found that groups fed with trace elements in proteinated form restricted to 50% Cu, 20% Fe, Zn, and Mn and on regular levels of Se had the same effect on feed conversion ratio.

Iqbal et al. (2011) reported that the feed conversion ratio of broiler chickens was not influenced by supplementation of zinc from different sources (organic and inorganic) and concentrations (40 and 80 mg/kg) in a six-week (0–6 weeks) feeding trial when combined with different sources and concentrations of copper. Ellen et al. (2012) conducted a feeding trial on broilers by supplementing diets with amino acid chelates of Copper, Zinc, Manganese, and Iron and found that there was no significant difference in feed efficiency. Osama et al. (2012) carried out an experiment on Arbor Acres female broiler chicks by feeding the diet supplemented with organic zinc, manganese, and copper and found that there was a significantly improved FCR in birds fed with a diet with a combination of 50 mg Zn, 60 mg Mn, and 8 mg Cu.

12.5.4 LIVABILITY

Hudson et al. (2005) observed that providing broilers with diets having added zinc from either organic or inorganic sources had no significant effect on mortality. Ellen et al. (2012) conducted a feeding trial on broilers by supplementing diets with amino acid chelates of copper, zinc, manganese, and iron and found no significant difference in mortality. Osama et al. (2012) carried out an experiment on Arbor Acres female broiler chicks by feeding a diet supplemented with organic zinc, manganese, and copper and found that mortality was not related to the dietary treatments.

12.6 TIBIA ZINC

When considering the accumulation of Zn in different tissues, bone was found to contain the highest level of accumulated Zn, while the liver and kidney were found to contain lower levels than the bone (Shan, 1993). Emmert & Baker (1995) confirmed that Zn could accumulate in bone, liver, and intestine and could subsequently be released for use during a period of Zn deficiency. They also indicated that chicks fed a diet containing the minimum requirement of Zn could not accumulate reserves of Zn that become available for use during a subsequent period of Zn depletion.

Ao et al. (2011) conducted an experiment to study the effect of feeding various concentrations and forms of zinc (Zn) on the tibia Zn status of broiler chicks. They observed that chicks fed a diet with 12 mg of organic Zn per kg had the same growth rate, but lower tibia Zn content compared with those fed either 40 mg of Zn as sulfate or 24 mg of Zn as organic. Hence, they concluded that lower than NRC or commercial levels of Zn provided by an organic source can be used in broiler diets and may better support the optimal growth of broiler chicks.

Stofanikova et al. (2014) evaluated the effect of diet with various Zn amounts on the mechanical properties of bones in broilers. There were significant changes in the mechanical properties of the bones, suggesting reduced weight-bearing capabilities in broilers fed a diet with a zinc content of 50 mg/kg of feed compared to chickens fed with an increased level of zinc (100 mg/kg of feed). In the group with lower Zn content, the incidence of various deformities was shown in 10% of the broilers. They concluded that Zn content in feed has a direct statistically remarkable result on the strength of bones in broiler chickens.

12.7 EFFECT OF ORGANIC ZINC ON CARCASS CHARACTERISTICS IN POULTRY

Rossi et al. (2007) investigated the effect of improved levels of organic Zn (Bioplex Zn) on growth production and carcass traits of broilers fed a plan consisting of organic Se (Sel-Plex). Nutrition treatment plans considered adding 0, 15, 30, 45, or 60 ppm organic Zn to a basal nutrition procedure with 60 ppm Zn using an inorganic source, and it is observed that organic Zn did not change the growth program of broilers but enhanced the resistance of skin to tearing, therefore improving carcass quality.

12.7.1 DRESSING PER CENT

Vladimir et al. (2010) carried out an experiment on broiler chickens fed with dietary supplementation of inorganic Cu, Zn, Fe, Mn, and sodium selenite. They found that groups fed with trace elements in proteinated form restricted to 50% Cu, 20% Fe, Zn, and Mn, and at regular levels of Se had the same effect on carcass yield. Ellen et al. (2012) conducted a feeding trial on broilers by supplementing diets with amino acid chelates of copper, zinc, manganese, and iron. They found that the dressing percentage was significantly higher in the group fed with 2.5 g/ton, 11.25 g/ton, 15.0 g/ton, and 18.75 g/ton of amino acid chelates of Cu, Zn, Mn, and Fe, respectively.

12.7.2 ORGAN WEIGHT

Liver: Osman & Ragab (2007) reported that broiler chicks fed with a diet supplemented with 0.3 g/kg Zn methionine had the highest percentage of liver. Iqbal et al. (2011) observed that broiler chickens fed with dietary supplements of zinc from different sources and concentrations had a significant effect on liver weight at lower levels of 40 mg/kg zinc from an organic source when mixed with various concentrations of copper from different sources.

12.7.3 HEART

Iqbal et al. (2011) observed that broiler chickens fed with dietary supplements of zinc from different sources and concentrations had a significant effect on heart weight at lower levels of 40 mg/kg zinc from an organic source when combined with different sources and concentrations of copper. Gizzard: Osman & Ragab (2007) reported that broiler chicks fed with a diet supplemented with 0.3 g/kg Zn methionine had the highest gizzard percentage.

12.8 NANO-ZINC ON BROILER PERFORMANCE

Mohammadi et al. (2015) studied the characteristics of various zinc complexes and nano-complexes on broiler production and performance. Broilers in six groups considered a basal nutrition plan (without zinc) and a basal diet supplemented with zinc-sulfate, zinc-methionine, zinc-nano-sulfate, zinc-nano-methionine, and zinc-nanomax with a concentration of 80 mg/kg of diet. They observed that the application of zinc in the structure of zinc-nano-methionine and zinc-nano-max at concentrations of 80 mg/kg of the fed program helped to enhance the growth plan fairly. However, the zinc-nano-sulfate nutrition plan reduced the growth program in broilers.

Ahmadi et al. (2014) observed that levels of zinc oxide nanoparticles from 60 to 90 mg/kg basal diet improved the antioxidant state and serum enzymes of broiler chickens and also stated that Zinc oxide nanoparticles, based on the current study, may be used as a new additive to meet zinc requirements in broiler nutrition.

12.8.1 ZINC AND METALLOTHIONEIN (MT) GENE EXPRESSION IN BROILERS

Metallothionein (MT) is a small, cysteine-rich protein that can bind Zn and other heavy metals with an acceptable connection. The single isoform of Metallothionein (MT) in the chicken was noticed in the liver, pancreas, kidney, and intestinal mucosa (Oh et al., 1979; McCormick, 1984; Sandoval et al., 1997; Cao et al., 2002). However, the absorption of tissue Metallothionein (MT) in chickens was highly related to Zn intake (Fleet et al., 1988), and long-term supplementation was limited to continue to raise the synthesis of Metallothionein (MT) in rats despite high-level exposure to dietary Zn (Reeves, 1995). This case has also been observed in avian species fed elevated Zn for three weeks (Sandoval et al., 1997; Cao et al., 2002). Tissue

MT, as well as bone Zn, could potentially be considered as an index of Zn bioavailability in chicks; however, only in short-term Zn supplementation trials with dietary Zn concentrations nearby the specification.

Shen et al. (2013) studied the effect of intravenously injected zinc (Zn) on tissue Zn absorption and pancreas (MT) gene expression in broilers. Chickens were injected with either a 0.9% (w/v) NaCl solution (control) or a saline solution supplemented with Zn sulfate, or one of the 3 organic Zn chelates with weak (Zn-AA W), ordinary (Zn-Pro M), or high (Zn-Pro S) chelation strengths at two dosages determined based on two Zn absorbability levels (6% and 12%). Chickens injected with the Zn-Pro S had lower bone Zn absorption than those injected with the Zn-Pro M or Zn-AA W on day 6 after injections. They observed that the injected Zn-Pro S was more or less acceptable for using bone Zn of broilers. The pancreas Zn absorption and pancreas Metallothionein gene expressions may not be sensitive enough to detect variation in the tissue utilization of injected Zn in broilers between organic and inorganic Zn.

KEYWORDS

- **absorption**
- **bioavailability**
- **broilers**
- **gene expressions**
- **metallothionein**
- **nano-Zn**
- **zinc**

REFERENCES

Abbas, A., Ali, R. F., Majid, T., & Mohammad, M. G. (2010). Effects of organic chelates of zinc, manganese, and copper in comparison to their inorganic sources on the performance of broiler chickens. *Journal of Animal and Plant Sciences, 6*(2), 630–636.

Abdallah, A. G., El-Husseiny, O. M., & Abdel-Latif, K. O. (2009). Influence of some dietary organic mineral supplementations on broiler performance. *International Journal of Poultry Science, 8*(3), 291–298.

Ahmadi, F., Ebrahimnezhad, Y., Maheri sis, N., & GhiasiGhalehkandi, J. (2014). The effects of zinc oxide nanoparticles on performance, digestive organs, and serum lipid concentrations in broiler chickens during the starter period. *International Journal of Biosciences, 3*(7), 23–29.

Ammerman, B. C., Baker, H. D., & Lewis, A. J. (1995). *Bioavailability of Nutrients for Animals: Amino Acids, Minerals and Vitamins.* Academic Press, Inc.

Ao, T., Pierce, J. L., Power, R., Pescatore, A. J., Cantor, A. H., Dawson, K. A., & Ford, M. J. (2009). Effects of feeding different forms of zinc and copper on the performance and tissue mineral content of chicks. *Poultry Science, 88*, 2171–2175.

Brooks, M. A., Grimes, J. L., Lloyd, K. E., Verissimo, S., & Spears, J. W. (2013). Bioavailability in chicks of zinc from zinc propionate. *Journal of Applied Poultry Research, 22*(2), 153–159.

Brown, T. F., & Zeringue, L. K. (1994). Laboratory evaluations of solubility and structural integrity of complexed and chelated trace mineral supplements. *Journal of Dairy Science, 77*, 181–189.

Burrell, A. L., Dozier III, W. A., Davis, A. J., Compton, M. M., Freeman, M. E., Vendrell, P. F., & Ward, T. L. (2003). Responses of broilers to dietary zinc concentrations and sources in relation to environmental implications. *British Poultry Science, 45*, 255–263.

Devrim, S., Aksu, T., Ozsoy, B., & Baytok, E. (2010). The effects of replacing inorganic with a lower level of organically complexed minerals (Zinc, Copper, and Manganese) in broiler diets on lipid peroxidation and antioxidant defense systems. *Asian-Australasian Journal of Animal Sciences, 23*(8), 1066–1072.

Dieck, H. T., Doring, F., Roth, H. P., & Daniel, H. (2003). Changes in rat hepatic gene expression in response to zinc deficiency as assessed by DNA arrays. *Journal of Nutrition, 133*, 1004–1010.

Edwards, H. M., & Baker, D. H. (1999). Bioavailability of zinc in several sources of zinc oxide, zinc sulfate, and zinc metal. *Journal of Animal Science, 77*, 2730–2735.

Ellen, B. B., Merca, F. E., Angeles, A. A., Acda, S. P., & Luis, E. S. (2012). Effects of supplementing diets with amino acid chelates of copper, zinc, manganese, and iron on the performance of broilers. *Philippine Journal of Veterinary and Animal Sciences, 38*(1), 1–10.

Fleet, J. C., Qureshi, M. A., Dietert, R. R., & McCormick, C. C. (1988). Tissue-specific accumulation of metallothionein in chickens as influenced by the route of zinc administration. *Journal of Nutrition, 118*, 176–182.

Gheisari, A. A., Rahimi-Fathkoohi, A., Toghyani, M., & Gheisari, M. M. (2011). Influence of feeding diets supplemented with different levels and sources of zinc, copper, and manganese on the mineral concentrations in tibia and performance of broiler chickens. *Asian Journal of Animal and Veterinary Advances, 6*(2), 166–174.

Hudson, B. P., Dozier, W. A., & Wilson, J. L. (2005). Broiler live performance response to dietary zinc Source and the influence of zinc supplementation in broiler breeder diets. *Animal Feed Science and Technology, 118*, 329–335.

Iqbal, Shariq, Deo, Chandra, Mandal, A. B., Tyagi, K. Praveen, & Wachat, V. B. (2011). Response of feeding different sources and concentrations of zinc and copper on the performance of broiler chickens. *Indian Journal of Poultry Science, 46*(3), 330–336.

Kopeć, W., Jamroz, D., Wiliczkiewicz, A., Biazik, E., Pudlo, A., Hikawczuk, T., & Korzeniowska, M. (2013). Influence of different histidine sources and zinc supplementation of broiler diets on dipeptide content and antioxidant status of blood and meat. *British Poultry Science, 54*(4), 454–465.

Mohammadi, V., Ghazanfari, S., Mohammadi-Sangcheshmeh, A., & Nazaran, M. H. (2015). Comparative effects of zinc-nano complexes, zinc-sulfate, and zinc-methionine on performance in broiler chickens. *British Poultry Science, 56*(4), 486–493.

Nollet, L., Huyghebaert, G., & Spring, P. (2008). Effect of different levels of dietary organic (Bioplex) trace minerals on live performance of broiler chickens by growth phases. *Journal of Applied Poultry Research, 17*, 109–115.

Oh, S. H., Nakaue, H., Deagen, J. T., Whanger, P. D., & Arscott, G. H. (1979). Accumulation and depletion of zinc in chick tissue metallothioneins. *Journal of Nutrition, 109*, 1720–1729.

Osama, M., El-Hussein, M., Samia, A., Hashish, R., Ali, A., Sohair, A., & Ahmad Olemy, A. (2012). Effects of feeding organic zinc, manganese, and copper on broiler growth, carcass characteristics, bone quality, and mineral content in bone, liver, and excreta. *International Journal of Poultry Science, 11*(6), 368–377.

Petrovic, V., Nollet, L., & Kovac, G. (2010). Effect of dietary supplementation of trace elements on the growth performance and their distribution in the breast and thigh muscles depending on the age of broiler chickens. *Acta Veterinaria, 79*, 203–209.

Poulsen, H. D., & Carlson, D. (2001). Bioavailability of zinc from different zinc sources. In Proc. 52nd Ann. Meeting Eur. Ass. Anim. Prod., Budapest, Hungary (p. 123). EAAP publications (Abstract).

Poulsen, H. D., & Larsen, T. (1995). Zinc excretion and retention in growing pigs fed increasing levels of zinc oxide. *Livestock Production Science, 43*, 235–242.

Rossi, P., Rutz, F., Anciuti, M. A., Rech, J. L., & Zauk, N. H. F. (2007). Influence of graded levels of organic zinc on growth performance and carcass traits of broilers. *Journal of Applied Poultry Research, 16*, 219–225.

Rucker, R. B., Lönnerdal, B., & Keen, J. L. (1994). Intestinal absorption of nutritionally important trace minerals. In L. R. Johnson (Ed.), *Physiology of the Intestinal Tract* (3rd ed., pp. 2183–2202). Raven Press.

Saenmahayak, B., Bilgili, S. F., Hess, J. B., & Singh, M. (2010). Live and processing performance of broiler chickens fed diets supplemented with complexed zinc. *Journal of Applied Poultry Research, 19*, 334–340.

Sahraei, M., Janmmohamadi, H., Taghizadeh, A., Moghadam, G. A., & Rafat, S. A. (2013). Estimation of the relative bioavailability of several zinc sources for broilers fed a conventional corn-soybean meal diet. *Journal of Poultry Science, 50*(1), 53–59.

Sandoval, M., Henry, P. R., Ammerman, C. B., Miles, R. D., & Littell, R. C. (1997). Relative bioavailability of supplemental inorganic zinc sources for chicks. *Journal of Animal Science, 75*, 3195–3205.

Shen, S. F., Wang, R. L., Lu, L., Li, S. F., Liu, S. B., Xie, J. J., & Luo, X. G. (2013). Effect of intravenously injected zinc on tissue zinc and metallothionein gene expression of broilers. *British Poultry Science, 54*(3), 381–390.

Štofaníková, J., Šály, J., Molnár, L., Sesztáková, E., & Vrabec, V. (2014). The mechanical properties of broiler chicken bones are affected by different dietary zinc levels. *African Journal of Biotechnology, 11*(20), 4681–4686.

Tronina, W., Kinal, S., & Lubojemska, B. (2007). Effect of various forms of zinc applied in concentrate mixtures for broiler chickens on its bioavailability as well as meat composition and quality. *Polish Journal of Food and Nutrition Sciences, 57*(4), 577–581.

Underwood, E. J., & Somers, M. (1969). Studies of zinc nutrition in sheep. *Australian Journal of Agricultural Research, 20*, 889–897.

Underwood, E. J., & Suttle, N. F. (1999). *The Mineral Nutrition of Livestock* (3rd ed.). CAB International.

Virgil, W. H., & Melvin, J. S. (1970). Minerals. In *Duk's Physiology of Domestic Animals* (8th ed., pp. 1463). Comstock Publishing Company.

Wang, X., Fosmire, G. J., Gay, C. V., & Leach, R. M. (2002). Short-term zinc deficiency inhibits chondrocyte proliferation and induces cell apoptosis in the epiphyseal growth plate of young chickens. *Journal of Nutrition, 132*(4), 665–673.

Wedekind, K. J., & Baker, D. H. (1990). Zinc bioavailability in feed-grade sources of zinc. *Journal of Animal Science, 68*, 684–689.

World Health Organization. (2001). E*nvironmental Health Criteria* 221. Zinc. WHO.

Zhao, J., Shirley, R. B., Vazquez-Anon, M., Dibner, J. J., Richards, J. D., Fisher, P., & Giesen, A. F. (2010). Effects of chelated trace minerals on growth performance, breast meat yield and footpad health in commercial meat broilers. *Journal of Applied Poultry Research, 19*, 365–372.

CHAPTER 13

Anti-Bacterial and Anti-Viral Properties of Phytobiotics in Poultry Nutrition

A. VARUN,[1] M. NITHYA QUINTOIL,[1] S. EZHILVALAVAN,[2] and PORTEEN KANNAN[2]

[1]*Rajiv Gandhi Institute of Veterinary Education and Research, Puducherry, Tamil Nadu, India*

[2]*Tamil Nadu Veterinary and Animal Sciences University, Chennai, Tamil Nadu, India*

ABSTRACT

Commercial broiler is one of the most widespread birds, and its maturation and development are dynamic processes. Currently, there is a need to completely avoid the use of antibiotics in poultry feed due to increasing consumer concerns about antibiotic residues in eggs and meat. As a result, alternatives to antibiotics such as phytobiotics, prebiotics, probiotics, and synbiotics are being considered in poultry diets to improve productivity, growth, and disease resistance. In this context, phytobiotics such as garlic (*Allium sativum*) and nilavembu (*Andrographis paniculata*) can be considered substitutes. The effects of nilavembu and garlic have been studied independently in different research studies. However, phytochemical screening of nilavembu and garlic has revealed the presence of similar phytochemical compounds that are common to both. Garlic has demonstrated antimicrobial activity, while nilavembu has antiviral properties. The combination of these two phytobiotics has indeed resulted in a synergistic effect on production and enhanced resistance in broilers.

Sustainable Agricultural and Food Systems Engineering: New Research, Practices, and Developments.
Jose Juan Buenrostro-Figueroa, Julio César Tafolla-Arellano, Leonardo Sepúlveda Torre, Porteen Kannan, A. K. Haghi (Eds.)

13.1 INTRODUCTION

Feed additives are commonly added to poultry feed to improve the nutritive value of ingredients and enhance broiler performance by increasing growth rate and improving feed conversion efficiency. Currently, there is a need to completely avoid the use of antibiotics in poultry feeding due to increasing consumer concern about poultry drug residues in meat and eggs. The production performance of birds has been increased by utilizing phytogenic substances, which are mainly responsible for stimulating and secreting digestive enzymes, leading to enhanced digestion and absorption (Recoquillay, 2006). Furthermore, the presence of active ingredients and phenolic compounds can reduce the number of intestinal pathogens, thus minimizing nutrient loss. Garlic is a proven herbal medicine with antibacterial, antiviral, antifungal, anticancer, antioxidant, immunomodulatory, anti-inflammatory, hypoglycemic, and hypocholesterolemic effects. Nilavembu is known as Mahatikta, literally meaning "king of bitters." The herb and its isolates are reported to possess antiviral, anti-inflammatory, hepatoprotective, astringent, and antipyretic properties. The phytochemical survey of garlic exhibits the presence of chemical compounds such as alkaloids, flavonoids, saponin, tannin, carbohydrates, cardiac glycosides, and glycosides (Pavni et al., 2011), whereas nilavembu contains triterpenoids, saponin, quinone, coumarin, protein, sugar, steroids, alkaloids, flavonoids, tannins, and gum (Salna et al., 2011). The effects of nilavembu and garlic on chickens have been studied separately in a few research studies. However, the phytochemical screening of garlic and nilavembu revealed the presence of many similar phytochemical compounds that are common to both.

13.2 NILAVEMBU (*ANDROGRAPHIS PANICULATA*)

Andrographis paniculata belongs to the family *Acanthaceae*. It is native to tropical Asian countries, with South India and Sri Lanka considered as the center of origin and diversity for native populations. Nilavembu is known in North-Eastern India as Maha-Tikta (referred to as the "king of bitters"). In Ayurvedic medicine, it is known as Kalmegh or Kalamegha (meaning "dark cloud"). In Tamil, it is called Nilavembu, which translates to "neem of the ground." There are about 28 species in the genus Andrographis, with *Andrographis paniculata* being the most popular. It is widely used in Ayurveda, Unani, and Siddha medicines, as well as in home remedies for various illnesses and traditional medicine in India. Nilavembu is reported to possess anti-inflammatory, hepato-protective, astringent, and antipyretic properties, and is helpful in the treatment of bronchitis, diarrhea, and influenza, mainly due to the presence of bioactive compounds in the herb and its isolates such as andrographolide, neoandrographolide, dehydroandrographolide, isoandrographolide, etc.

Abhishek et al. (2010) reported that nilavembu (*Andrographis paniculata* Nees) has a wide range of medicinal and pharmacological applications. It is most extensively used in traditional systems of medicine, exhibiting anti-inflammatory, antibacterial, antioxidant, antiparasitic, antispasmodic, antidiabetic, anticarcinogenic, antipyretic, hepatoprotective, nematocidal properties, etc.

13.2.1 PHYTOCHEMICAL PROPERTIES OF NILAVEMBU

Salna et al. (2011) studied the phytochemical analysis of Nilavembu and found the presence of different types of compounds like triterpenoids, tannins, alkaloids, steroids, and saponin, which could be extensively used to cure various ailments. Qualitative analysis of *Andrographis neesiana* plant parts showed the presence of various compounds of therapeutic importance, including tannins, saponin, phenolic compounds, glycosides, mucilases, steroids, and triterpenoids, whereas the absence of carbohydrates, proteins, and amino acids has been reported for the first time in India (Alagesaboopathi & Sivakumar, 2011). The preliminary phytochemical analysis of Nilavembu chooranam and nilavembu tablet revealed the presence of glycosides, tannins, and alkaloids by thin-layer chromatography (Balakrishnan et al., 2013).

13.2.2 ACTIVE PRINCIPLES OF NILAVEMBU

The bioactive principle present in nilavembu which gives a bitter taste is andrographolide. It is a diterpenoid and is mainly responsible for its anti-inflammatory effects. Apart from the anti-inflammatory effect, andrographolide also exhibits immunomodulatory effects by effectively enhancing the cytotoxic T cells, natural killer (NK) cells, phagocytosis, and antibody-dependent cell-mediated cytotoxicity (ADCC). Collectively, all these properties of andrographolide form the base for the use of this supernatural compound to restrain virus replication and virus-induced pathogenesis. Effects of andrographolide on the immune system, virus entry, and propagation include inhibiting the H1N1-induced Retinoic acid-inducible gene-1-like receptor signaling pathway, suppressing Hepatitis C virus genome replication by promoting IFNa response, increasing CD4+ T cell count, inhibiting Chikungunya virus RNA copy, etc. The effect of nilavembu was most extensively carried out in humans against various viral diseases, but in chickens, there is a scanty report on the action of nilavembu on production performance and anti-viral properties. The antibacterial action of nilavembu versus Influenza has been studied by (Chen et al., 2010), Herpes simplex virus-1 (Wiart et al., 2005), and HIV (Calabrese et al., 2000). Therefore, this chapter will discuss the production performance and anti-viral properties of nilavembu against common viral diseases in chickens.

13.2.3 EFFECT OF NILAVEMBU ON BODY WEIGHT

Mathivanan et al. (2006) investigated the effect of *Andrographis paniculata* at 2 g/kg feed on the productive performance of commercial broilers from 0–42 days and they found a highly significant ($P<0.05$) difference in the body weight (1875.07± 38.88 g) in birds fed with *Andrographis paniculata* when compared to the control group (1785.52 ± 36.47 g). Malahubban et al. (2013) evaluated the effect of different dosages of *Andrographis paniculata* in the diet of broiler chicken on growth performance. The dietary treatments were as follows: basal diet (control), basal diet + 0.2% *A. paniculata* (2 g/kg), basal diet + 0.4% *A. paniculata* (4 g/kg), and basal diet + 0.8% *A. paniculata* (8 g/kg). Birds who received 8 g of *A. paniculata*/kg of feed had a significantly ($P<0.05$) higher body weight (1320.8 ± 57.57 g) when compared to the control group (1218.0 ± 48.39 g). Similarly, Laing et al. (2013) investigated the effects of *Andrographis paniculata* and *Zingiber cassumuna* mixture in day-old commercial broiler chicken (Arbor Acres) for a period of 42 days. The basal experimental diet was supplemented with *Andrographis paniculata* and *Zingiber cassumunar* mixture at 0, 0.05, 0.10, 0.15, and 0.20%, respectively. They reported that broilers fed with 0.05% *Andrographis paniculata* and *Zingiber cassumunar* mixture showed a significantly ($P<0.05$) improved body weight when compared to the control group. Improved body weight could be simply because of the existence of bioactive substances that will suppress the growth of pathogenic microorganisms. Supplementation of nilavembu causes reduced pH and decreased intestinal thickness which leads to improved digestion and absorption.

13.2.4 EFFECT OF NILAVEMBU ON FEED CONSUMPTION

Mathivanan et al. (2006) observed the effect of *Andrographis paniculata* supplementation at 1, 2, and 3 g/kg feed on the production performance of commercial day-old broiler chickens for 42 days. Results proved that the consumption of feed was much lower ($P<0.01$) in the birds fed with *A. paniculata* at 2.0 and 3.0 g/kg of diet when compared with the control group. This could be because of the bitter taste of nilavembu, which mainly influences feed consumption. However, Malahubban et al. (2013) supplemented different dosages of *Andrographis paniculata* (2 g, 4 g, and 8 g/kg of feed) in the diet of day-old male broiler chickens. They showed that cumulative feed consumption was not statistically influenced by the supplementation of *A. paniculata* at different levels.

13.2.5 EFFECT OF NILAVEMBU ON IMMUNE PERFORMANCE

Suriani et al. (2015) evaluated the effect of the ethanolic extract of *A. paniculata* against the Newcastle disease virus. They found that the ethanolic extract of *A. paniculata* at the extraction of 15 μg/ml had better antiviral activity and was non-toxic to the embryo. Al-Hadid (2016) evaluated the antiviral effect of the medicinal plant and reported that the HA titer of 3 to 5.6 for plant extracts of 20–80% mortality in the virus neutralization study. Similarly, Arify et al. (2018) reported that the aqueous extract of nilavembu at a higher concentration (100 mg/ml) and the ethanol extract of nilavembu (10 mg/ml and 100 mg/ml) were found to have better antiviral properties against Velogenic strains of Newcastle disease in broiler chickens.

13.2.6 ANTI-BACTERIAL EFFECT OF NILAVEMBU

Das et al. (2013) found that nilavembu had shown a 16 mm zone of inhibition versus *Staphylococcus aureus*. Padmalochana & Rajan (2014) studied the comparative effect of three different solvent-derived leaf extracts of *A. paniculata* and they showed that the ethanol extract of nilavembu had a 12 mm zone of inhibition for *Klebsiella sp.*

13.3 GARLIC (*ALLIUM SATIVUM*)

Garlic (*Allium sativum*) is the king of medicinal plants, which has growth-promoting effects and improves the production performance of broiler chicks (Tollaba & Hassan, 2003). It has:

- Anti-inflammatory;
- Antibacterial;
- Anti-protozoal;
- Anti-viral;
- Antioxidant;
- Anti-fungal;
- Anti-cancer;
- Immunomodulatory;
- Hypoglycemic; and
- Cardiovascular properties.

Moreover, it boosts the immune system, improves body weight gain, enhances the digestibility of ingredients, decreases bad cholesterol, and also enhances the meat standard factors (Rehman & Munir, 2015). Allicin is the principal compound in aqueous garlic extract or raw garlic that is responsible for this effect. When a raw garlic bulb is split or crushed, the enzyme alliinase converts alliin (a non-protein amino acid within the intact garlic) into allicin (allyl 2 propenethiosulfinare or diallylthiosulfinate). Other important sulfur-containing compounds present in garlic homogenates are diallymeghylthiosulfonae, I-propenyl allyl thiosulfonate, and y-L-glutamyl-S-alkyl-L-cysteine (Bannerjee et al., 2002). Sulfur-containing compounds such as alliin, diallyl sulfides, and allicin present in garlic are responsible for this (Amagase et al., 2001).

- Antioxidant;
- Antithrombotic;
- Anti-cancerous;
- Anti-bacterial;
- Anti-fungal;
- Anti-parasite;
- Anti-viral; and
- Vasodilator characteristics.

Yu-Yan & Liu (2001) studies showed that garlic weakened the hepatic activities of lipogenic and cholesterogenic enzymes such as malic enzyme, fatty acid synthase, glucose-6-phosphate dehydrogenase, and cholesterol-7-alpha-hydroxylase, as well as 3-hydroxy-3-methyl-glutaryl-coA reductase, while it increases the excretion of cholesterol. This hepato-protective activity of garlic might be attributed to its excellent antioxidant potential to prevent liver damage, thereby limiting the bioavailability of these biomarkers in the blood.

13.3.1 PHYTOCHEMICAL PROPERTIES OF GARLIC

Phytochemical screening of the extracts from the various developing stages of *Allium sativum* pods, shoots, and roots indicated the presence of flavonoids, monosaccharides, glycosides, saponins, terpenoids, and free-reducing sugars. This study revealed the presence of glycosides and terpenoids in developing pods and flavonoids, and glycosides in developing roots (Pavni et al., 2011). The crude extract of *A. sativum* bulb showed higher protein concentrations; medium concentrations of saponins, reducing sugars, oils, and steroids, while lower concentrations of flavonoids and acidic compounds. Tannins and resins were absent in the extract (Ameh et al., 2013). The presence of tannins, flavonoids, and phlobatannins in the plant extract, whereas steroids, glycoside, and non-reducing sugar were absent in the plant material as reported by Abdullahi et al. (2014).

13.3.2 EFFECT OF GARLIC ON BODY WEIGHT AND BODY WEIGHT GAIN

Ologhobo et al. (2008) showed that supplementation of raw garlic paste and sun-dried garlic powder in the diet of starter and finisher of broiler at 1%, 2%, and 3% had no remarkable effect on body weight and body weight gain. According to Anvar et al. (2012), supplementation of garlic and thyme extract at 0.3% and 0.6% to broiler chicks for a period of 42 days showed that there was no remarkable difference in body weight boost over the analyzed groups.

Onyimonyi et al. (2012) conducted an experiment in broiler chickens to evaluate the growth properties of dried garlic powder. The dietary treatment included the inclusion of 0%, 0.25%, 0.50%, and 0.75% garlic powder. Results showed that the effect of feeding varying dietary levels of garlic powder at 0.75% had a significantly ($P<0.05$) higher final body weight of 2,780 g, which differed significantly from the birds in the control, 0.25, and 0.50% treated groups (2,240, 2,350, and 2,460 g), respectively. The average daily gain followed the same trend as that of body weight. This might be due to enhanced villus and goblet cell numbers in the duodenum, jejunum, and ileum of birds supplemented with garlic. As a result of these intestinal morphological changes, the entire absorptive process in the birds is better activated.

Oleforuh et al. (2014) evaluated the effect of supplementation of garlic and ginger in the form of powder (14 g/kg feed) and liquid (50 ml/liter of drinking water) to commercial broiler chickens for a period of 42 days and showed that the birds fed with the powder form of garlic showed significantly ($P<0.05$) better final body weight (2475.00 g) and daily body weight gain (336.45 g) than those fed the water-based infusion (2150.00 g and 289.29 g, respectively). This may be due to Allicin in garlic that increases the conduct of the intestinal flora thereby improving digestion and enhancing the utilization of energy, leading to improved growth. Furthermore, garlic enhances pancreatic enzyme activity and provides a better space for absorption of nutrients.

Makwana et al. (2015) conducted an experiment on commercial broiler (Cobb 400) chickens to study the effect of incorporating garlic powder on growth performance. Basal diets were supplemented with garlic at 0%, 0.1%, and 0.5%. Total body weight gain was significantly ($P<0.01$) higher in birds supplemented with 0.1% garlic powder when compared to 0.5% garlic supplementation and the control group. Better weight gain in garlic-fed birds might be due to the action of Allicin that constrains the growth of pathogenic bacteria by intervening with bacterial cell metabolism.

13.3.3 EFFECT OF GARLIC ON SERUM PARAMETERS

Ologhobo et al. (2008) showed that supplementation of raw garlic paste and sun-dried garlic powder in the broiler starter and finisher diets at 1%, 2%, and 3% had significantly ($P<0.05$) decreased the levels of total cholesterol, triglycerides, and low-density lipoprotein concentrations. The significant ($P<0.05$) reduction observed in total cholesterol was probably due to the effect of garlic on the activation of lipogenic and cholesterogenic enzymes such as malic enzyme, fatty acid synthase, glucose-6-phosphate dehydrogenase, and 3-hydroxy-3-methyl-glutaryl-coenzyme A (HMG CoA reductase).

Rahimi et al. (2011) observed that supplementation of three herbs: thyme (*Thymus vulgaris*), coneflower (*Echinacea purpurea*), and garlic (*Allium sativum*) on serum biochemistry parameters of male broiler chickens up to 42 days of age. The results showed that supplementation of garlic significantly reduced the serum levels of triglycerides, low-density lipoprotein, and total cholesterol but significantly ($P<0.05$) increased the level of high-density lipoprotein. There was no significant difference observed in triglyceride levels among all the treatment groups and the control group. Jimoh et al. (2012) supplemented garlic at 0.0, 0.5, 1.0, 1.5, 2.0, and 2.5 g/kg of diet to broiler chicks for up to the 7th week of age. The results showed that garlic supplementation at 2.0 and 2.5 g/kg significantly ($P<0.05$) decreased the serum concentration of triglycerides. Total cholesterol and low-density lipoprotein were not significantly influenced, but high-density lipoprotein levels in the broiler blood slightly increased in all the groups fed with the supplemented diet. This shows that garlic powder supplementation in the experimental diet exhibited anti-lipogenic and anti-cholesterogenic effects on the birds.

13.3.4 ANTI-BACTERIAL EFFECT OF GARLIC

Newall et al. (1996) reported that garlic contains 33 sulfur-containing compounds, several enzymes, 17 amino acids, and minerals. They also found that sulfur compounds had an antioxidant effect. Zakaria (2003) studied the antibacterial effect of aqueous garlic extract on gram-positive and gram-negative bacteria and reported that around 15–29% of the antibacterial activity of garlic was lost when stored for six days. Mukhtar & Ghori (2012) conducted an experiment to investigate the antibacterial activity of garlic, cinnamon, and turmeric by the disc diffusion method. They showed that garlic had inhibitory activity against *E. coli* and also found that the aqueous extract of garlic was more useful than the ethanolic extract. Verma et al. (2015) evaluated the antibacterial properties of garlic and showed that the aqueous garlic extract at a 25% concentration had the maximum effect on *Staphylococcus aureus.* Abiy & Berhe (2016) evaluated the antibacterial effect of garlic against the standard isolation of *S. aureus* and *E. coli*. They found that garlic can be used as an effective antibacterial agent and also reported that 20% of the *E. coli* membrane was made up of lipids, hence decreasing the permeability of the extracts, whereas *Staphylococcus aureus* has only 2% lipid content.

13.4 COMBINED EFFECT OF GARLIC AND NILAVEMBU ON GROWTH PERFORMANCE

Arify et al. (2019) conducted an experiment to study the effect of feeding nilavembu (*Andrographis paniculata*) and garlic (*Allium sativum*) to broiler chickens (Vencobb 400) for 6 weeks on different production types and the cost management of feeding garlic and nilavembu. The analyzed groups were fed the following diets containing: nilavembu and garlic at various levels, such as control (T1), 5 g garlic + 1 g nilavembu (T2), 10 g garlic + 1 g nilavembu (T3), 5 g garlic + 2 g nilavembu (T4) and 10 g garlic + 2 g nilavembu (T5), respectively.

The results showed that there was a remarkable (P<0.01) distinction in body weight gain and feed conversion ratio. No remarkable dissimilarity was noted in feed used and sustainability. The cost management of the broiler chicken fed with different levels of nilavembu and garlic enhanced net profit per kg live weight in the analyzed group with 10 grams of garlic and two grams of nilavembu (T5). The mixed effect of garlic and nilavembu showed an environment for better absorption of nutrients and enhanced the utilization of energy, thereby increasing the growth rate of chickens. A better feed conversion ratio in the garlic-supplemented group might be due to decreased growth and colonization of various pathogenic microorganisms in the gut, resulting in enhanced feed efficiency. Hence, the combination of garlic and nilavembu leads to positive additive effects in broilers, which are responsible for better production performances.

Similarly, they also reported that birds receiving 10 g of garlic + 2 g of nilavembu (T5) had a remarkable increase in lymphocyte count and lower cholesterol and triglyceride levels (Arify et al., 2018). This could be due to the high levels of bioactive saponins in garlic that form complexes with cholesterol and inhibit the intestinal absorption of both internal and external cholesterol. These important saponins in garlic also have the ability to slow down key enzymes in the cholesterol and lipid biosynthetic pathways. Meanwhile, allicin inhibits the action of hydroxymethyl glutaryl-CoA reductase, which is a very effective enzyme in the synthesis of cholesterol and lipids.

13.5 ANTI-BACTERIAL EFFICACY OF GARLIC AND NILAVEMBU EXTRACTS

Arify et al. (2017) studied the zone of inhibition for different concentrations of garlic extract and found that a 30% concentration of garlic had the highest zone of inhibition (23 mm) against *Staphylococcus aureus*. Garlic extract at a concentration of 100% had the highest inhibition zone of 14 mm against *E. coli* and 12 mm against *Klebsiella* sp. On the other hand, Nilavembu exhibited an inhibition zone of 16 mm for *Staphylococcus aureus* at 50 mg/ml, but a lower inhibition zone of 12 mm for *Klebsiella* sp. at 100 mg/ml. The highest activity against the *Staphylococcus aureus* isolate was obtained with amoxicillin (32 mm), followed by gentamicin (20 mm). Garlic extract had a 23 mm zone, which was almost similar to the zone of inhibition of gentamicin antibiotic.

13.6 LETHALITY EVALUATION OF NILAVEMBU AND GARLIC EXTRACTS IN EMBRYONATED CHICKEN EGGS

Arify et al. (2017) studied the lethality assay of garlic and nilavembu extracts in embryonated chicken eggs and they found that in the lethality assay, 100% embryo mortality was observed in Group IV (aqueous garlic extract of 200 mg/ml) alone, and no fatality was noted in other groups.

13.7 ANTIVIRAL ADVANTAGE OF NILAVEMBU AND GARLIC BY VIRUS NEUTRALIZATION TEST (VNT)

Suriani et al. (2015) reported that the ethanolic extract of nilavembu, even at a small concentration of 15 µg/ml, had greater antiviral effectiveness. In order to study the benefits of the aqueous extract of nilavembu, a higher concentration was considered in this research so that the aqueous extract can be easily prepared and applied in the field environment.

The ethanolic extract of nilavembu has greater antiviral activity than the aqueous extract. It should be noted that a higher concentration (100 mg/ml) could be selected whenever the aqueous extract of nilavembu is considered in the environment.

Arify et al. (2017) studied the antiviral efficacy of garlic and nilavembu extracts by virus-neutralization test (VNT). In the virus neutralization study, no mortality was observed in Groups 1, 3, 4, 5, 6, and NC group. In Groups 2 and 7, 40% and 80% embryo mortality was observed, whereas Groups 8 and PC group had 100% embryo mortality. Lesions such as occipital and diffused hemorrhage were noticed in chicken embryos. No HA activity was found in Groups 1, 3, 4, 5, and 6, whereas HA titers of 8, 128, 128, and 128 were reported for Groups 2, 7, 8, and PC group, respectively.

13.8 CONCLUSION

It can be concluded that the supplementation of garlic and nilavembu at various concentrations has proven to give better results. The combinations of these two phytobiotics have brought about a synergistic effect on the production performances, immunity, and serum biochemical parameters in broilers. Hence, the combination of these phytobiotics (garlic and nilavembu) could potentially replace the use of antibiotics in broilers, which could lead to antimicrobial resistance.

Note: NC: Negative control; PC: Positive control; G1: Nilavembu aqueous extract 100 mg/ml; G2: Nilavembu aqueous extract 10 mg/ml; G3: Nilavembu 50% ethanol 100 mg/ml; G4: Nilavembu 50% ethanol 10 mg/ml; G5: Nilavembu 100% ethanol 100 mg/ml; G6: Nilavembu 100% ethanol 10 mg/ml; G7: Garlic aqueous extract 100 mg/ml; G8: Garlic aqueous extract 10 mg/ml.

KEYWORDS

- **antibacterial**
- **antiviral**
- **aqueous extract**
- **broilers**
- **garlic**
- **immunity**
- **nilavembu**

REFERENCES

Abdullahi, D. K., Michael, O. O., & Indabawa, I. (2014). Antibacterial activities and phytochemical screening of Aloe vera (*A. babardensis*), garlic (*A. sativum*), and ginger (*Z. officinale*). *Journal of Emerging Trends in Engineering and Applied Sciences, 5*(3), 172–178.

Abhishek, N., Tewari, S. K., & Lehri, A. (2010). Biological activities of Kalmegh (*Andrographis paniculata* Nees) and its active principles – A review. *Indian Journal of Natural Products Resources, 1*(2), 125–135.

Abiy, E., & Berhe, A. (2016). The anti-bacterial effect of garlic (*Allium sativum*) against clinical isolates of *Staphylococcus aureus* and *Escherichia coli* from patients attending Hawassa Referral Hospital, Ethiopia. *Journal of Infectious Diseases, 2*(18), 1–5.

Alagesaboopathi, C., & Sivakumar, R. (2011). Phytochemical screening studies on the leaves and stem of Andrographis neesiana wight – An endemic medicinal plant from India. *World Applied Sciences Journal, 12*(3), 307–311.

Al-Hadid. (2016). Evaluation of antiviral activity of different medicinal plants against Newcastle disease virus. *American Journal of Agricultural and Biological Sciences, 11*(4), 157–163.

Ameh, G. I., Eze, S. C., & Omeje, F. U. (2013). Phytochemical screening and antimicrobial studies on the methanolic bulb extract of (*Allium sativum* L.). *African Journal of Biotechnology, 12*(14), 1665–1668.

Anvar, A. B., Dastar, J., Nejad, J. G., Sung, K., Lohakare, J., & Forghani, F. (2012). Effects of garlic and thyme extracts on growth performance and carcass characteristics of broiler chicks. *Journal of Animal Science and Technology, 54*(3), 185–190.

Arify Toriyali, S., Ezhil Valavan, K., Manimaran, A., Sundaresan, A., & Varun, A. (2018). Effect of supplementation of garlic (*Allium sativum*) and nilavembu (*Andrographis paniculata*) on hematological and serum biochemical parameters of commercial broiler chicken. *International Journal of Current Microbiology and Applied Sciences, 7*(10), 3649–3656.

Arify Toriyali, S., Ezhil Valavan, Varun, A., Manimaran, K., & Sundaresan, A. (2019). Effect of garlic (*Allium sativum*) and nilavembu (*Andrographis paniculata*) on growth performance and cost-effectiveness of broiler chicken. *Indian Journal of Animal Sciences, 89*(11), 1242–1245.

Arify, T., Jaisree, S., Manimaran, K., Valavan, S., & Sundaresan, A. (2018). Antiviral effects of garlic (*Allium sativum*) and nilavembu (*Andrographis paniculata*) against the velogenic strain of Newcastle disease virus – An *in-ovo* study. *International Journal of Livestock Research, 8*(10), 157–164. doi: 10.5455/ijlr.20170814051902.

Balakrishnan, C., Senthilkumar, M., & Viswanath, M. (2013). Pharmacognosy and chromatographic analysis of nilavembu chooranam and nilavembu tablet, a Siddha polyherbal formulation. *International Journal of Pharmaceutical and Biological Sciences, 4*(2), 180–186.

Banerjee, S. K., Maulik, M., Mancahanda, S. C., Dinda, A. K., Gupta, S. K., & Maulik, S. K. (2002). Dose-dependent induction of endogenous antioxidants of garlic. *Life Sciences, 70*(12), 1509–1518.

Calabrese, C., Berman, S. H., Babish, J. G., Ma, X., Shinto, L., Dorr, M., & Standish, L. J. (2000). A phase I trial of andrographolide in HIV positive patients and normal volunteers. *Phytotherapy Research, 14*(5), 333–338.

Chen, Y., Wang, D. H. U. Y., Wang, G. Z., Zhao, J., Fan, X., & Xing, Y. (2010). Astragalus polysaccharide and oxymatrine can synergistically improve the immune efficacy of the Newcastle disease vaccine in chickens. *International Journal of Biological Macromolecules, 46*, 425–428.

Das, M. P., Banerjee, A., & Anu. (2013). Comparative study on in vitro antibacterial and antifungal properties of five medicinal plants of West Bengal. *Asian Journal of Plant Science and Research, 3*(2), 107–111.

Jimoh, A. A., Olorede, B. R., Abubakar, A., Fabiyi, J. P., Ibitoye, E. B., Suleiman, N., & Garba, S. (2012). Lipid profile and hematological indices of broiler chickens fed garlic (*Allium sativum*)-supplemented diets. *Journal of Veterinary Advances, 2*(10), 474–480.

Laing, D., Wongtangtintharn, S., Tungjarernkul, B., Sirilaophaisan, S., & Khajarern, J. (2013). Effects of Andrographis paniculata and *Zingiber cassumunar* mixture on productive performance and carcass quality of broiler chickens. *International Conference on Agriculture and Biotechnology, 60*, 7.

Makwana, A. R., Rinkesh, B., Raval, P., Chauhan, H., Kulkarni, C. A., Srivastava, K., & Bharat, B. R. (2015). Effects of garlic (*Allium sativum*) supplementation on growth performance, carcass characteristics and economics of broilers. *Indian Journal of Animal Research, 5*(4), 843–848.

Malahubban, M., Alimon, A. R., Sazili, A. Q., & Fakurazi, S. (2013). Effects of *Andrographis paniculata* and *Orthosiphon stamineus* supplementation in diets on growth performance and carcass characteristics of broiler chickens. *International Journal of Agriculture & Biology*, 1814–9596.

Mathivanan, R., Edwin, S. C., & Viswanathan, K. (2008). Effect of *Andrographis paniculata* supplementation on growth and feed conversion efficiency of broilers. *Indian Journal of Poultry Science, 43*(2), 185–188.

Mathivanan, R., Edwinand, S. C., & Viswanathan, K. (2006). *Panchagavya* and *Andrographis paniculata* as alternatives to antibiotic growth promoters on broiler production and carcass characteristics. *International Journal of Poultry Science, 5*(12), 1122–1155.

Mukhtar, S., & Ghori, I. (2012). Antibacterial activity of aqueous and ethanolic extracts of garlic, cinnamon, and turmeric against *Escherichia coli* and *Bacillus subtilis*. *IJABPT, 2*(3).

Newall, C. A., Anderson, L. A., & Phillipson, J. D. (1996). Herbal medicines: A guide for health care professionals (2nd ed.). Pharmaceutical Press.

Oleforuh, O. V. U., Chukwu, G. C., & Adeolu, A. I. (2014). Effect of ground ginger and garlic on the growth performance, carcass quality and economics of production of broiler chickens. *Global Journal of Bio-Science and Biotechnology, 3*(3), 225–229.

Ologhobo, A. D., Adebiyi, F. G., & Adebiyi, O. A. (2008). Effect of long-term feeding of raw and sundried garlic (*Allium sativum*) on performance and lipid metabolism of broiler chicks. Conference on International Research on Food Security, Natural Resource Management and Rural Development.

Onyimonyi, A. E., Chukwuma, P. C., & Igbokwe, C. (2012). Growth and hypocholesterolemic properties of dry garlic powder (*Allium sativum*) on broilers. *African Journal of Biotechnology, 11*(11), 2666–2671.

Padmalochana, K., & Rajan, M. S. D. (2014). Phytochemical analysis and antimicrobial activity of king of bitter *Andrographis paniculata* against pathogenic microbes. *International Journal of Institutional Pharmacy and Life Sciences, 4*(6), 135–143.

Pavni, K., Esha, B., Neha, J., Tushar, A., Shrey, K., Suchit, A., & Neeraj, W. (2011). Phytochemical screening of developing garlic and effect of its aqueous extracts on the viability of cardiac cell line. *Journal of Pharmaceutical Research, 4*(3), 902–904.

Rahimi, S., Teymouri, Z. Z., Karimi, M. A. T., Omidbaigi, R., & Rokni, H. (2011). Effect of three herbal extracts on growth performance, immune system, blood factors, and intestinal selected bacterial population in broiler chickens. *Journal of Agriculture Science Technology, 13*, 527–539.

Recoquillay, F. (2006). Active plant extracts show promise in poultry production. *Poultry International,* 28–30.

Rehman, & Munir, M. T. (2015). Effect of garlic on the health and performance of broilers. *Open Access Journal Veterinaria, 3*(1), 32–39.

Salna, K. P., Sreejith, K., Uthiralingam, M., Mithu, A. P., John, M. C., & Albin, T. F. (2011). A comparative study of phytochemicals investigation of *Andrographis paniculata* and *Murraya koenigii*. *Journal of Pharmaceutics and Pharmaceutical Sciences, 3*(3).

Suriani, M. N., Jamaliah, H., Normah, M., Goon, S. M. N., Iswadi, M. I., & Mona, R. (2015). Antiviral Effect of *Andrographis paniculata* Ethanolic Extract against Newcastle Disease Virus. *Veterinary Research Institute, Perak, Malaysia.*

Verma, S., Singh, J., Kaur, S., & Garg, A. (2015). Anti-bacterial Effect of Garlic (*Allium sativum* L) extract on different pathogenic and non-pathogenic Bacteria. *Research Journal of Pharmaceutical, Biological and Chemical Sciences, 6*(3), 1103–1107.

Zakaria, Y. A. A. (2003). Effect of storage and temperature of aqueous garlic extract on the growth of certain pathogenic bacteria. *Journal of Al Azhar University-Gaza, 6*(2), 11–20.

Index

C

D

E

G

I

J

K

L

O

P

T

U